Springer Proceedings in Physics **417**

Indexed by Scopus

The series Springer Proceedings in Physics, founded in 1984, is devoted to timely reports of state-of-the-art developments in physics and related sciences. Typically based on material presented at conferences, workshops and similar scientific meetings, volumes published in this series will constitute a comprehensive up to date source of reference on a field or subfield of relevance in contemporary physics. Proposals must include the following:

- Name, place and date of the scientific meeting
- A link to the committees (local organization, international advisors etc.)
- Scientific description of the meeting
- List of invited/plenary speakers
- An estimate of the planned proceedings book parameters (number of pages/articles, requested number of bulk copies, submission deadline).

Please contact:

For Americas and Europe: Dr. Zachary Evenson; zachary.evenson@springer.com
For Asia, Australia and New Zealand: Dr. Loyola DSilva; loyola.dsilva@springer.com

Ge Luo
Editor

Proceedings of the Second Global Forum on Space Information for Sustainable Development

GFSISD 2024, 25–28 September, Guangzhou, China

Springer

Editor
Ge Luo
China Association of Remote Sensing Application
Beijing, China

ISSN 0930-8989 ISSN 1867-4941 (electronic)
Springer Proceedings in Physics
ISBN 978-981-96-2489-8 ISBN 978-981-96-2490-4 (eBook)
https://doi.org/10.1007/978-981-96-2490-4

This Springer imprint is published by the registered company Springer Nature Singapore Pte Ltd.
The registered company address is: 152 Beach Road, #21-01/04 Gateway East, Singapore 189721, Singapore

Contents

Space Optical Instruments and Application

Moving Target Detection and Motion State Estimation Based on Multispectral Remote Sensing Data for Space Applications

Wenyi Zhang(✉), Xiangyu Dai, Xiaoyi Yin, Qingwei Chu, and Guangzuo Li

Aerospace Information Research Institute, Chinese Academy of Sciences, Beijing, China
wyzhang@aircas.ac.cn

Abstract. The internal spatial configuration of push-broom multispectral imagers causes different observation directions for various detectors, leading to parallax for the same object across images in different spectral bands. This parallax can be utilized to infer detection information and estimate parameters for moving targets. However, most existing methods rely on traditional paradigms, which suffer from low detection performance and robustness. This paper introduces the first multispectral dataset specifically designed for detecting airborne moving aircraft targets. To address the challenge of detecting small airborne moving targets in multispectral images, we propose modifications to the YOLOv8 network. By introducing skip connections in the neck section, we facilitate information transfer between multi-scale feature layers, thereby enhancing the network's ability to detect small targets. Experimental results demonstrate that the improved model significantly enhances detection accuracy and robustness in the task of target detection within multispectral images. Finally, the rainbow effect in multispectral images is utilized to estimate the parameters of the detected aircraft targets, and the results show that the proposed parameter estimation method offers high reliability.

Keywords: Moving Target Detection · Motion State Estimation · Multispectral Remote Sensing

1 Introduction

High-resolution visible optical remote sensing images provide only spatial information and lack temporal information. They are typically suitable for detecting stationary aircraft but are not applicable for detecting airborne targets in motion. A push-broom multispectral imager uses the motion of the instrument itself to drive the detector array for scanning and imaging [1]. The different observation directions of the detectors within the imager cause them to observe the same ground object at different times, which leads to the position of a moving target differing across images in different spectral bands, as shown in Fig. 1. According to the characteristics of the push-broom multispectral image, the detection of the flying target by fully combining the time and space information, and the estimation of the motion state of the flying target by using the relationship between time and displacement, can achieve the quantitative description of the flying target.

G. Luo (Ed.): GFSISD 2024, SPPHY 417, pp. 3–14, 2025.
https://doi.org/10.1007/978-981-96-2490-4_1

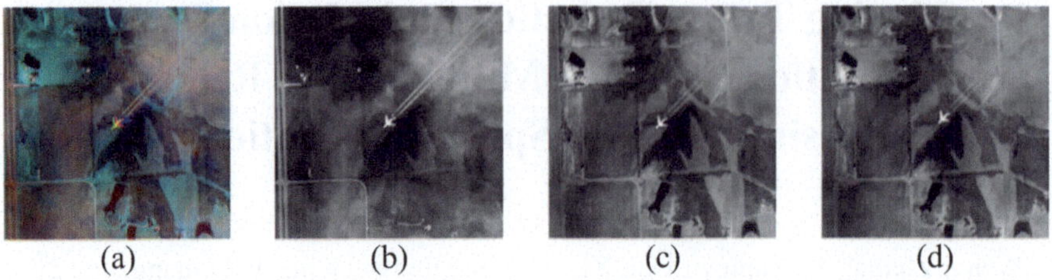

Fig. 1. Rainbow effect of moving targets in multispectral images

Currently, the main methods for detecting moving targets are mainly classified into: frame differencing [2, 3], background subtraction [4, 5], and optical flow [6, 7]. Traditional methods often rely on strong prior assumptions, such as the static nature of the background or the continuity of motion, which are often difficult to satisfy in practical applications. As the resolution of remote sensing images continues to improve and scene complexity increases, the traditional moving target detection methods are not sufficient to cope with the complex scene and detailed features. In recent years, deep learning-based methods have made significant progress in the field of computer vision and have gradually been applied to target detection in remote sensing images [8].

To address the above problems, this paper proposes a novel deep learning-based method for detecting moving aircraft targets, aimed at overcoming the limitations of traditional methods in complex scenarios. Firstly, in this paper, a novel multispectral dataset is constructed specifically for the detection of airborne moving aircraft. The establishment of this dataset not only fills a gap in existing datasets but also provides abundant training data for future research. Given the small target characteristics of aircraft in multispectral images, the YOLOv8 network is improved. In the feature extraction stage, a skip connection mechanism is introduced to ensure effective information transfer between feature layers of different scales. This improvement enables the network to better capture the features of small targets, thereby enhancing detection accuracy and robustness. Finally, the speed and azimuth of the aircraft are estimated using the rainbow effect of the multispectral image. In order to accurately estimate the azimuth of the aircraft, we propose an angle estimation algorithm based on principal component analysis (PCA). Using the shape characteristics of the aircraft, the azimuth of the aircraft is estimated more accurately, and the speed estimation accuracy of the aircraft is improved at the same time.

2 Prior Work

Since the launch of high-resolution satellites like QuickBird and WorldView, some scholars have studied the methods of detecting moving targets such as aircraft and ships from a single remote sensing image. Etaya et al. [9] discovered the phenomenon of inconsistent imaging positions for moving targets between panchromatic and multispectral images in QuickBird images. They investigated this and proposed the idea of estimating motion parameters by using the changes in the imaging positions of moving targets caused by the imaging intervals of the panchromatic and multispectral line array sensors in the focal plane of the satellite. The idea was then further studied and developed, resulting in a series of theories and methods for obtaining target motion parameters from a single remote sensing image [10–13]. These methods require the satellite images to be composite images obtained from different sensors, such as panchromatic and multispectral sensors, and are not applicable to images acquired by a single sensor.

Meng et al. [14] proposed a target tracking algorithm for time-series images that includes a three-step process of motion target detection, target modelling and target matching. Liu et al. [15] developed a method for detecting flying aircraft using the imaging time difference between bands in Sentinel-2 multispectral images. The method detects flying aircraft by the difference in spectral reflectance in images of different bands and estimates the speed. However, the speed estimation results in this work are not precise, as the estimated speed only maintains a linear relationship with the actual speed, and the method is unable to calculate the direction of the aircraft. Peder Heiselberg and Henning Heiselberg [16] conducted statistical analysis of the spectral reflectance across different backgrounds in the various bands of Sentinel-2 multispectral images to remove background noise, thereby enabling the extraction of aircraft targets and speed estimation. This method estimates direction based on the contrail, but wind direction and speed can impact the formation and propagation of the contrail. Under unstable atmospheric conditions, the contrail may become distorted or even disappear. Eric et al. [17] used the frame difference method to determine the position of a moving aircraft in each band of a multispectral image and to calculate the velocity of the aircraft target. Their study demonstrates the great potential of multispectral data in the field of motion state estimation.

3 Method

3.1 Model

Since the first generation of the YOLO network was proposed, many researchers have continuously improved and iterated on it, making its performance increasingly powerful [18–21]. The YOLOv8 is the standout performance model in the YOLO series. The model consists of three components: the backbone, the neck, and the detection head. The backbone part is the feature extraction module, which is responsible for extracting multi-scale and multi-level features from the input image, mainly including a series of Conv, C2f, and SPPF modules. The neck plays a crucial role in the object detection, primarily responsible for feature fusion between feature maps at different scales, generating feature maps used for object classification and localization. The detection head receives feature maps from the neck and converts these feature maps into the final output, which are the positional coordinates, confidence levels, and category labels of the target. The YOLOv8 detection head enhances the overall performance by decoupling the classification and localization tasks, allowing each to be optimized separately.

However, YOLOv8 still faces some challenges in detecting airborne aircraft. Detection is made more difficult by the fact that aircraft usually fly at high altitudes and have a small target size relative to the background. Although YOLOv8 performs well in most object detection tasks, there is still room for improvement when it comes to detecting small objects, particularly in the dynamic and variable aerial environment. The background is highly variable, including cloud cover, terrain and weather factors, which can easily be confused with the target, leading to false or missed detections. Therefore, to better adapt to the requirements of detecting small airborne targets, this study introduces improvements to the YOLOv8 model and subsequently introduces YOLOv8-CLSC.

In the process of feature extraction, semantic information is gradually enriched and spatial information describing image details is gradually reduced as the feature map is continuously downsampled. However, detailed information is crucial for detecting small targets, especially the contours and structures of aircraft that can easily be overlooked against a high-altitude background, as shown in Fig. 2. To address this issue, this paper introduces a cross-layer skip connection (CLSC) mechanism in the YOLOv8 model. The feature maps obtained from the first stage are downsampled by a factor of four and then fused with the feature maps from the third stage to achieve cross-stage feature fusion between the first and third stages. Similarly, the feature maps from the first stage are downsampled by a factor of eight and then fused with the feature maps from the fourth stage to achieve cross-stage fusion between the first and fourth stages. Specifically, the skip connections fuse the high-resolution spatial information from the early feature extraction layers with the semantic information from the later feature layers, thereby generating multi-scale feature maps. These feature maps contain more detailed information that contributes to the detection of small targets, allowing the model to locate and identify objects more accurately.

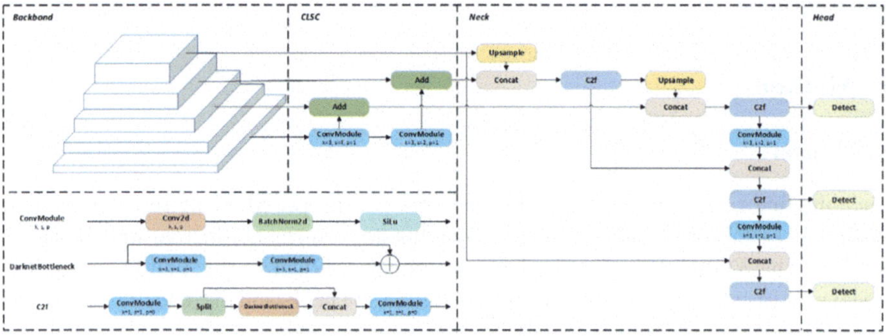

Fig. 2. Structure of YOLOv8-CLSC

3.2 Motion State Estimation

In the high-resolution multi-mode satellite, a CCD imaging array is mounted perpendicular to the direction of motion of the platform, as shown in Fig. 3. As the satellite moves forward, the array captures a strip of image data, with a width corresponding to the field of view observed by the array.

During the imaging process, assume that the incident angles of two bands along the along-track direction are ω_1 and ω_2, respectively. Since the field of view angles are small, the time difference between the two bands can be approximated as the ground

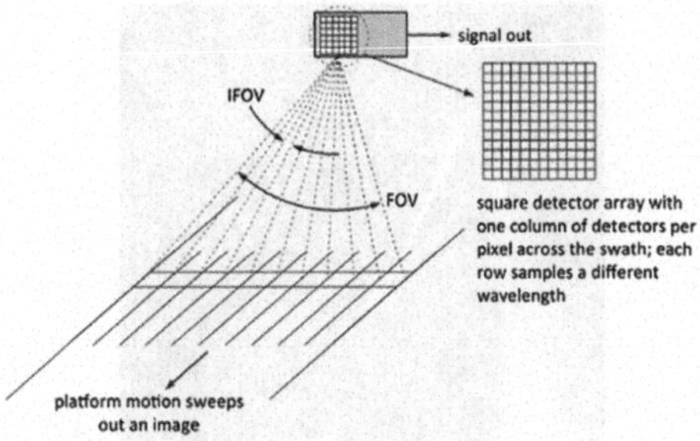

Fig. 3. Schematic of linear array push-broom mode imaging

distance corresponding to the difference in field of view angles between the two bands divided by the speed of the satellite in orbit, expressed as:

$$\Delta t = \frac{H\Delta\omega}{v} \tag{1}$$

where H represents the average orbital altitude of the satellite, and v is the velocity of the satellite along the orbit.

Assuming that the pixel coordinates of the moving target in the remote sensing image at time t_1 are (x_1, y_1), and the pixel coordinates at time t_2 are (x_2, y_2). Then the displacements can be expressed as $\Delta x = x_2 - x_1,\ \ \Delta y = y_2 - y_1$. Subsequently, the approximate velocity of the moving target can be calculated as:

$$v_0 = \frac{\sqrt{(\Delta x)^2 + (\Delta y)^2} \cdot r}{\Delta t} \tag{2}$$

where r represents the spatial resolution. In addition, the motion azimuth of the moving target can be calculated from the pixel displacement, with the azimuth defined as 0° for north and increasing clockwise up to 360°. When $\Delta x = 0$, the azimuth is defined as 90° ($\Delta y > 0$), or 270° ($\Delta y < 0$). When $\Delta y = 0$, the azimuth is defined as 180° ($\Delta x > 0$), or 0° ($\Delta x < 0$). Otherwise, the azimuth θ (in degrees) of the moving target is determined as follows:

$$\theta = \begin{cases} 180 - \alpha,\ \Delta x > 0,\ \Delta y > 0 \\ 180 + \alpha,\ \Delta x > 0,\ \Delta y < 0 \\ 360 - \alpha,\ \Delta x < 0,\ \Delta y < 0 \\ \alpha,\ \Delta x\langle 0,\ \Delta y\rangle 0 \end{cases} \tag{3}$$

where $\alpha = \tan^{-1}\frac{\Delta y}{\Delta x}$.

The speed and azimuth can be calculated using Eqs. (2) and (3), but the results do not represent the true speed and azimuth of the aircraft. Due to the movement of the satellites, the direction in which the aircraft moves between the different bands is not in the same direction as the contrail of the aircraft, as shown in Fig. 4.

Since the continuous movement of the satellite during imaging, the velocity calculated directly from the image is the superposition of the aircraft velocity and the satellite velocity, called the apparent velocity, as shown in Fig. 5.

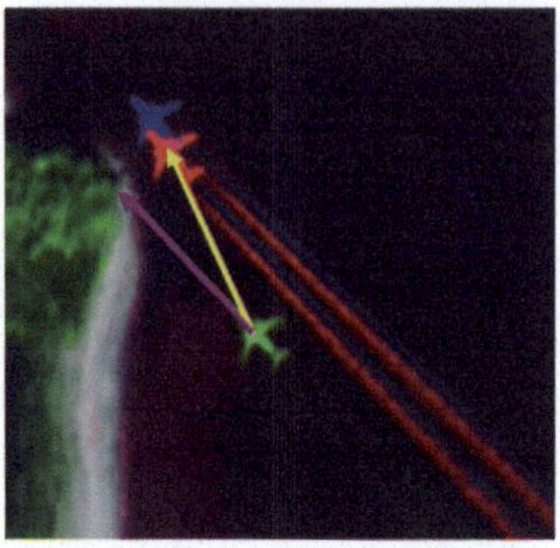

Fig. 4. Three-band fusion image of multispectral image, yellow line: apparent velocity direction; Purple line: true velocity direction (color figure online).

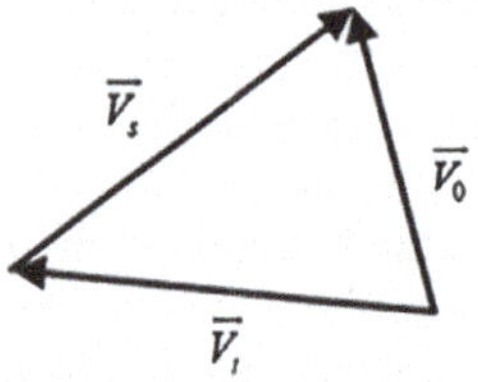

Fig. 5. Schematic of satellite velocity, target apparent velocity, and target true velocity.

Where $\overrightarrow{V_0}$ represents the apparent velocity estimated directly from the image, $\overrightarrow{V_t}$ represents the actual velocity of the target, and $\overrightarrow{V_s}$ represents the speed of the satellite. Based on the geometric relationship, the following can be derived:

$$V_t = \frac{\sin(\theta_0 - \theta_s)}{\sin(\theta_t - \theta_s)} \cdot V_0 \tag{4}$$

where θ_0 denotes the apparent velocity angle, θ_s denotes the satellite trajectory angle, and θ_t denotes the actual flight angle of the target. Therefore, to calculate the true velocity of the target, it is necessary to first determine θ_0, θ_s and θ_t. Among these, θ_0 has already been calculated. Due to the curvature of the Earth, θ_s is related to the latitude of its position, and is expressed as follows:

$$\cos\theta_s = \frac{\cos i}{\cos\phi} \tag{5}$$

where i denotes the inclination angle of the satellite orbit and ϕ denotes the latitude. Generally, the direction of contrail is the same as the true direction of flight. However, there are specific conditions for the generation of contrails, and the method is completely ineffective in the absence of contrails. For θ_t, we propose a more universally applicable method to obtain it. By analyzing the morphology of the target, true angle information is extracted based on the characteristics of the aircraft target in the spectral image. Specifically, an aircraft consists of a fuselage and wings, and the combination of these parts roughly resembles a cross shape, with the fuselage typically being larger than the wings. Therefore, Principal Component Analysis (PCA) is well-suited for analyzing aircraft pixels, and the direction of the first principal component can be considered as the direction of aircraft flight.

4 Experiments

4.1 Experiment Settings

Dataset. This section employs our dataset for experimentation. We screened a large number of multispectral images and ultimately obtained 153 scenes, manually annotating 155 aircraft. Each aircraft was then sliced across different spectral bands, followed by a secondary screening process to remove severely blurred images. In the end, we obtained 924 images with a size of 256 × 256 pixels and a resolution of 2 m. The dataset is divided into training set and validation set in the ratio of 8:2. The training set includes 726 images and the validation set includes 182 images.

Evaluation Metrics. We evaluated the performance of the model by four metrics: Precision (P), Recall (R), $mAP_{50–95}$ and F1 score. Combining false positives (FP), true positives (TP), false negatives (FN), and true negatives (TN), P and R were defined as follows:

$$P = \frac{TP}{TP + FP}, R = \frac{TP}{TP + FN} \tag{6}$$

After setting the IoU threshold, the Precision-Recall curve can be plotted using precision and recall. The area under the curve represents the Average Precision (AP). The $mAP_{50–95}$ provides a more stringent evaluation of model performance compared to a single IoU threshold (e.g., mAP_{50}) and is an important metric for assessing the overall performance of object detection models. The formula is as follows:

$$AP = \int_0^1 P(R)dR, mAP_{50-95} = \frac{1}{N}\sum\nolimits_{i=1}^{N} AP_{IoU=0.5+0.05(i-1)} \tag{7}$$

where $N = 10$, corresponding to IoU thresholds of 0.50, 0.55, 0.60,…, 0.95 of 10 values.

The F1 score provides a comprehensive measure by balancing the Precision and Recall of a model. It is derived by calculating the harmonic mean of Precision and Recall with the following formula:

$$F1 = \frac{2 \times P \times R}{P + R} \tag{8}$$

The value of F1 score is between 0 and 1, with larger values indicating better performance of the model.

Training Settings. All experiments are performed under CUDA 11.8 on NVIDIA GTX 3060 GPU. We use AdamW optimizer with the learning rate of 0.01, the betas of (0.9, 0.09), the weight decay of 0.0005. The epoch is set to 500, and the batch size is set to 8. Leave the rest of the configuration as the default configuration of the original YOLOv8 model.

4.2 Experimental Result

Target Detection Result. In this study, we not only compared the performance of YOLOv8 and YOLOv8-CLSC but also evaluated the superiority of the proposed network by comparing it with the latest version of the YOLO series, YOLOv10. The results are shown in Table 1.

Table 1. Performance of YOLOv8, YOLOv10 and YOLOv8-CLSC.

Model	P	R	mAP_{50-95}	F1
YOLOv8	0.989	0.983	0.740	0.986
YOLOv10	0.993	0.978	0.718	0.985
YOLOv8-CLSC	0.998	0.984	0.750	0.991

As shown in Table 1, our network achieved the best performance. Compared to the original YOLOv8 network, our method improved the P, R, mAP_{50-95}, and F1 evaluation metrics by 0.9%, 0.1%, 1.0%, and 0.5%, respectively. Additionally, we visualized some of the detection results, as shown in Fig. 6. The figure demonstrates that, compared to the other two networks, the proposed method exhibits superior detection performance for small targets that are difficult to distinguish.

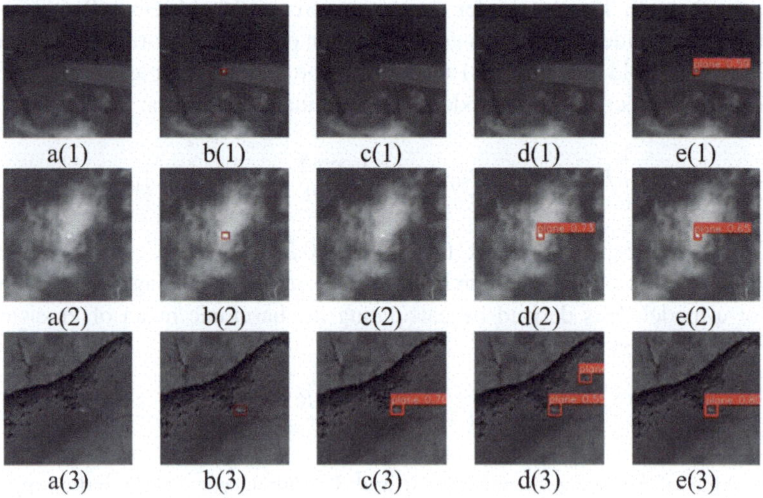

Fig. 6. Visualization of target detection results. Column a represents the original image, Column b represents the ground truth, Column c represents YOLOv8, Column d represents YOLOv10, Column e represents YOLOv8-CLSC.

State Estimation Result. In this experiment, four multispectral images from the high-resolution multi-mode satellite were used. Each band of these images has a resolution of

2 m and contains a total of 7 aircraft in flight. After detecting the aircraft in each band of the multispectral images using the YOLOv8-CLSC, the position of target in each band image can be obtained. The moving distance of the target can be calculated according to Eq. (2). The structure of the high-resolution multimode satellite and its on-board sensors is shown in Fig. 7. It is equipped with a 5-band TDICCD sensor and a 4-band TDICCD sensor. The former includes the panchromatic band and multispectral bands 1–4, while the latter includes multispectral bands 5–8. The field of view (FOV) for each band are shown in Table 2.

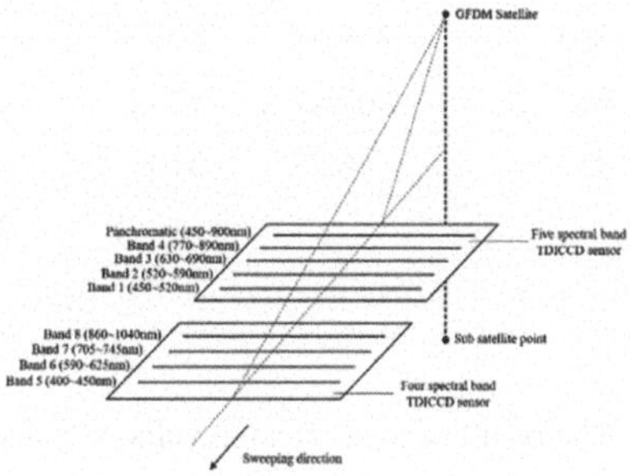

Fig. 7. Schematic diagram of the five-band and four-band TDICCD sensors of the high-resolution multimode satellite and their corresponding bands.

Table 2. Field of view of each band.

Band	Band4	Band3	Band2	Band1	Band8	Band7	Band6	Band5
FOV (deg)	0.300	0.320	0.340	0.365	0.620	0.645	0.665	0.690

Substituting the data from Table 2 into Eq. (1), the imaging time difference for each band can be calculated and the results are shown in Table 3.

Table 3. The time difference of each band.

Band	Band5	Band6	Band7	Band8	Band1	Band2	Band3	Band4
Time (s)	0	0.0373	0.0671	0.1044	0.4848	0.5220	0.5519	0.5817

Substituting the above results into Eqs. (2) and (3) can calculate the apparent velocity of the aircraft target movement, and the results are shown in Table 4.

As can be seen from the above results, the uncorrected velocities and azimuths exhibit significant errors. By applying Eq. (4) to correct the above results, the outcomes are presented in Table 5. The results demonstrate a significant reduction in the errors of the corrected velocity and azimuth.

Table 4. The result of the apparent velocity calculation.

	Estimated velocity (km/h)	Estimated azimuth (deg)	True velocity (km/h)	True azimuth (deg)	Velocity difference (%)	Azimuth difference (deg)
1	1193.23	50.42	942.67	64.24	26.58	13.82
2	498.58	265.69	761.17	234.8	34.50	30.89
3	695.20	273.46	842.66	250.49	17.50	22.97
4	506.36	272.36	796.36	232.94	36.42	39.42
5	315.57	325.28	318.54	321.13	0.93	4.15
6	723.84	92.97	807.47	129.6	10.36	36.63
7	1022.28	333.62	766.73	310.00	33.33	23.62
Mean difference					22.80	24.50

Table 5. The result of the corrected velocity calculation.

	Estimated velocity (km/h)	Estimated azimuth (deg)	True velocity (km/h)	True azimuth (deg)	Velocity difference (%)	Azimuth difference (deg)
1	1002.91	60.75	942.67	64.24	6.39	3.49
2	580.35	245.54	761.17	234.8	23.76	10.74
3	732.68	260.60	842.66	250.49	13.05	10.11
4	741.63	232.81	796.36	232.94	6.87	0.13
5	234.67	299.97	318.54	321.13	7.55	21.16
6	783.41	124.53	807.47	129.6	8.43	5.07
7	698.64	307.82	766.73	64.24	10.99	2.18
Mean difference					11.00	12.63

5 Conclusion

In this paper, we propose a deep learning-based method for airborne moving target detection and motion state estimation, which addresses the limitations of existing methods in small target detection and complex scenes. By establishing the first multispectral dataset for detecting moving airborne targets and improving the YOLOv8 network, the proposed method significantly enhances the detection performance of small targets in multispectral images. Additionally, this study introduces a reliable aircraft motion parameter estimation method utilizing the rainbow effects in multispectral images. Experimental results demonstrate that the proposed approach not only achieves notable improvements in detection performance but also shows high reliability in parameter estimation accuracy. Future work will analyse the sources of motion state estimation errors in depth and apply the method to other dynamic target detection scenarios.

References

1. Heiselberg, H.: Aircraft and ship velocity determination in Sentinel-2 multispectral images. Sensors **19**(13), 2873 (2019)
2. Cheng, Y.H., Wang, J.: A motion image detection method based on the inter-frame difference method. Appl. Mech. Mater. **490**, 1283–1286 (2014)
3. Wan, M., Gu, G., Cao, E., et al.: In-frame and inter-frame information based infrared moving small target detection under complex cloud backgrounds. Infrared Phys. Technol. **76**, 455–467 (2016)
4. Xin, Y., Jie, H., Dong, L., et al.: A self-adaptive optical flow method for the moving object detection in the video sequences. Optik **125**(19), 5690–5694 (2014)
5. Han, P., Du, J., Zhou, J., et al.: An object detection method using wavelet optical flow and hybrid linear-nonlinear classifier. Math. Probl. Eng. **2013**(9), 1–14 (2013)
6. Akula, A., Khanna, N., Ghosh, R., et al.: Adaptive contour-based statistical background subtraction method for moving target detection in infrared video sequences. Infrared Phys. Technol. **63**, 103–109 (2014)
7. Kaushal, M., Khehra, B.S.: BBBCO and fuzzy entropy based modified background subtraction algorithm for object detection in videos. Appl. Intell. **47**(4), 1008–1021 (2017)
8. Girshick, R., Donahue, J., Darrell, T., et al.: Rich feature hierarchies for accurate object detection and semantic segmentation. In: Proceedings of the IEEE Conference on Computer Vision and Pattern Recognition, pp. 580–587 (2014)
9. Etaya, M., Sakata, T., Shimoda, H., et al.: An experiment on detecting moving objects using a single scene of QuickBird data. J. Remote Sens. Soc. Jpn. **24**(4), 357–366 (2003)
10. Yamazaki, F., Liu, W., Vu, T.T.: Speed detection for moving objects from digital aerial camera and QuickBird sensors. In: 5th International Workshop on Remote Sensing Applications to Natural Hazards (2007)
11. Xiong, Z., Zhang, Y.: An initial study of moving target detection based on a single set of high spatial resolution satellite imagery. In: Proceedings of the ASPRS 2006 Annual Conference, pp. 1–5 (2006)
12. Zhang, Y., Xiong, Z.: Moving vehicle detection using a single set of QuickBird imagery – an initial study. In: Proceedings of the ISPRS Commission VII Mid-Term Symposium "Remote Sensing: From Pixels to Processes", Enschede, The Netherlands, pp. 8–11 (2006)
13. Xiong, Z., Zhang, Y.: An initial study on vehicle information extraction from single pass QuickBird satellite imagery. Photogram. Eng. Remote. Sens. **74**(11), 1401–1411 (2008)
14. Meng, L., Kerekes, J.P.: Object tracking using high resolution satellite imagery. IEEE J. Sel. Top. Appl. Earth Observ. Remote Sens. **5**(1), 146–152 (2012)
15. Liu, Y., Xu, B., Zhi, W., et al.: Space eye on flying aircraft: from Sentinel-2 MSI parallax to hybrid computing. Remote Sens. Environ. **246**, 111867 (2020)
16. Heiselberg, P., Heiselberg, H.: Aircraft detection above clouds by sentinel-2 MSI parallax. Remote Sens. **13**(15), 3016 (2021)
17. Keto, E., Watters, W.A.: Velocity Analysis of Moving Objects in Earth Observation Satellite Images Using Multi-Spectral Push Broom Scanning (2024). arXiv:2406.13710
18. Redmon, J., Farhadi, A.: Yolov3: An Incremental Improvement (2018). arXiv:1804.02767
19. Bochkovskiy, A., Wang, C.Y., Liao, H.Y.M.: Yolov4: Optimal Speed and Accuracy of Object Detection (2020). arXiv:2004.10934

20. Wang, C.Y., Bochkovskiy, A., Liao, H.Y.M.: YOLOv7: trainable bag-of-freebies sets new state-of-the-art for real-time object detectors. In: Proceedings of the IEEE/CVF Conference on Computer Vision and Pattern Recognition, pp. 7464–7475 (2023)
21. Reis, D., Kupec, J., Hong, J., et al.: Real-Time Flying Object Detection with YOLOv8 (2023). arXiv:2305.09972

Discovering Hidden Caves: Innovative Applications of UAV and LiDAR Technology in Vegetated Areas

Congyuan Zhang[1], Liangsheng Ge[2,3], Nan Luo[1], Xingwu Liu[1], Wangchuan Guo[1], Sipeng Han[1,3](✉), and Yaxi Li[1]

[1] Research Center of Applied Geology of China Geological Survey, Chengdu 610000, China
hansipeng@mail.cgs.gov.cn

[2] Command Center of Integrated Natural Resources Survey, China Geological Survey, Beijing 100000, China

[3] Key Laboratory of Geological Survey and Evaluation of Ministry of Education, China University of Geosciences, Wuhan 430074, China

Abstract. The global distribution of karst landscapes is extensive, and in regions with suitable climates, a large number of natural caves have developed. However, these caves are often concealed by vegetation, making them difficult to detect. To address the challenges of locating and detecting caves in karst regions under vegetative cover, this study integrates Unmanned Aerial Vehicle (UAV) remote sensing and Light Detection and Ranging (LiDAR) technology to explore a method for identifying hidden caves in vegetated areas using both aerial and ground-based approaches. The findings demonstrate that the penetration capabilities of LiDAR dense point clouds in vegetated areas can effectively identify depression structures in the terrain, which are indicative of concealed caves that are not easily detectable by the naked eye. Additionally, as natural underground spaces, caves can be readily converted for purposes such as civil defense or storage, significantly reducing labor and material costs. To this end, a centimeter-level precision three-dimensional (3D) model of the caves was developed, and the concept of "Cubic Space Volume" was proposed to simulate the effectively usable space within the caves, providing valuable practical reference. This study not only offers a new approach for cave detection and exploration but also enriches the application of remote sensing technology in cave exploration research.

Keywords: Caves · UAV · LiDAR · Vegetated Areas

1 Introduction

Karst topography is a unique geomorphological feature on the Earth's surface formed by the chemical dissolution of soluble rocks (carbonate rocks) and is widely distributed across continents globally. According to previous studies, the modern global karst area is approximately 20.3 million km^2, accounting for 15.2% of the ice-free land surface [1].

G. Luo (Ed.): GFSISD 2024, SPPHY 417, pp. 15–26, 2025.
https://doi.org/10.1007/978-981-96-2490-4_2

Carbonate rocks, as typical soluble rocks, provide a rich material basis for cave development. In regions with suitable climates, the number of natural caves developed through prolonged water erosion is vast, varying greatly in scale and size, and preserving numerous irreplaceable and non-renewable geological relics and cultural resources. Caves, as naturally formed underground spaces, hold significant research and utilization value. Therefore, the exploration and survey of caves are of great importance. However, cave development areas are often covered by dense vegetation, and traditional optical remote sensing images cannot penetrate this vegetation, making it difficult to locate the sudden structural changes at cave entrances from optical remote sensing images [2]. Ground surveys are also challenging to conduct due to the steep terrain and dense vegetation in mountainous areas.

With the advancement of Unmanned Aerial Vehicle (UAV) and LiDAR technology, new insights have been provided for cave detection. LiDAR technology measures the distance to a target object by emitting laser beams and receiving the reflected signals, thus generating high-precision 3D point cloud data. Compared to traditional optical remote sensing technologies, mounting LiDAR sensors on UAV platforms can penetrate vegetation to capture terrain information beneath the canopy [2], flexibly and efficiently obtaining high-resolution topographic data over large areas. This approach is particularly suitable for regions with complex terrains and dense vegetation and can operate around the clock [3], enabling the detection of concealed caves in densely vegetated karst areas. Furthermore, ground-based LiDAR scanning technology can capture the intricate structural information inside caves, constructing high-precision 3D models of caves, thereby providing a reliable basis for scientific research and spatial analysis of caves. Therefore, this paper aims to establish an integrated aerial and terrestrial technology method for the detection of concealed caves, combining UAV and LiDAR technologies.

2 Related Work

In practical scenarios, point cloud data often exhibit sudden distance changes, such as in the interspersed growth of trees of varying heights and undulating terrains. However, these features are distinctly different in spatial scale from cave targets. The most noticeable characteristic distinguishing cave targets from typical ground objects (like buildings) is their concave geometric structure [4]. Xin, W. et al. considering the unique geometric structure of cave-like targets, simulated point clouds for two typical cave targets, residential cave dwellings and natural caverns. They based their rapid target identification on sudden distance changes between laser points and the LiDAR sensor, exploring the feasibility of using UAV LiDAR technology for cave target recognition and extraction [4]. However, their point cloud data were computer-generated and did not involve real-world data collection or experimentation, which creates a certain discrepancy with actual conditions. Chunlei, X. et al. also noted that UAV LiDAR technology could play a significant role in micro-relief morphology surveys over large areas. It can effectively identify subsurface collapse structures in vegetated areas due to its strong vegetation penetration capability, effectively eliminating the influence of vegetation height on ground elevation measurements, and they studied methods for identifying the extent of surface collapses in mining areas [2]. However, their focus on identifying surface collapses in mining areas differs from the cave structures discussed in this paper. Ullman, M. et al. used UAV and LiDAR systems to scan Har Sifsof Cave (HSC) in northern Israel, collecting integrated point cloud data of the surface topography and the underground cave. The cave is accessed through a vertical shaft, with its entrance nearly invisible at the surface [5]. However, their focus was on comprehensive 3D mapping of the cave, without detailing the point cloud characteristics and identification capabilities of UAV-mounted LiDAR sensors for vertical cave entrances. Ferreira, C.F. et al. examined the capability of UAV-based DEM using automatic target detection algorithms to detect dolines in karst regions, but the results differed from those obtained with LiDAR data, as the sensors used were unable to accurately detect the subsurface morphology beneath vegetation [6]. In the realm of cave spatial analysis, Vassilakis, E. and colleagues utilized LiDAR technology to scan a cave in Koutouki, Peania, Greece, and calculated the complete volume of the cave using the elevation difference between the cave ceiling and the cave floor [7]. However, this paper focuses on the actual usable space within caves, intending to use an alternative method to represent cave volume to cater to specific developmental and utilization approaches of caves.

The distinctive feature of this paper lies in its utilization of UAV-mounted LiDAR sensors for accurately reflecting surface features under vegetation cover, pinpointing the location of unknown caves in karst regions under vegetative cover, and extracting and identifying the structural characteristics of cave entrances. Coupled with ground-based laser scanning, this approach establishes a cave 3D model with centimeter-level precision, offering new insights for the potential spatial utilization of caves. This methodology allows for precise mapping and modeling of cave structures, facilitating enhanced understanding and strategic planning for their use.

3 Study Area

China is one of the countries where karst development is most typical, with a karst distribution area of approximately 2.55 million km^2, accounting for 26.5% of the land surface area [1]. According to research data, the exposed carbonate rock regions in China have an estimated cave density of about 0.8 per km^2, with the total number of caves exceeding 500,000 [8], and many caves remain undiscovered. The southwestern region of China, with its extensive carbonate rock distribution and karst topography, features numerous karst caves and other karstic features. This paper selected three caves with distinct characteristics in this region for data collection and experimentation (Fig. 1).

The three caves, each with different developmental characteristics, were used for various experimental purposes. Cave 1 and Cave 2 were utilized for UAV LiDAR scanning experiments at oblique and vertical angles, respectively, to verify the capability to recognize cave entrance structures under vegetative cover. Cave 1 is developed on a hillside next to a road, with a small entrance that is difficult to detect from both ground and aerial views; Cave 2 is located in a mountain valley and is larger in scale, with its entrance covered by tall vegetation. Cave 3, a tourist cave with some development and utilization, was used for constructing 3D cave models and cave space analysis. Therefore, these three caves were selected for the experiments.

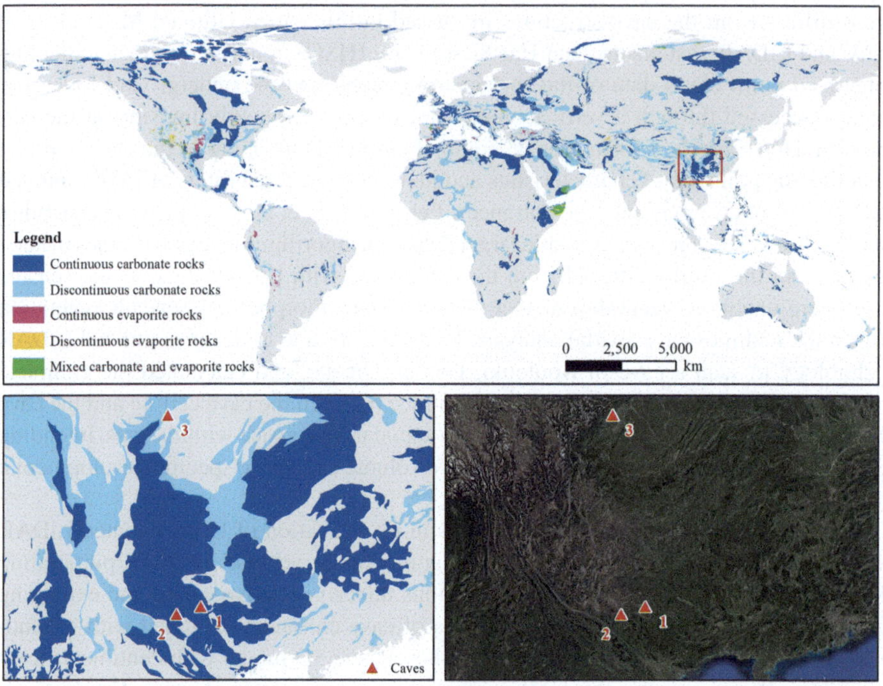

Fig. 1. Cave locations in the study area and Google imagery (modified from Chen, Z. [9]).

4 Methods

UAV and LiDAR technology, as emerging terrestrial observation tools, integrate 3D laser scanning, global positioning, and inertial navigation into a single system [10]. Their active imaging approach enables all-weather, continuous measurements. In this study, we used the DJI Matrice 300 RTK UAV, equipped with the Zenmuse L1 LiDAR sensor, to collect experimental data in karst areas (Fig. 2a). The Zenmuse L1 LiDAR has an effective scanning range of approximately 190 m to 450 m and is equipped with a 20-million-pixel, 1-inch sensor optical camera that captures color information, enabling the point cloud to be assigned true colors. Subsequently, LiDAR 360 point cloud processing software was used to extract surface depression structures under vegetative cover based on the point cloud profile, identifying them as potential cave locations, followed by field investigations and verification. For the collection of internal 3D point clouds within caves, the GVI LiGrip V100 handheld rotary laser scanner was used (Fig. 2b). This device, with an effective scanning range of 100 m, features a simple design, compact body, and convenient operation, making it particularly suitable for recording complex spatial 3D data of caves. The specific technical workflow is as follows (Fig. 3).

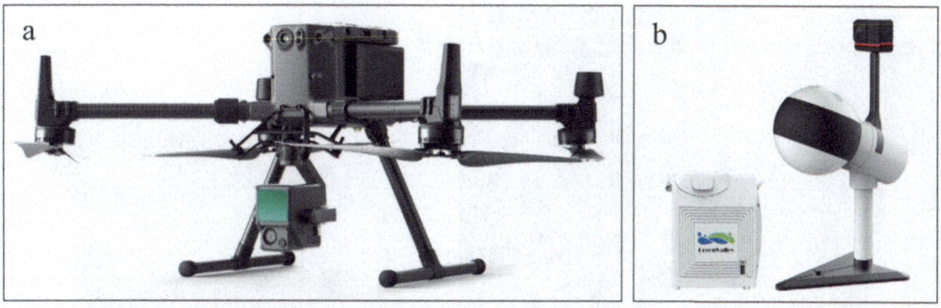

Fig. 2. DJI Matrice 300 RTK UAV (equipped with Zenmuse L1 LiDAR sensor) and GVI LiGrip V100 handheld rotary laser scanner.

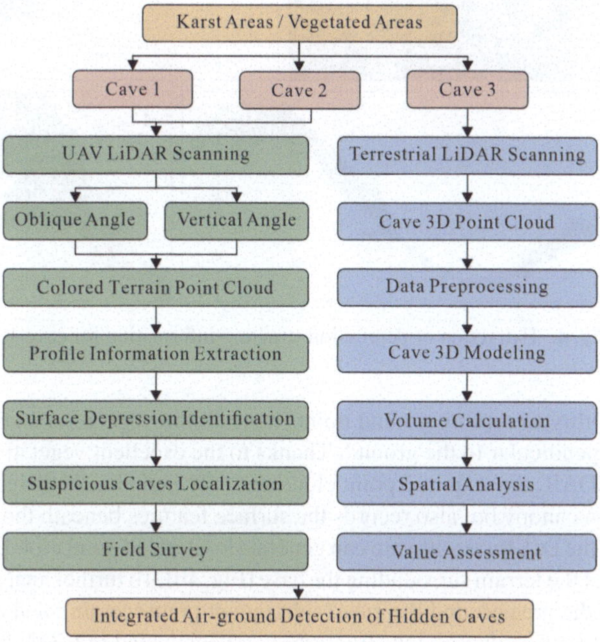

Fig. 3. Flowchart.

5 Data Acquisition, Processing and Results

5.1 Oblique Angle Scanning Experiment

The entrance of Cave 1 is approximately 4 m wide, 2.5 m high, and has a depth of 12 m, with stalactites developed inside. It is located on the eastern side of the road on a steeply sloped hillside, with dense vegetation covering the entrance. The author marked the cave's planar projection outline on orthoimagery, but the cave's features are completely obscured by vegetation, making it difficult to detect (Fig. 4a).

During data collection, the UAV flight altitude was set to 100 m, with the LiDAR sensor tilted at a 30-degree angle relative to the horizontal ground. This setup significantly

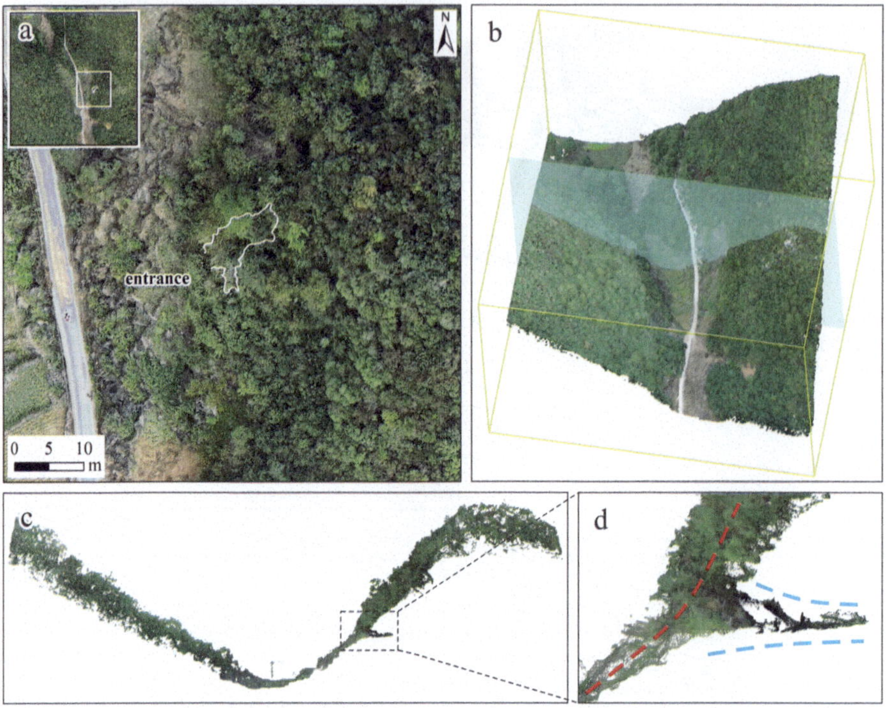

Fig. 4. Extraction of depression structure information for cave 1.

enhances the ability to capture ground points beneath the vegetation canopy compared to scanning perpendicular to the ground. Thanks to the excellent vegetation penetration capability of LiDAR, the acquired point cloud not only includes the height information of the vegetation canopy but also records the surface features beneath the canopy. After data collection, the DJI Terra software can generate a colorized point cloud, providing 3D visual features of the terrain surrounding the cave (Fig. 4b). To further analyze the surface morphology of the area around the cave, subsequent preprocessing and extraction and identification of surface depression structures were conducted in LiDAR 360 software. The point cloud preprocessing includes noise removal, filtering, and adjusting point cloud density. These steps are crucial for ensuring the accuracy of subsequent analyses.

To highlight the 3D topographic information recorded by the UAV LiDAR point cloud, a section of the point cloud profile is shown to display the features of the cave's depression structure (Fig. 4c). It is clearly visible that there is a distinct surface depression on the right-hand slope, where the point cloud not only successfully penetrated the surface vegetation but also reached into the subsurface space. Based on this, we inferred the presence of a cave at this location (Fig. 4d). A red dashed line indicates the normal surface morphology, and a blue dashed line marks the point cloud entering the cave interior.

Subsequently, through field investigation, a terrestrial 3D laser scanner was used to collect point cloud data of the cave interior, clearly recording the cave's complex morphological features (Fig. 5a), including the development of stalactites, and the cave entrance contour was extracted (Fig. 5b). This provides data support for further research on the cave. By observing at a larger spatial scale, the overall morphology and development direction of the cave can be identified, while at a smaller scale, the internal microstructures of the cave can be analyzed in greater detail. The development scale of this cave is relatively small, and the entire equipment setup and scanning process took no more than 1 h.

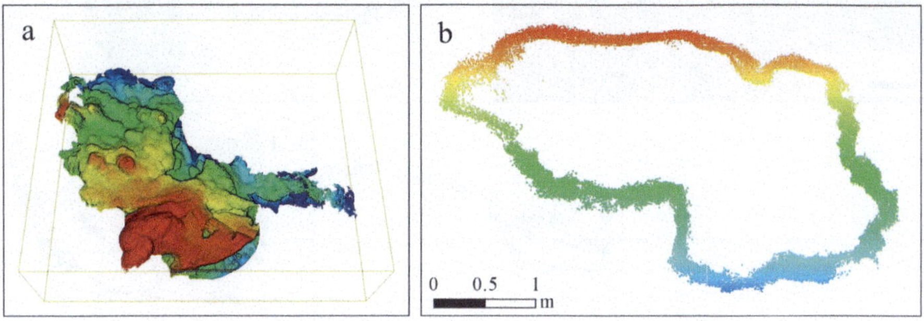

Fig. 5. 3D point cloud and entrance outline of cave 1.

5.2 Vertical Angle Scanning Experiment

Cave 2 is located within a mountain valley, with its entrance heavily covered by vegetation, providing strong concealment (Fig. 6a). The entrance is approximately 4 m wide and 10 m high, while the depth of the cave is unknown. An underground river flows through it, and the interior space of the cave is elongated, possessing natural attributes suitable for civil defense and storage. Due to the complex environment inside the cave and the lack of adequate lighting, for safety reasons, we only explored up to a depth of 60 m.

The UAV equipped with a LiDAR sensor was set to fly at an altitude of 80 m, scanning perpendicularly to the horizontal ground, and collected surface point cloud data (Fig. 6b). We employed the same method to extract features of surface depressions beneath the vegetation, which clearly revealed a significant surface depression at the slope of the mountain, where the point cloud penetrated the vegetation to reach the cave entrance (Fig. 6c). A red dashed line indicates the normal surface morphology, while a blue dashed line marks the point cloud at the depressed part of the cave entrance (Fig. 6d). However, since the cave extends horizontally and the LiDAR sensor scanned vertically, forming a 90-degree angle with the direction of the cave, the laser beams could not enter the cave interior. Although the vertical scan accurately captures the surface depressions and effectively identifies the entrance location, it has limitations in detecting the horizontal outline of cave entrances. This highlights that the scanning angle is a critical factor when detecting underground structures.

The 3D point cloud obtained from terrestrial laser scanning of the cave interior shows that Cave 2 has developed more typically (Fig. 7a). The cross-section of the cave

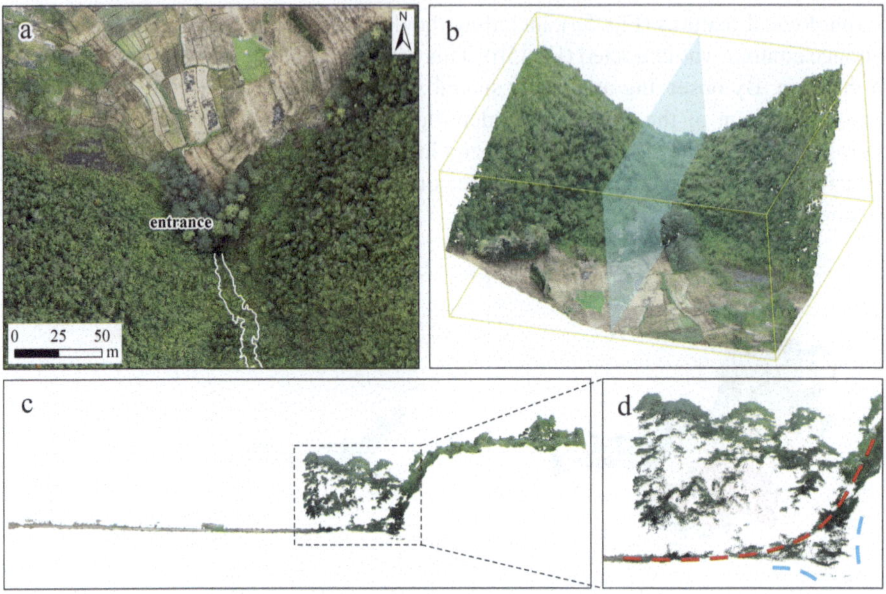

Fig. 6. Extraction of depression structure information for cave 2.

entrance is inclined (Fig. 7b). Due to the long-term erosive action of an underground river, Cave 2's depth and scale far exceed those of Cave 1., offering greater research value in terms of scientific study and spatial utilization. However, due to the complex environment inside the cave, a complete point cloud of the cave could not be collected.

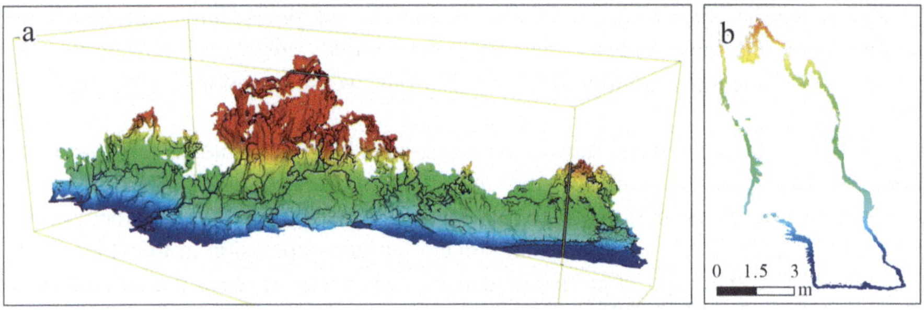

Fig. 7. 3D point cloud and entrance outline of cave 2.

5.3 Spatial Analysis

This paper focuses on the spatial utilization value of caves, which as naturally formed underground spaces, hold immense potential for applications such as civil defense and storage. In special circumstances, caves can be directly used as shelters or for other emergency purposes, thus avoiding the time and economic costs associated with constructing dedicated civil defense structures. This aspect of research is not only significant for the field of public safety but also offers a new perspective on the multifunctionality of karst caves. Caves as natural underground civil defense spaces possess the following distinctive characteristics:

(1) Structural stability: Caves typically form in soluble rocks, such as limestone or dolomite, which are generally very robust, providing good structural stability.
(2) Environmental stability: The temperature inside caves varies little, offering a natural "constant temperature" characteristic, and being underground, they are less affected by surface climate changes and human activities.
(3) Concealment: Caves are usually located underground, making them difficult to detect from the surface. Many cave entrances are small or hidden among cliffs, forests, or other natural terrain features, making them hard to find.

Cave 3 has an entrance approximately 10 m high and 7 m wide, with a depth of about 120 m (Fig. 8a). The erosional space exceeds 10,000 m^3, and at the end of the cave, there is a circular karst window formed by erosion and collapse, with well-developed features, including a rich landscape of stalactites, stalagmites, and columns. It has already been developed for tourism, with a high safety factor, facilitating comprehensive data collection. The author collected complete 3D point cloud data of Cave 3 using 3D laser scanning, and combined field photographs with point cloud data, constructing a centimeter-accurate real-scene 3D model of the cave (Fig. 8b), which visually displays the structural characteristics of the cave.

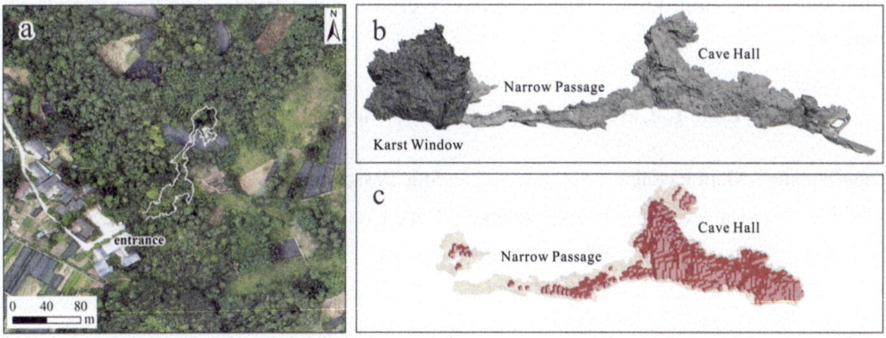

Fig. 8. 3D model of Cave 3 and cubic space volume calculation.

When planning the specific utilization of cave spaces for purposes such as military use, civil defense, or material storage, there is typically a preference for using areas with regular shapes. Due to the complexity of cave structures, such as narrow passages and irregular cave walls, the actual usable volume in certain areas is often limited. For example, a spacious cave hall may accommodate more people and equipment, whereas narrow passages restrict space utilization (Fig. 8c). Therefore, the usable space in a cave often constitutes only a portion of its total volume. When calculating cave volume, it is important to distinguish between the total volume and the actual usable space to provide more valuable reference information.

The method for calculating the usable volume of cave interiors is inspired by the concept of "voxels". A voxel, short for "volume pixel", is a method of representing data in 3D digital space, combining the notions of "volume" and "pixel". Similar to the resolution in a 2D image, a voxel can be considered a pixel in 3D space, where each voxel is a cube or rectangular prism in 3D space, serving as the basic unit of a 3D image or model [11]. Based on this concept, to make the calculation of usable volume more aligned with real-world conditions for cave space utilization, we propose the concept of "Cubic Space Volume", which involves calculating the actual usable volume by counting the number of regular-shaped cubes that can fit within the cave (Fig. 8c).

The calculation of the usable volume for the non-open-air portions of the cave, excluding the karst window at the cave's end, yielded a total volume of 7,797 m^3 using traditional methods. When applying the concept of "cubic space" and setting the cube size to a 0.5 m edge length, the number of cubes that can fit within the cave is 44,928, resulting in an effective usable volume of 5,616 m^3. When the cube size is set to a 1 m edge length, the number of cubes is 4,086, with an effective usable volume of 4,086 m^3. For a cube size of 1.5 m edge length, the number of cubes is 858, resulting in an effective usable volume of 2,896 m^3. Lastly, for a cube size of 2 m edge length, the number of cubes is 252, yielding an effective usable volume of 2,016 m^3 (Table 1). It is evident that as the edge length of the regular cube increases, the cave's effective usable volume significantly decreases. The difference between the total volume and the usable volume reflects the space lost due to the complex structure of the cave. Therefore, in future spatial planning and utilization, depending on the specific military or civil defense purposes, or the size of the stored materials, different "Cubic Space Volume" can be selected as a reference.

Table 1. Cave "Cubic Space Volume" calculation record table.

Cube/Square Edge Length (m)	Main Passage		Side Passage		Total Usable Volume (m^3)
	Cube Count	Usable Volume (m^3)	Cube Count	Usable Volume (m^3)	
0		7566		231	7797
0.5	44256	5532	672	84	5616
1	4065	4065	21	21	4086
1.5	856	2889	2	7	2896
2	252	2016	0	0	2016

6 Discussion

This study conducted experiments on the use of UAV-mounted LiDAR sensors for the detection of concealed caves under vegetative cover in karst regions. The LiDAR sensor demonstrated excellent penetration capability in surface mapping within densely vegetated areas. When identifying cave entrances, the fixed scanning angle of the LiDAR sensor used in this study was a limitation. The results showed that when the sensor's field of view was parallel to the direction of cave extension, it effectively identified surface depression structures. Therefore, the authors suggest that using a LiDAR sensor capable of 360-degree scanning would likely provide an optimal angle for clearly identifying surface depressions, thereby improving the detection capability of cave entrances. Future research could explore multi-angle scanning methods to obtain more comprehensive 3D point cloud data of surface cave entrances. Moreover, the point cloud density and intensity vary with the performance of different instruments. Higher-performance LiDAR sensors can produce higher-density point clouds, offering better vegetation canopy penetration and more efficient surface depression structure identification. In regions where cave development is typical and caves are densely distributed, it may be possible to study automated methods for extracting surface depression features or to explore real-time point cloud processing technologies to enhance detection efficiency [12].

The methodology provided in this study presents a complete technical process for cave exploration, mapping, 3D model construction, and value analysis. While the application cost may be high in areas where caves are sparsely distributed, it has significant practical value in regions with dense cave distributions or in other similar scenarios involving dense target applications.

7 Conclusions

This study innovatively applies UAV and LiDAR technology to the detection and identification of concealed caves under vegetative cover in karst regions. By integrating terrestrial 3D laser scanning, it achieves efficient detection and precise modeling of caves in vegetated areas, thereby enriching the research on remote sensing technology in cave exploration.

The identification of cave entrance features in vegetated areas using UAV LiDAR sensors has demonstrated both feasibility and effectiveness. The 3D laser point cloud can penetrate through gaps in the vegetation canopy and reach the cave interior. By analyzing the depression structures in the surface point cloud, it is possible to locate concealed caves that are difficult to detect with the naked eye, providing a new approach for cave exploration and detection. Caves, as natural underground spaces, offer significant value for civil defense and storage due to their stable structures and environments. With simple modifications, they can be effectively utilized, greatly reducing the cost of constructing such facilities. The concept of "cubic space volume" proposed in this paper offers a more practical reference for the spatial utilization of caves. Terrestrial 3D laser scanners have shown outstanding performance in mapping the internal structures of caves, enabling the construction of centimeter-precision 3D models of cave spaces. Their high precision and convenience make them ideal tools for cave research.

Funding. This research was funded by the Science and Technology Innovation Foundation of Command Center of Integrated Natural Resources Survey (KC20240004, KC20230006).

References

1. Goldscheider, N., et al.: Global distribution of carbonate rocks and karst water resources. Hydrogeol. J. **28**, 1661–1677 (2020)
2. Chunlei, X., Zhaocheng, G., Xiongwei, Z., Shengwei, L., Boxuan, S.: Typical applications of airborne LiDAR technique in geological investigation. Remote Sens. Nat. Resour. **28**, 136–143 (2016)
3. Panghe, Q., Yuliang, T., Chunhui, W., Ruimin, F.: Application and development of laser altimetry in satellite surveying and mapping. Spacecraft Recov. Remote Sens. **44**, 102–111 (2023)
4. Xin, W., Sheng, L., Tao, L.: Air borne LiDAR point cloud simulation and fast recognition of cave target. J. Geomat. Sci. Technol. **31**, 278–282 (2014)
5. Ullman, M., et al.: Formation processes and spatial patterning in a late prehistoric complex cave in northern Israel informed by SLAM-based LiDAR. J. Archaeol. Sci. Rep. **47**, 103745 (2023)
6. Ferreira, C.F., Hussain, Y., Uagoda, R., Silva, T.C., Cicerelli, R.E.: UAV-based doline mapping in Brazilian karst: a cave heritage protection reconnaissance. Open Geosci. **15** (2023)
7. Vassilakis, E., Konsolaki, A.: Quantification of cave geomorphological characteristics based on multi source point cloud data interoperability. Z. Geomorphol. **63**, 265–277 (2022)
8. Weihai, C.: An outline of speleology research progress. Geol. Rev. **52**, 783–792 (2006)
9. Chen, Z., et al.: World Karst Aquifer Map (WHYMAP WOKAM). BGR, IAH, KIT, UNESCO (2017)
10. Kovanič, Ľ., Topitzer, B., Peťovský, P., Blišťan, P., Gergeľová, M.B., Blišťanová, M.: Review of photogrammetric and Lidar applications of UAV. Appl. Sci. **13** (2023)
11. Lee, B.: A study of 3D model voxelization method for artificial intelligence learning. J. Auton. Intell. **7** (2023)
12. Chao, L., Chuanrong, L., Jian, H., Wenjing, H., Jingyu, N., Fanrong, M.: Application of EDMA3 in real-time LiDAR point cloud calculation. Spacecraft Recov. Remote Sens. **37**, 57–65 (2016)

Geoinformatics for Sustainable Ecosystems and Society

Key Technology and Engineering Application of PolSAR Data Processing for China C/S/L Band SAR Satellites

Lili Yan(✉)

China Centre for Resources Satellite Data and Application, Haidian District, Fengxian East Road 5, Beijing, China
yanlili116@163.com

Abstract. In recent years, the China SAR (Synthetic Aperture Radar, SAR) satellites, including Gaofen-3 satellite (GF3), Environmental disaster reduction 2E satellite (HJ2E) and Land Exploration 1A satellite (LT1A), covering three different frequency bands (C-band, S-band, L-band), were launched. They all have polarimetric SAR imaging mode, which can apply to ecological civilization construction and sustainable development. To form standard PolSAR (Polarimetric Synthetic Aperture Radar, PolSAR) application products according to the characteristics of domestic polarized SAR satellites and apply them to natural environment monitoring and disaster detecting is of great significance. This paper mainly resolves key technology of PolSAR data processing, explores a set of flow generation technology of domestic polarized SAR satellite products (PolSAR classification and PolSAR orthorectified products), and verifies the accuracy of the products in engineering application. An automatic PolSAR data processing system of China C/S/L band SAR satellites is constructed. Considering polarization characteristics of three different bands satellites, PolSAR orthorectified product generation and land classification mapping under rare control points were carried out on three domestic SAR satellites in three study areas, and the corresponding accuracy was evaluated. The engineering application results show that the orthoimage rectification accuracy of China PolSAR satellites in the three study areas is 1.5 pixel, 1.9 pixel and 1.7 pixel respectively, all of which are better than 2 pixel. The overall accuracy of land classification mapping obtained is 85.25%, 80.15% and 87.02%, respectively. The Kappa coefficients are 0.75, 0.72 and 0.81, which can meet the application accuracy of natural environment monitoring and disaster detecting. It provides a scientific reference for the further application and research of domestic PolSAR satellites.

Keywords: Polarimetric SAR · PolSAR orthorectified product · PolSAR classification mapping

1 Introduction

Synthetic Aperture Radar (SAR) actively emits electromagnetic waves and receives signals reflected from terrestrial targets. Its electromagnetic wave penetration properties allow for the acquisition of ground data unaffected by weather, clouds, and other factors.

G. Luo (Ed.): GFSISD 2024, SPPHY 417, pp. 29–44, 2025.
https://doi.org/10.1007/978-981-96-2490-4_3

This fills the data gap issues encountered by optical remote sensing satellites under adverse conditions, serving as a crucial foundation for Earth observation [1–3].With the development and rise of Polarimetric Synthetic Aperture Radar (PolSAR) [4, 5], utilizing polarization characteristics for land classification has become a significant technical approach. By using the four distinct polarization modes shared by full polarimetric data to detect deep subsurface objects, polarization information is extracted based on the varying scattering characteristics of ground objects for the verification of land objects classification, which has been widely applied in various fields [6–8].

For a long time, the SAR satellites data used have relied heavily on foreign countries. Compared with mature SAR satellite data from abroad, the application capabilities of domestically produced polarimetric SAR satellites in radiation calibration, geometric correction, and polarization information utilization urgently need to be improved [9, 10]. Many domestic scholars have conducted varying degrees of SAR remote sensing observation and application demonstration research around relevant areas requirements using domestic satellite data, but most of them are based on single technology applications and have not systematically completed the production process application of SAR satellite products [11–15]. This article uses fully polarized data from domestic SAR satellites, including GF3 C-band satellite, HJ2E S-band satellite, and LT1A L-band satellite, to evaluate the geometric accuracy of polarimetric SAR satellites in rare control points. And the systematic application technology process of domestic polarimetric SAR satellite product production is explored using image classification methods for land classification research. Overall, it improves the technology and application level of polarimetric SAR data for domestic SAR satellites, and provides reference for promoting the in-depth application and research of domestic polarimetric SAR satellites in various fields such as resources and disasters.

2 Characteristics of Domestic SAR Satellite Data

2.1 Radiation Calibration

GF3 is the first domestically manufactured C-band multi polarization SAR satellite, and its radiometric calibration accuracy meets the strict requirements for geophysical parameter measurement [11]. Subsequently, domestically manufactured SAR satellites HJ2E in the S band and LT1A in L band, which were launched successively, were also able to meet the calibration accuracy requirements in terms of radiation quality [16, 17]. But compared with foreign countries, foreign radiation calibration technology is mature, and the calibration accuracy can reach 0.4 dB, even up to 0.2 dB [11]. At the same time, obtaining images with a large time span and inconsistent radiation information requires image dodging and uniform color processing to achieve radiation consistency.

2.2 Geometric Positioning

After the launch of GF3, the positioning error was summarized to verify that its positioning accuracy on system level after calibration can reach about 3 m [12, 18]. The launch time of HJ2E satellite is still short, and there is relatively few research on its geometric positioning accuracy. LT1A is the first domestically launched L-band fully polarized SAR satellite, with a direct positioning accuracy of 10 m and an indirect positioning accuracy of 2.4 pixels in its geometric positioning accuracy evaluation [17], but related research is still few.

Based on the radiation and geometric characteristics of domestic SAR satellites data, this paper evaluates the geometric positioning accuracy of GF3 satellite, HJ2E satellite, and LT1A satellite, produces orthoimage product, and develops application.

3 Key Technology

The domestic SAR satellite products are systematically generated using fully polarized Single Look Complex (SLC) images from GF3, HJ2E, and LT1A satellite. The key technical process flow is shown in Fig. 1.

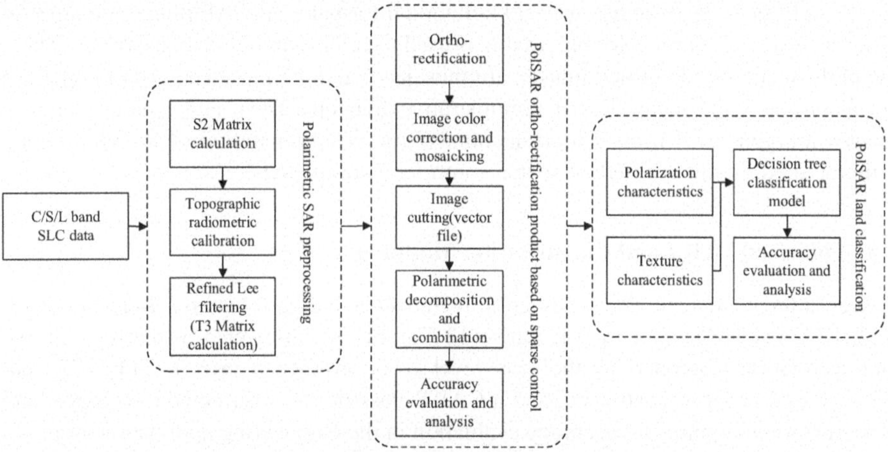

Fig. 1. Technical flow chart.

3.1 Polarization SAR Preprocessing

Firstly, the SAR raw data is processed by calculating the Sinclair matrix (S2 matrix). Then the terrain radiometric correction method for mountainous areas [19] is applied to complete the terrain radiometric correction process. Calculate the polarization coherence matrix (T3 matrix) of the S2 image after radiometric correction, and then perform Refined Lee filtering on the T3 matrix to complete the preprocessing of polarimetric SAR.

3.2 PolSAR Image Orthorectified Product Generation Based on Sparse Control

SAR adopts a side view imaging technology in slant range direction, and the original SAR images obtained are often affected by terrain undulations, resulting in geometric distortions such as perspective shrinkage, overlay, and shadows [19]. In order to facilitate the interpretation and measurement of land features, it is necessary to eliminate these geometric distortions present in the original SAR images through orthorectification. The geometric positioning model is the foundation of image orthorectification, and the classic range Doppler (R-D) model [20–22] was used as the geometric positioning model in the application. However, in a wide area with complex terrain, it is difficult to obtain control points, so very few control points are needed to achieve geometric calibration of the image. This paper achieves orthorectification under rare control points, thereby realizing the generation of large-scale SAR orthorectified products.

The engineering application uses the basic base map and elevation data as the correction basis. Rare and obvious feature corresponding ground points evenly distributed on the image to be corrected and the basic base map are selected as the correction control points. After the residual of the correction control points and the residual of the registration control points meet the tolerance requirements, the original SAR image can be orthorectified using the R-D geometric positioning model.

To address the issue of inconsistent radiation information in SAR images obtained at different times, uniform color processing is used to achieve radiation consistency. Then, the multi scene orthorectified images are mosaiced, and the mosaiced orthoimages are cut according to the vector files of administrative divisions. The pseudo color composite images are generated using Cloud decomposition. The generation of PolSAR image orthorectified products based on sparse control are completed.

3.3 Polarized SAR Land Classification Mapping

Based on the T3 matrix, the polarization features (including 6 features T11, T12, T13, T22, T23, and T33) and texture features (12 features including energy, inertia, entropy, and correlation generated by the gray level co-occurrence matrix in T11, T22, and T33) of land cover are extracted, and a CART classifier is constructed to achieve land classification mapping and accuracy evaluation in the engineering application area.

CART (Classification And Regression Tree), also known as Classification and Regression Tree [23], was proposed by Breiman in 1984 as a decision tree algorithm. It is a binary tree structure with a training dataset consisting of test variables and target variables for cyclic analysis [24, 25].

As CART is a typical supervised classification algorithm, it is necessary to determine the classification threshold based on high-quality training samples before constructing the decision tree [24]. The training sample structure of CART decision tree is shown in Eq. (1):

$$\begin{gathered} L := \{X_1, X_2, \ldots, X_m, Y\}, \\ X_1 : (x_1, x_2, \ldots x_{1n_1}), \ldots X_m = (x_{m1}, x_{m2}, \ldots x_{mn1}) \\ Y = (y_1, y_2, \ldots y_t) \end{gathered} \tag{1}$$

In the formula, X and Y are the attribute vector and label vector, respectively. The decision tree is a classification tree or a regression tree based on whether the vector value Y is discrete data or continuous data. This study conducted engineering applications on land cover classification, therefore the model was constructed using a classification decision tree.

The classification tree uses the Gini Index minimum criterion for feature selection to generate a binary tree, to determine the optimal attribute features and filter threshold. A Gini value equal to 0 indicates that the data types are consistent, and the larger it is, the greater the uncertainty of the sample data. The Gini Index formula is as follows:

$$Gini = 1 - \sum_{i} p_i^2 \tag{2}$$

In the formula, p_i represents the ratio of the sample size to the total sample size in a certain category, with smaller Gini indicating uniform sample distribution and larger Gini indicating uneven distribution.

3.4 Accuracy Evaluation

Apply existing accuracy indicators to evaluate the accuracy of the generated PolSAR orthoimage products and PolSAR land classification products.

(1) Accuracy evaluation index for orthorectification

Residual calculation: Calculate the directional difference between the known value and the corrected value.

Mean square error of point position:

$$m = \sqrt{\Delta X^2 + \Delta Y^2} \tag{3}$$

In the formula, $\triangle X$ and $\triangle Y$ represent the residuals in the X and Y directions, respectively.

Mean square error:

$$\Delta m = \sqrt{\frac{\sum_{i=1}^{n} (\Delta X)^2}{n}} \quad \Delta m = \sqrt{\frac{\sum_{i=1}^{n} \Delta Y^2}{n}} \tag{4}$$

In the formula, $\triangle m$ is mean square error, and n is the number of control points.

(2) Accuracy evaluation index for classification

The evaluation indicators for classification results can quantify the level of algorithm performance, among which the Confusion Matrix is a commonly used accuracy verification method for classification. It uses the land cover category truth value and classification value to solve and calculate [26, 27]. The expression form of the Confusion Matrix is shown in Table 1. The main accuracy evaluation indicators used are Overall Accuracy (OA) and Kappa coefficient.

TP in the table refers to correctly classified pixels; FP refers to pixels that belong to the same category but are not recognized; FN refers to misclassified pixels; TN refers to

Table 1. Confusion matrix.

Confusion matrix		Predictive value	
		Positive	Negative
True value	Positive	TP	FN
	Negative	FP	TN

pixels that do not belong to and are not classified into that category. If the total number of samples is N, then

$$N = TP + FP + FN + TN \tag{5}$$

Overall accuracy (OA): The ratio of the total number of correctly classified pixels to the total number of samples, as shown in Eq. (6).

$$OA = \frac{TP + TN}{TP + TN + FN + FP} \tag{6}$$

Kappa coefficient: It is used to measure the accuracy of classification, which is an accuracy indicator for evaluating classification quality based on confusion matrix. Its value range is −1 to 1. The closer it is to 1, the better the classification effect is. Its calculation is shown in Eq. (7).

$$Kappa = \frac{p_o - p_e}{1 - p_e} \tag{7}$$

In this formula,

$$P_{\mathrm{o}} = \frac{TN + TP}{N} = OA \tag{8}$$

$$P_{\mathrm{e}} = \frac{\sum_{i=1}^{c} a_i * b_i}{n^2} \tag{9}$$

Among them, a_i represents the number of correctly classified samples for each class, b_i represents the number of predicted samples for each class, and n is the total number of samples.

4 Data in Engineering Application

The data used in engineering application is the C-band GF3 satellite with a spatial resolution of 8 m, the S-band HJ2E satellite with a spatial resolution of 5 m, and the L-band LT1A satellite with a spatial resolution of 6 m. The polarization mode is the fully polarized SLC image as the data source. Changfeng County in Anhui Province is selected as the research area of GF3 satellite. Longyou County in Zhejiang Province is selected as the research area of HJ2E satellite. Deqing County in Zhejiang Province is selected as the research area of LT1A satellite. The above three study areas are shown in Fig. 2.

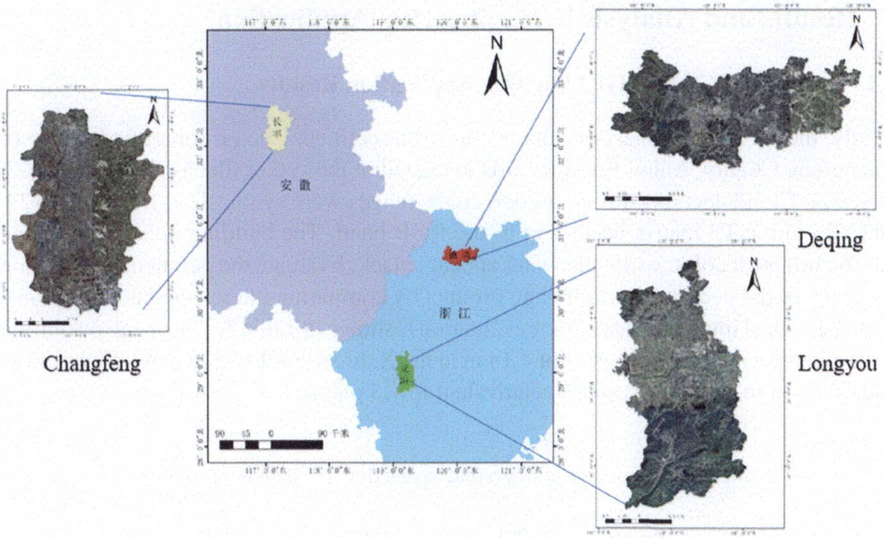

Fig. 2. Study area.

Fig. 3. Orthoimage of Changfeng.

5 Results and Analysis in Engineering Application

5.1 Analysis of C-band GF3 Satellite Application Results

Firstly, the GF3 fully polarized image was orthorectified. The orthorectified image of Changfeng County, Anhui Province was extracted in the vector file, as shown in Fig. 3. Based on Cloud decomposition, pseudo color images were synthesized from T11, T22, and T33 in the T3 matrix according to the RGB band. The building area in the image has the brightest color, while the water appears black. Evaluate the geometric correction accuracy of the generated orthoimage product by comparing the selected control points with the optical image. The accuracy evaluation is shown in Table 2. The image resolution is 10 m, with a mean square error of 4.14 m in the X direction, 14.35 m in the Y direction, and 14.94 m in the point position, equivalent to 1.5 pixels.

Table 2. Accuracy of orthographic correction in Changfeng (unit: m)

Points	Rectified coordinates X	Rectified coordinates Y	Known coordinates X	Known coordinates Y	DX	DY	DS
1	117.260717464	32.169070188	117.260717464	32.168842452	0.000	25.327	25.327
2	117.233869704	32.184271704	117.233849001	32.184147484	1.949	13.815	13.952
3	117.093758895	32.02229693	117.093692644	32.022197554	6.237	11.052	12.69
4	117.143483838	31.970310364	117.143533526	31.970260676	−4.678	5.526	7.24
5	117.254193959	32.04826758	117.25426021	32.048151642	−6.237	12.894	14.323
6	117.087797339	32.162506028	117.087813902	32.162373527	−1.559	14.736	14.818
7	117.081286564	32.202459396	117.081299814	32.202326895	−1.247	14.736	14.789
8	117.144655049	32.261452146	117.144628549	32.261359395	2.495	10.315	10.613
9	117.22308889	32.327879109	117.22308889	32.32769692	0.000	20.262	20.262
10	117.232000171	32.390862404	117.231983608	32.390812716	1.559	5.526	5.742
11	117.154021912	32.4564937	117.154088163	32.456394324	−6.237	11.052	12.69
12	117.338538469	31.986320873	117.338588157	31.986188372	−4.678	14.736	15.46
13	117.289139133	32.01593521	117.289155696	32.015753021	−1.559	20.262	20.322
14	117.198169541	32.315527006	117.198086728	32.315527006	7.796	0.000	7.796
Mean square error of X		4.14					
Mean square error of Y		14.35					
Mean square error of Point		14.94					

Then, the CART decision tree was used to classify the orthorectified images of Changfeng County into four categories: water, forest, building, and farmland. The PolSAR classification results are shown in Fig. 4. From the orthoimage, it can be observed that the building area corresponds to the brightest color area, while the water body corresponds to the black area. The accuracy of the decision tree classification results was evaluated by constructing a confusion matrix, as shown in Table 3. According to the data analysis in the table, the classification accuracy of buildings and water bodies is high, while forest land and farmland have similar scattering characteristics and weak C-band penetration, resulting in lower classification accuracy. The overall accuracy of decision tree classification reaches 85.25%, with a Kappa coefficient of 0.75, which can meet the needs of land classification in engineering application.

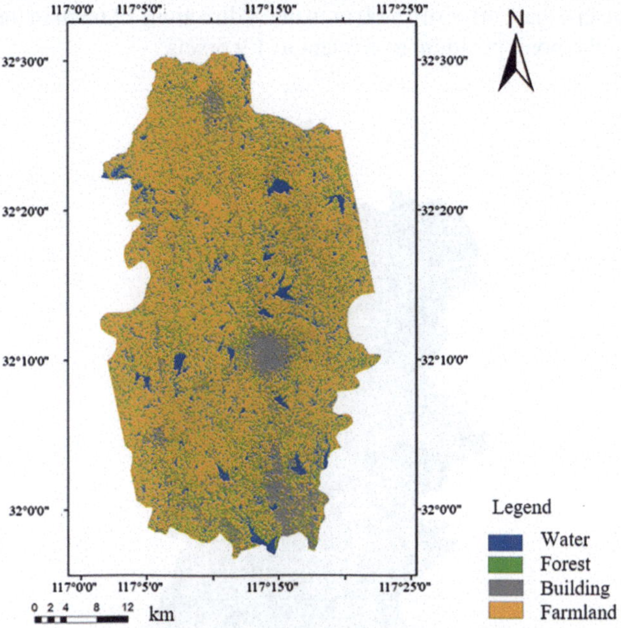

Fig. 4. PolSAR classification results in Changfeng.

Table 3. PolSAR classification accuracy of Changfeng.

Categories	Water	Forest	Building	Farmland
Water	99.86	0.28	0.22	0.35
Forest	0	72.17	1.93	7.33
Building	0.01	3.95	91.54	7.14
Farmland	0.13	23.6	6.31	85.18
Overall	100	100	100	100
Overall Accuracy	85.25%			
Kappa coefficient	0.75			

5.2 Analysis of S-band HJ2E Satellite Application Results

Similar to the GF3 decision tree classification engineering application, the HJ2E satellite was used for land classification in Longyou County. Firstly, the orthoimages of Longyou County were processed, and pseudo color images were synthesized as shown in Fig. 5. It can be seen that the terrain south and north of Longyou County is complex with multiple mountains and hills, while the central region is flat. The accuracy evaluation of orthorectified images in Longyou County is shown in Table 4. The image resolution is 10 m, with a mean square error of 15.00 m in the X direction, 12.63 m in the Y direction, and 19.61 m in the point position, equivalent to 1.9 pixels.

Fig. 5. Orthoimage of Longyou.

The S-band classification results of Longyou County were obtained through CART decision tree classification, as shown in Fig. 6. Forest land is mainly distributed in the southern and northern areas of Longyou County, while building areas and farmland are mostly located in the central flat areas, which is consistent with the classification results. Longyou County has complex terrain, and the PolSAR classification accuracy is shown in Table 5. The overall classification accuracy OA reaches 80.16%, and the Kappa coefficient is 0.72. The S-band HJ2E satellite can meet the classification requirements in areas with complex terrain.

Table 4. Accuracy of orthographic correction in Longyou (unit: m)

Points	Rectified coordinates X	Rectified coordinates Y	Known coordinates X	Known coordinates Y	DX	DY	DS
1	119.0564065	28.98304135	119.0566948	28.98303693	−28.053	0.491	28.058
2	119.0884884	28.9864117	119.0886058	28.9865113	−11.42	−11.08	15.911
3	119.0907012	29.04117011	119.0905298	29.0412573	16.675	−9.699	19.291
4	119.0687514	29.10109298	119.0687345	29.10121157	1.642	−13.192	13.294
5	119.1005679	29.13155936	119.1005221	29.13173018	4.463	−19.001	19.518
6	119.1370586	29.17937451	119.1374234	29.17931797	−35.498	6.289	36.05
7	119.2919613	29.26874427	119.2918075	29.26876381	14.964	−2.174	15.121
8	119.3314922	29.20839296	119.331456	29.20849501	3.525	−11.351	11.886
9	119.2810674	29.17923689	119.2811278	29.17935419	−5.875	−13.047	14.309
10	119.1882482	29.10103848	119.1881933	29.10117461	5.338	−15.142	16.056
11	119.1832691	29.04558402	119.1832716	29.04570231	−0.242	−13.158	13.16
12	119.2927293	29.05988194	119.2928292	29.06004927	−9.72	−18.613	20.998
13	119.2917903	29.03485601	119.2917163	29.03499251	7.193	−15.183	16.8
Mean square error of X		15.00					
Mean square error of Y		12.63					
Mean square error of Point		19.61					

5.3 Analysis of L-band LT1A Satellite Application Results

Firstly, obtain the L-band orthorectified image of Deqing County, and synthesize a pseudo color image similar to the above engineering applications, as shown in Fig. 7. The terrain in the western part of Deqing County is mountainous, gradually transitioning from west to east to plain terrain. The L-band has strong penetration characteristics and clear differentiation of land cover categories. The accuracy of orthorectification has been verified, as shown in Table 6. The analysis accuracy evaluation table shows that the image resolution is 10 m, with a mean square error of 14.64 m in the X direction, 9.15 m in the Y direction, and 17.26 m in the point position, equivalent to 1.7 pixels.

The PolSAR classification of the L-band Deqing County image is shown in Fig. 8. A portion of the water body is incorrectly identified as forest land due to the complex terrain in the western region, which is mostly mountainous. The SAR side view imaging characteristics make it prone to geometric distortion and differences in scattering mechanisms in mountainous areas, resulting in the water body being identified as forest land. The PolSAR classification accuracy is shown in Table 7, with an overall classification accuracy OA of 87.02% and a Kappa coefficient of 0.81. The L-band LT1A satellite can meet the classification requirements in complex terrain areas and achieve good classification mapping products.

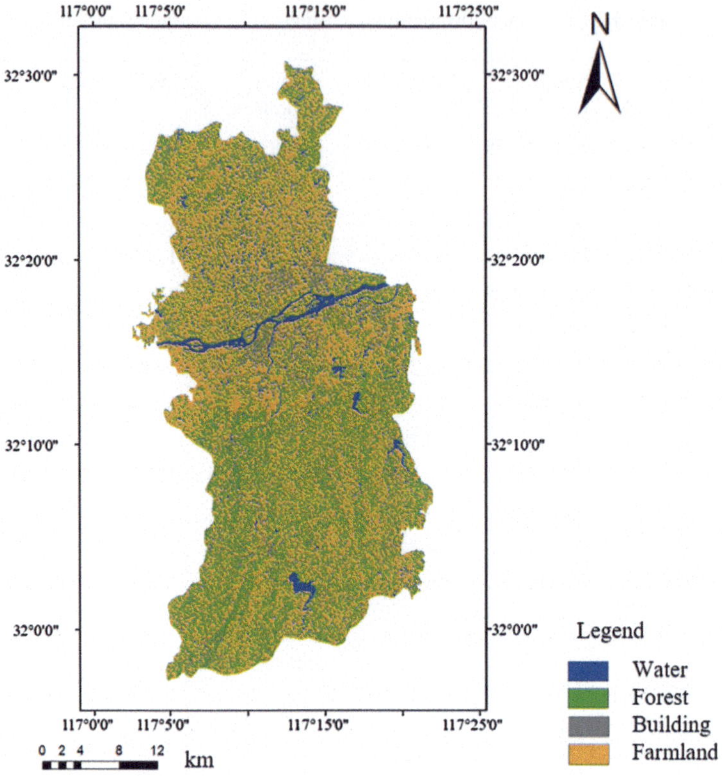

Fig. 6. PolSAR classification results in Longyou.

Table 5. PolSAR classification accuracy of Longyou.

Categories	Water	Building	Forest	Farmland
Water	98.96	0.20	0.06	0.61
Building	0.10	59.91	9.23	4.81
Forest	0.16	11.99	61.34	10.93
Farmland	0.78	27.91	29.38	83.65
Overall	100	100	100	100
Overall accuracy	80.15%			
Kappa coefficient	0.72			

Fig. 7. Orthoimage of Deqing.

Table 6. Accuracy of orthographic correction in Deqing (unit: m)

Points	Rectified coordinates X	Rectified coordinates Y	Known coordinates X	Known coordinates Y	DX	DY	DS
1	119.7834857	30.47926556	119.7832877	30.47930113	18.978	−3.956	19.386
2	119.8070565	30.45901062	119.8069308	30.45904466	12.043	−3.786	12.624
3	119.931239	30.4879235	119.9310829	30.48814637	14.967	−24.788	28.956
4	119.9848804	30.51438648	119.9848066	30.51434117	7.071	5.04	8.683
5	119.9509747	30.54944172	119.9508861	30.5495064	8.489	−7.194	11.127
6	120.0451075	30.62694314	120.0448516	30.6268708	24.529	8.046	25.815
7	120.0962598	30.55697933	120.0961342	30.55701365	12.042	−3.818	12.633
8	120.1262082	30.68784421	120.1260623	30.68781639	13.987	3.094	14.325
9	120.1816088	30.66800518	120.1813905	30.6679792	20.928	2.89	21.127
10	120.197857	30.54770478	120.1976762	30.54770922	17.336	−0.494	17.343
11	120.1652031	30.45699367	120.1651128	30.45699586	8.661	−0.244	8.664
12	120.1362395	30.50892705	120.1361511	30.50899185	8.473	−7.207	11.123
13	120.2219083	30.45929743	120.2219667	30.45942131	−5.591	−13.779	14.87
14	120.2762437	30.62254029	120.2760658	30.62263485	17.058	−10.518	20.039
Mean square error of X		14.64					
Mean square error of Y		9.15					
Mean square error of Point		17.26					

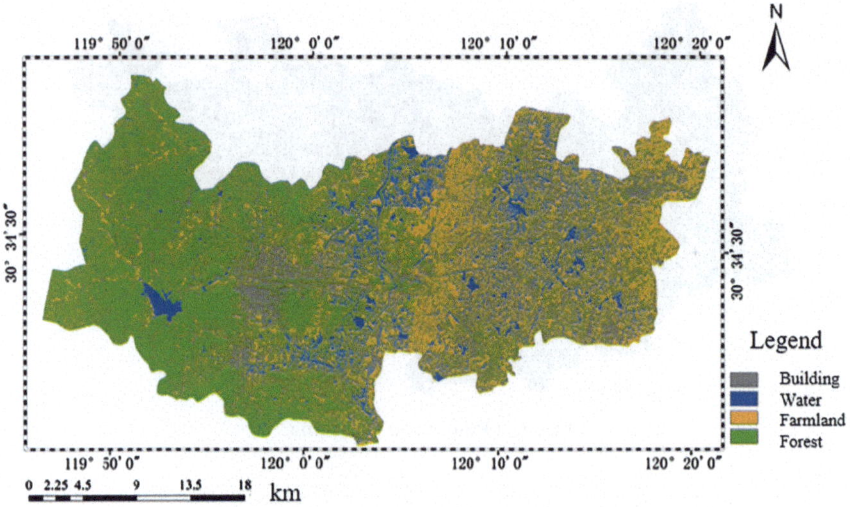

Fig. 8. PolSAR classification results in Deqing.

Table 7. PolSAR classification accuracy of Deqing.

Categories	Water	Building	Forest	Farmland
Water	77.05	0.00	0.00	0.00
Building	0.03	96.36	0.06	1.55
Forest	22.92	0.00	99.16	9.54
Farmland	0.00	3.64	0.77	88.91
Overall	100.00	100.00	100.00	100.00
Overall accuracy	87.02%			
Kappa coefficient	0.814			

6 Conclusion

The C-band GF3 satellite, the S-band HJ2E satellite, and the L-band LT1A satellite are three domestically manufactured polarimetric SAR remote sensing satellites. This paper presents key technology of domestic C/S/L band polarized SAR satellites products (PolSAR classification and PolSAR orthorectified products), explores an automatic PolSAR data processing system of China C/S/L band SAR satellites, and applies in engineering experiments. The results and theoretical analysis in engineering application are as follows:

(1) The correction accuracy of the PolSAR orthoimage products generated by three domestically manufactured polarimetric SAR satellites in different frequency bands under rare control points is 1.5 pixels for the C-band in Changfeng County, 1.9 pixels for the S-band in Longyou County, and 1.7 pixels for the L-band in Deqing County.

All three satellites can meet the positioning accuracy requirement of 2 pixels for orthoimage products.

(2) Applying the CART decision tree classification algorithm to land classification research under three polarimetric SAR satellites, the PolSAR classification mapping accuracy in the three county areas can reach 80% level, and the effective classification results with Kappa coefficients greater than 0.7 can meet the application accuracy of land classification related work.

(3) The three domestic PolSAR satellites data can meet the accuracy requirements of PolSAR classification mapping and PolSAR orthorectified product in engineering application. The systematic and automatic processing system of China C/S/L band PolSAR satellites is conducive to improve the application of domestic PolSAR satellites in ecological civilization construction, sustainable development, resources, environment, and to promote the research and development of domestically manufactured PolSAR satellites.

References

1. Zhang, T., Wang, G., Qi, J., Dai, H.: Application practice of satellite remote sensing service in national space planning and supervision. Satell. Appl. **9**, 33–37 (2022)
2. Lee, H., Liu, J.G.: Analysis of topographic decorrelation in SAR interferometry using ratio coherence imagery. IEEE Trans. Geosci. Remote Sens. **39**(2), 223–232 (2001)
3. Zhao, C., Liu, X., Gao, Y., et al.: Early identification of high-elevation geohazards in the lower Yarlung Zangbo River based on the SAR/InSAR technology. J. Ceomech. **28**(6), 981–994 (2022)
4. Cloude, S.R., Papathanassiou, K.P.: Polarimetric SAR interferometry. IEEE Trans. Geosci. Remote Sens. **36**(5), 1551–1565 (1998)
5. Cloude, S.R.: Polarization Applications in Remote Sensing. Oxford University Press, New York (2009)
6. Yang, J., Liu, L.: Feature extraction of spatial and temporal change of land in functional areas based on weighted full-polarization SAR image classification. Bull. Survey. Mapp. **7**, 138–142 (2022)
7. Zhao, J.: Application of Land Classification based on Polarization Characteristics of Domestic L-SAR Satellite. China University of Mining and Technology, Xuzhou (2023)
8. Hua, W., Wang, C., Sun, N., et al.: Multi-scale contrastive learning method for PolSAR image classification. J. Appl. Remote Sens. **18**(1), 014502 (2024)
9. Zhang, Q., Liu, Y.: Overview of Chinese first C band multi-polarization SAR satellite GF-3. Aerospace **3**, 22–31 (2017)
10. Wan, J., Zang, J., Liu, S.: Fusion and classification of SAR and optical image with consideration of polarization characteristics. Acta Opt. Sin. **37**(6), 292–301 (2017)
11. Deng, M., Chen, Z., Zhang, G., et al.: Verification of absolute radiometric calibration accuracy of domestic C-band satellite. Chin. Space Sci. Technol. (2024)
12. Li, X., Jiang, B., Wang, T., et al.: Geometric positioning accuracy improvement method without ground control points for global orthorectification of GF-3. Acta Geodaetica et Cartographica Sinica **52**(11), 1929–1940 (2023)
13. Li, T., Tang, X., Li, S., et al.: Classification of basic deformation products of L-band differential interferometric SAR satellite. Acta Geodaetica et Cartographica Sinica **52**(5), 769–779 (2023)
14. Liu, X., Liu, L., Shao, Y., et al.: Water detection in urban areas from GF-3. Sensors **18**(4), 1299 (2018)

15. Li, J., Wang, Y., Ma, W., et al.: Research on cloud removal based on combination of domestic HISEA-1 SAR and optical image. Remote Sens. Inf. **39**(1), 66–72 (2024)
16. Chen, J., Wang, G., Wang, D., et al.: Overview of environmental disaster reduction No. 2 05 satellite. Space Int. **12**, 33–36 (2022)
17. Cai, S., Zhang, Y., Kang, Y., et al.: Preliminary evaluation of the LT-1 geolocation accuracy and radiation characteristics. Sci. Survey. Mapp. **49**(2), 165–174 (2024)
18. Zhang, W., Zhang, B., Liu, R., et al.: A five parameter transformation model for GF-3 SAR image precise geometric correction. Sci. Technol. Eng. **23**(20), 8726–8734 (2023)
19. Zhang, C., Wang, R., Zhao, C., et al.: Study on terrain radiant correction of high spatial resolution image of mountainous area. J. Zhejiang Univ. (Agric. Life Sci.) **32**(6), 693–698 (2006)
20. Lebrl, F.: Radargrammetry for Image Interpretation. International Institute for Aerial Survey and Earth Sciences (ITC) (1978)
21. Zhao, B., Qi, X., Song, H., et al.: An accurate range model based on the fourth-order Doppler parameters for geosynchronous SAR. IEEE Geosci. Remote Sens. Lett. **11**(1), 205–209 (2013)
22. Dai, C., Deng, M., Dai, M., et al.: Geometric calibration of helicopter-borne wide-angle SAR imagery based on the range Doppler model. Remote Sens. Lett. **13**(10), 1040–1047 (2022)
23. Utkin, L.V., Kovalev, M.S., Coolen, F.P.A.: Imprecise weighted extensions of random forests for classification and regression. Appl. Soft Comput. **92**, 106324 (2020)
24. Zhang, J., Liu, N., Song, W., et al.: A fully polarimetric radar image classification method based on feature selection. Sci. Survey. Mapp. **47**(6), 127–134 (2022)
25. Zhang, J., Wang, J., Hu, W., et al.: Research on land use classification based on object-oriented CART decision tree. Geospat. Inf. **21**(1), 113–118 (2023)
26. Ji, H., Ren, Z., Rao, Z.: Discriminant analysis of millet from different origins based on hyperspectral imaging technology. Spectrosc. Spect. Anal. **39**(7), 2271–2277 (2019)
27. West, R.D., LaBruyere, T.E., III., Skryzalin, J., et al.: Polarimetric SAR image terrain classification. IEEE J. Sel. Top. Appl. Earth Observ. Remote Sens. **12**(11), 4467–4485 (2019)

Spatiotemporal Analysis Methods Based on Ground-Based Intelligent Multi-parameter Environmental Health Remote Sensing Diagnostic Instrument

Yu Zhang[1,2], Chunxiang Cao[1,2](✉), Min Xu[1,2], and Shaohua Wang[1,2]

[1] State Key Laboratory of Remote Sensing Science, Aerospace Information Research Institute, Chinese Academy of Sciences, Beijing 100101, China
caocx@aircas.ac.cn
[2] University of Chinese Academy of Sciences, Beijing 100094, China

Abstract. This research addresses key challenges in spatial information mining, particularly related to the complexities and heterogeneity of geospatial data generated by diverse sensors and platforms. To enhance data collection and analysis, we introduce the Ground-Based Intelligent Multi-Parameter Environmental Health Remote Sensing Diagnostic Instrument (GB-EHRSDI). By leveraging Swin Transformer for cloud segmentation and PatchTST for time-series forecasting, we utilize edge computing to process multi-dimensional environmental data efficiently. This approach significantly improves the correlation analysis between cloud cover, rainfall, cloud motion, wind speed, and vertical cloud distribution, advancing weather prediction capabilities. Additionally, the GB-EHRSDI highlights the current limitations in Internet of Things (IoT)-based big data analysis techniques and underscores the need for future work to integrate more efficient data acquisition methods, IoT architecture, and Artificial Intelligence (AI) technologies to further enhance localized weather forecasting and environmental monitoring.

Keywords: Spatial Information Mining · Geospatial Data · Environmental Monitoring · GB-EHRSDI

1 Introduction

Spatial information mining plays a crucial role in fields such as environmental monitoring, urban planning, and disaster prediction, where geospatial data are utilized to identify patterns and perform in-depth analysis [1, 2]. However, current methods encounter significant challenges in processing the increasingly complex, heterogeneous, and dynamic nature of spatial data, which are generated by diverse sensors, platforms, and observation techniques [3]. These challenges become especially prominent when attempting to correlate phenomena such as cloud cover and rainfall, cloud motion and wind speed/direction, as well as acquiring cloud vertical distribution with meteorological conditions. Moreover, the lack of efficient methods to acquire and integrate multi-dimensional data for big data analysis further complicates efforts to improve the reliability of weather predictions.

G. Luo (Ed.): GFSISD 2024, SPPHY 417, pp. 45–56, 2025.
https://doi.org/10.1007/978-981-96-2490-4_4

Understanding the relationship between cloud cover and meteorological variables such as rainfall, temperature, and humidity is complicated by the intricate interactions among atmospheric components. For instance, empirical studies show that variations in cloud cover directly influence surface temperature and humidity, particularly in tropical regions, where cloud formation and dissipation play a significant role in local climate regulation [4]. Moreover, research has indicated that increases in cloud cover are often accompanied by changes in rainfall patterns, especially at lower altitudes, where cloud-rainfall correlations exhibit a significant relationship [5].

Similarly, tracking cloud motion and accurately correlating it with wind speed and direction remain a challenge, particularly across various altitudes and geographic locations. Studies have shown that cloud motion is highly dependent on wind patterns, and deviations in wind speed can affect the spatial distribution of clouds [6]. Furthermore, vertical cloud distribution, essential for understanding atmospheric stratification and meteorological processes, is often poorly represented in traditional models, making it difficult to predict weather patterns based on cloud dynamics alone [7]. These limitations hinder precise environmental monitoring and localized weather forecasting, both of which are critical for disaster prevention and mitigation.

The advent of the Internet of Things (IoT) has helped address some of these challenges by enabling real-time data processing through edge computing, allowing for quicker response times in weather forecasting and disaster prevention. IoT systems also facilitate large-scale analysis and contribute to comprehensive big data frameworks for climate prediction by transmitting environmental data to cloud platforms [8]. However, current IoT systems often lack the comprehensive multi-altitude and multi-dimensional capabilities necessary for fully integrating cloud cover, motion, and vertical distribution data. This limitation hampers the development of more accurate localized weather forecasting and highlights the need for more advanced and integrated data systems to enhance predictive accuracy [9, 10].

To address these challenges, this study introduces the **Ground-Based Intelligent Multi-parameter Environmental Health Remote Sensing Diagnostic Instrument** (GB-EHRSDI). Tested in diverse locations, including Beijing, the Daihai Basin, and Lhasa, the GB-EHRSDI gathers comprehensive multi-altitude data. With advanced algorithms such as semantic segmentation and time-series models, it enables precise tracking and analysis of cloud cover, motion, and vertical distribution.

The primary contributions of this study are as follows:

(1) Introducing a ground-based intelligent remote sensing platform capable of collecting and processing multi-component environmental data.
(2) Demonstrating the efficacy of the system across multiple geographical locations for reliable meteorological monitoring and cloud vertical distribution.
(3) Validating the method's effectiveness through experiments on cloud-meteorological analysis, cloud motion-wind correlations, and localized weather forecasting.

2 Methodology

The purpose of this study is to utilize the GB-EHRSDI for meteorological monitoring and environmental data analysis. GB-EHRSDI is an end-to-end solution fully developed in-house, which is based on the IoT architecture and equipped with edge computing

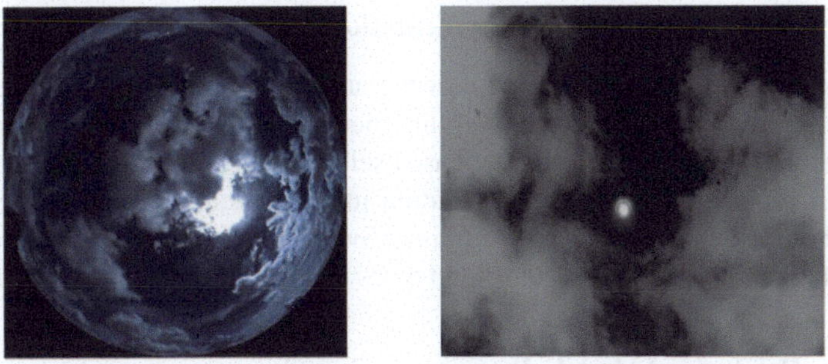

Fig. 1. Infrared and Visible Light All-Sky Cloud Image

capabilities. The instrument integrates multiple sensors, to enable comprehensive real-time data collection, processing, and display.

Table 1. Parameters of GB-EHRSDI

Parameter	Range	Resolution	Parameter	Range	Resolution
Temperature	−40 °C to 85 °C	0.1 °C	NH_3	0–100 ppm	1 ppm
Humidity	0–100% RH	0.001	H_2S	0–100 ppm	1 ppm
Wind Speed	0–60 m/s	0.01 m/s	CO	0–1000 ppm	1 ppm
Wind Direction	0–360°	0.1°	O_3	0–10 ppm	0.1 ppm
P_atm	300–1100 hPa	0.1 hPa	SO_2	0–20 ppm	0.1 ppm
Rainfall	0–200 mm/h	0.2 mm	NO_2	0–20 ppm	0.1 ppm
Noise	0–150 dB	0.1 dB	eCO_2	400–5000 ppm	1 ppm
UV Sensor	0–200 W/m^2	0.1 W/m^2	TVOC	0–50000 $\mu g/m^3$	50 $\mu g/m^3$
PM2.5	0–1000 $\mu g/m^3$	15 $\mu g/m^3$	HCHO	0–2000 $\mu g/m^3$	1 $\mu g/m^3$
PM1.0	0–1000 $\mu g/m^3$	15 $\mu g/m^3$	CO_2	0–2000 ppm	1 ppm
PM10	0–1000 $\mu g/m^3$	15 $\mu g/m^3$	Negative Ions	0–500,000/cm^3	20 ions/cm^3

The GB-EHRSDI system integrates infrared and visible light all-sky cameras with 22 sensors (shown in Table 1) to capture detailed cloud and meteorological data continuously across various altitudes and times of the day (shown in Fig. 1). Its core innovation lies in combining edge computing with IoT, enabling real-time, localized data processing and analysis close to the source, which is crucial for time-sensitive applications like weather forecasting. Simultaneously, large datasets are transmitted to cloud platforms for big data analysis, supporting long-term monitoring of cloud cover, motion, and vertical distribution over wide geographical areas. The system also incorporates a web-based display interface, enabling multi-scale spatiotemporal information to be displayed in real-time across various locations.

2.1 Swin Transformer for Cloud Segmentation

We use Swin Transformer [11], a state-of-the-art transformer model, for segmenting cloud cover from the images and classifying them into different cloud types. Its hierarchical feature extraction and window-based self-attention mechanism make it highly effective for processing large, high-resolution images, such as those from Sentinel-2 or other satellite sources. We use a pre-trained Swin Transformer model based on ImageNet for segmenting and classifying cloud types from all-sky visible and infrared cloud images.

Loading the Pre-trained Model. We select the Swin Transformer model pre-trained on the ImageNet-1K dataset. For this work, we choose the Swin-B (Base) variant due to its balance of accuracy and computational efficiency. This model serves as the foundation for our cloud segmentation task.

Data Preprocessing. The all-sky cloud images, which include both visible light and infrared data, are first resized to 224×224 pixels to fit the model's input requirements. Following this, the pixel values are normalized to ensure consistency with the pre-trained model's expectations (mean = [0.485, 0.456, 0.406], std = [0.229, 0.224, 0.225]).

Fine-Tuning the Model. Instead of retraining the entire model, we apply fine-tuning by freezing most of the pre-trained weights and only updating the parameters in the final classification layers. This allows us to efficiently adapt the model to the specific task of cloud segmentation, where it learns to identify cloud types such as cirrus and cumulus from the all-sky data.

Inference and Results. Once fine-tuned, the model is used to generate pixel-level segmentation maps for cloud types. These results are further analyzed in three key aspects:

Vertical Cloud Distribution. By utilizing the Swin Transformer to segment and classify both cloud cover and cloud types across various altitudes, the system provides a comprehensive view of the vertical distribution of clouds. This analysis reveals patterns in atmospheric stratification, offering key insights into how cloud classifications and cloud cover change with altitude.

Cloud-Meteorological Correlation. A regression analysis is conducted to examine the relationship between segmented cloud cover and key meteorological variables, including rainfall, temperature, and humidity. This method quantifies how cloud cover affects these parameters over time, particularly across different altitudes.

Cloud Motion and Wind Speed. We apply cross-correlation to extract cloud motion trajectories. The relationship between cloud motion and wind speed/direction is then assessed by calculating the Pearson correlation coefficient, providing valuable insights into the dynamics of cloud movement and its correlation with wind patterns.

2.2 PatchTST for Weather Forecasting

In addition to segmentation and correlation analysis, the system uses the PatchTST [12] model to forecast weather-related parameters such as temperature, rainfall, and wind speed. By analyzing historical time-series data collected from sensors, the model predicts future conditions, to help to generate localized weather forecasts. PatchTST is specifically designed for multi-step and multi-scale time-series forecasting using a sliding window approach.

Model Configuration and Data Preprocessing. We employ the PatchTST-Large (PatchTST-L) model, pre-trained on the ETDataset, to forecast weather-related parameters such as temperature, rainfall, and wind speed. This model consists of 4 transformer layers with a hidden size of 512 and 8 attention heads. It uses a patch length of 16, which divides the input time series into patches of 16 time steps, capturing both short-term and long-term temporal dependencies. The historical data from our sensors, including temperature, rainfall, and wind speed, are normalized using min-max scaling. A sliding window approach is applied, where 96 time steps (representing 96 h) are used to predict the next 24 time steps (24 h into the future).

Fine-Tuning Procedure. To adapt the pre-trained model to our specific dataset, we fine-tune the PatchTST-L model by freezing most of the pre-trained layers and training only the final fully connected layers. This allows the model to retain its general forecasting capabilities while learning patterns specific to our localized data. The fine-tuning is conducted over 50 epochs with a learning rate of $1e{-4}$, ensuring efficient convergence without overfitting.

Ranger Method. To further enhance model performance, we incorporate the Ranger optimizer. By introducing different sequences and observing prediction accuracy, we ensure that the model better generalize to unseen data, leading to more robust forecasting results.

Forecasting and Evaluation. Once fine-tuned, the model is used to generate multi-step forecasts for key parameters such as temperature, rainfall, and wind speed. The model's performance is evaluated using mean squared error (MSE) and mean absolute error (MAE). It achieves an MSE of 0.015 for rainfall prediction and an MAE of 0.07 for wind speed, demonstrating its accuracy in forecasting future weather conditions.

Note that, although the Ranger method enhances forecasting accuracy in PatchTST for weather parameters, its application in cloud segmentation tasks, such as those performed with Swin Transformer, has limited direct impact. The two models serve different purposes: PatchTST focuses on time-series weather forecasting, while Swin Transformer excels in cloud segmentation and related atmospheric analyses. Thus, we approach these tasks independently to better align with their respective goals.

3 Experiment

This section outlines the experimental design used to analyze all-weather atmospheric data using two different models: Swin Transformer for cloud segmentation and PatchTST for weather forecasting. The experiments are based on 24-h continuous data collected from multiple altitudes in Beijing, Daihai, and Lhasa. The experiments are designed to demonstrate the independent performance of each model in its respective task.

3.1 Swin Transformer Cloud Segmentation Experiment

Data Collection and Preprocess. The dataset used for this experiment consists of 24-h continuous infrared and visible light all-sky cloud images. Images are captured at 5-min intervals, resulting in 288 images per day for each modality (visible and infrared). This dataset covers varying times of the day, providing a full temporal and spatial representation of cloud behavior. The images are resized to both 224×224 pixels to match the input size of the Swin Transformer model. Normalization is applied, scaling the RGB channels of visible light images to a range of [0, 1], using ImageNet normalization mean (mean = [0.485, 0.456, 0.406]) and standard deviation (std = [0.229, 0.224, 0.225]). Similarly, infrared images are normalized based on their respective channel statistics. To improve generalization, random horizontal flips and color jittering are applied as data augmentation techniques during training.

Implementation Details. We use a pre-trained Swin Transformer Base model with 12 transformer blocks, pre-trained on the ImageNet dataset. For segmentation tasks, the final classification head is replaced with a multi-class segmentation head. The first 8 layers of the model are frozen, allowing only the last 4 layers to be fine-tuned using our cloud dataset. The training process is conducted with a batch size of 16 for 50 epochs, using the AdamW optimizer with a learning rate of 1×10^{-4}, which decays using a cosine annealing schedule.

The cross-entropy loss function is employed for pixel-wise classification. The loss for a single pixel i is defined as:

$$\text{CrossEntropy}(p_i, y_i) = -\sum_{c=1}^{C} y_i^{(c)} \log p_i^{(c)}, \tag{1}$$

where $p_i^{(c)}$ is the predicted probability of pixel i belonging to class c, $y_i^{(c)}$ is the ground truth (1 if the pixel belongs to class c, otherwise 0), and C is the total number of classes. Class weighting is applied to prioritize underrepresented cloud types (e.g., cirrus clouds) to address class imbalance.

Accuracy Assessment. The accuracy of the Swin Transformer is evaluated using Intersection over Union (IoU) and F1-Score, assessing the overlap between predicted segmentation masks and ground truth. The IoU is defined as:

$$\text{IoU} = \frac{TP}{TP + FP + FN}, \tag{2}$$

where TP, FP, and FN represent the number of true positives, false positives, and false negatives, respectively. F1-Score is calculated as:

$$\text{F1 - Score} = 2 \times \frac{\text{Precision} \times \text{Recall}}{\text{Precision} + \text{Recall}}. \tag{3}$$

Precision and Recall are defined as:

$$\text{Precision} = \frac{TP}{TP + FP}, \quad \text{Recall} = \frac{TP}{TP + FN}, \tag{4}$$

where *TN* represent the number of true negatives. Additionally, Classification Accuracy is introduced to evaluate the overall ratio of correctly predicted observations:

$$\text{Accuracy} = \frac{TP + TN}{TP + TN + FP + FN}. \tag{5}$$

The model's ability to segment various cloud types (e.g., cirrus, cumulus, stratus) is assessed both quantitatively and visually. Additionally, the relationship between segmented cloud coverage and rainfall patterns, as well as cloud motion vs. wind speed, is analyzed using graphical representations.

3.2 PatchTST Weather Forecasting Experiment

Data Collection and Preprocess. The meteorological data for this experiment are collected over a 24-h period from Beijing, Daihai, and Lhasa. Each location record key parameters at 5-min intervals, resulting in 288 data points per day for each parameter, including temperature, rainfall, wind speed, wind direction, humidity, and pressure. Each parameter is normalized to a range of [0, 1] using min-max scaling, according to the following formula:

$$X_{\text{norm}} = \frac{X - X_{\text{min}}}{X_{\text{max}} - X_{\text{min}}}, \tag{6}$$

where X_{norm} is the normalized value, X is the raw sensor data, X_{min} and X_{max} are the minimum and maximum values of the dataset. A sliding window approach is used to structure the data, where 96 time steps (8 h) are used to predict the next 24 time steps (2 h). The dataset is split into training (70%), validation (15%), and test (15%) sets.

Implementation Details. We utilize a PatchTST-Large model with 4 transformer layers, each with a hidden size of 512 and 8 attention heads, pre-trained on the ETDataset. The first 3 layers are frozen, and only the final layer is fine-tuned using our meteorological dataset. The training is conducted over 50 epochs with a batch size of 32 and a learning rate of 1×10^{-4}. A cosine annealing schedule is applied to adjust the learning rate dynamically. The model is optimized using the mean squared error (MSE) loss function:

$$\text{MSE} = \frac{1}{n}\sum_{i=1}^{n}(y_i - \hat{y}_i)^2, \tag{7}$$

where y_i represents the actual value and $\hat{y}_i$ is the predicted value. We use the Ranger optimizer, which combines RAdam and Lookahead, to enhance convergence and accuracy, especially with noisy time-series data.

Accuracy Assessment. The forecasting accuracy of the PatchTST model is evaluated using MSE and Mean Absolute Error (MAE)

$$\text{MAE} = \frac{1}{n}\sum_{i=1}^{n}|y_i - \hat{y}_i|. \tag{8}$$

MSE is used to penalize larger errors more heavily, making it suitable for evaluating fluctuations in weather parameters. MAE provides an interpretable measure of average prediction error. Additionally, the performance of the model with the AdamW optimizer is used to enhance its performance on forecasting stability and accuracy.

4 Results

From left to right, Fig. 2 shows segmented examples of Cirrus, Cumulus, Hybrid, and Stratus clouds, generated using the Swin Transformer model. These results demonstrate the Swin Transformer model's superior ability to differentiate between various cloud types. The model excels in segmenting fine details of cloud formations, ensuring a high degree of accuracy in classifying cloud types. This segmentation and classification enable more effective calculations of cloud cover and cloud dynamics, as illustrated by the metrics in Table 2, where key performance indicators such as IoU, F1 Score, Precision, and Recall further confirm the model's strong performance.

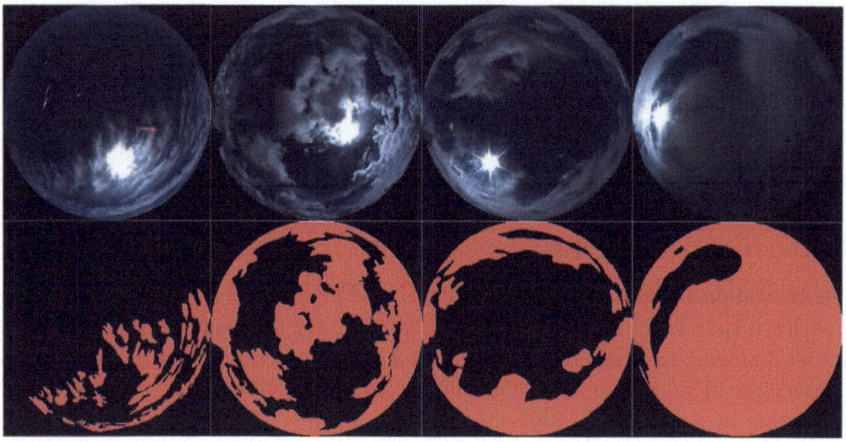

Fig. 2. Semantic Segmentation Results of Cloud Types

As shown in Fig. 3, the segmented and classified cloud cover data reveals significant variations in the vertical distribution of clouds across different altitudes. Lhasa is dominated by cumulus and stratus clouds, exhibiting a clear pattern typical of high-altitude regions. As the altitude decreases, cirrus clouds become more frequent in Daihai and Beijing, reflecting the changing atmospheric conditions. In terms of cloud cover, Lhasa has the greatest cloud amount, followed by Daihai, with Beijing showing the least cloud cover overall.

The regression analysis presented in Fig. 4 demonstrates key correlations between cloud cover and meteorological parameters. A positive correlation is observed between cloud cover and rainfall, where increased cloud cover generally leads to higher rainfall. Conversely, a negative correlation is found between cloud cover and temperature,

Table 2. Quantitative Evaluation of Cloud Segmentation and Classification Performance

Cloud Type	IoU (%)	F1-Score (%)	Precision (%)	Recall (%)	Classification Accuracy (%)
Cirrus	65.0	77.6	79.2	76.1	85.0
Cumulus	68.5	80.5	81.7	79.5	87.5
Hybrid	63.0	75.8	77.0	74.7	82.0
Stratus	66.8	78.9	80.3	77.5	84.0

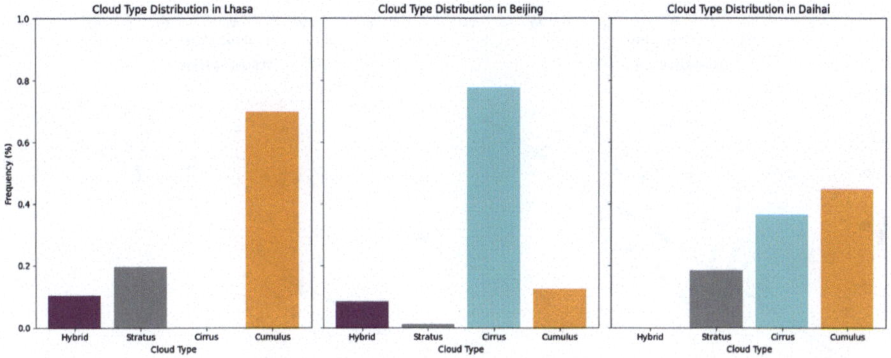

Fig. 3. Vertical Distribution of Cloud Types

indicating that higher cloud cover tends to reduce temperatures, likely due to the obstruction of solar radiation. Additionally, a positive correlation exists between cloud cover and humidity, as increased cloud cover corresponds to rising humidity levels. Lastly, the analysis reveals a positive relationship between rainfall and humidity, with higher humidity often coinciding with precipitation events.

The extracted cloud motion vectors and wind vectors are visualized in Fig. 5, offering a clear representation of the correlation between these two parameters. The analysis of the Pearson correlation coefficient between cloud motions and wind speed yields a value of 0.71, indicating a strong positive correlation. This suggests that both wind speed and wind direction significantly influence cloud motions. As wind speed increases, cloud motions tend to follow closely, with cloud direction aligning with wind direction.

Finally, we present the prediction performance of PatchTST for key meteorological parameters in Beijing in Table 3. The model demonstrates strong predictive capabilities, as reflected in the low MSE and MAE values across various parameters. Specifically, the model shows excellent performance in predicting humidity and temperature, with the lowest MSE values of 0.062 and 0.067, respectively. The prediction of wind speed and wind direction also performs well, with MSE values of 0.071 and 0.088, indicating reliable forecasting accuracy for cloud and wind dynamics.

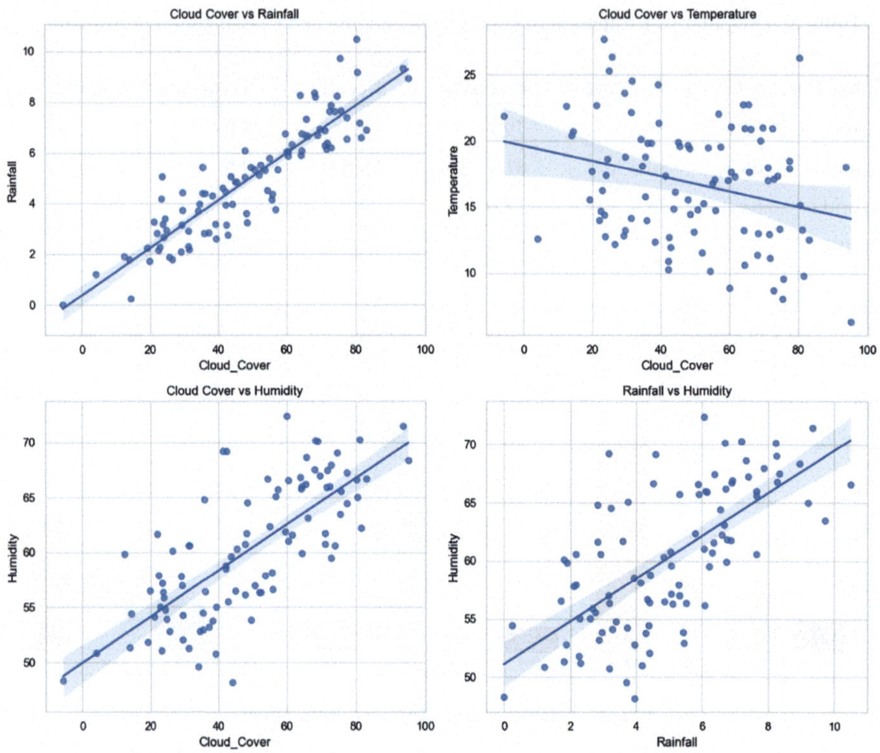

Fig. 4. Correlation Between Cloud Cover and Meteorological Parameters

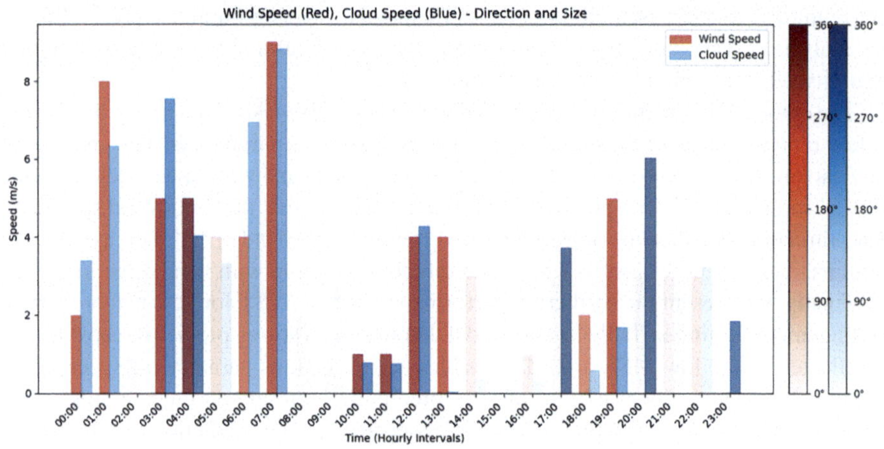

Fig. 5. Correlation Between Cloud and Wind Motions

Table 3. Quantitative Evaluation of PatchTST Prediction Performance

Metrics	Rainfall	Temperature	Humidity	Wind Speed	Wind Direction
MSE	0.083	0.067	0.062	0.071	0.088
MAE	0.048	0.041	0.039	0.045	0.051

5 Discussion

Additionally, the distinct cloud distribution patterns observed in Lhasa and Daihai reflect the influence of geographical features on cloud formation. The high-altitude environment of Lhasa, surrounded by mountains, promotes the formation of stratus clouds, which tend to dominate lower and middle atmospheric layers [13]. In contrast, Daihai, though also encircled by mountains, exhibits a higher prevalence of cirrus clouds at greater altitudes. These results suggest that terrain and elevation play significant roles in shaping the vertical distribution of clouds.

The analysis demonstrates clear correlations between cloud cover and key meteorological factors. Increased cloud cover is associated with higher rainfall and humidity, indicating that denser clouds contribute to greater atmospheric moisture and precipitation. Additionally, a negative correlation between cloud cover and temperature shows that clouds act as insulators, reducing surface heating by blocking sunlight. These findings underscore the role of cloud cover in influencing local weather patterns and highlight its predictive value for meteorological forecasting. Furthermore, wind speed and direction are well-established factors in cloud dynamics, particularly affecting cloud movement and the organization of cloud systems [14]. The data from the GB-EHRSDI system supports this by providing empirical evidence that wind parameters significantly influence cloud motion, reinforcing their importance in meteorological modeling and forecasting.

The strong performance of PatchTST in predicting rainfall, temperature, and other meteorological parameters across diverse regions demonstrates its potential for real-world applications. Its low prediction error rates highlight its robustness as a forecasting tool. Deployed within an IoT-based multi-station setup, PatchTST enables real-time localized weather forecasts, which can greatly aid agricultural planning and disaster prevention by providing accurate, site-specific predictions. This system can enhance response times and decision-making for weather-dependent industries, particularly in rural and agricultural sectors.

6 Conclusion

In conclusion, this study demonstrates the effectiveness of GB-EHRSDI in capturing, analyzing, and interpreting cloud dynamics and related meteorological data across diverse geographical regions. By integrating advanced models like Swin Transformer for cloud segmentation and PatchTST for time-series forecasting, we have successfully identified significant correlations between cloud cover and key meteorological parameters, including rainfall, temperature, and humidity. Additionally, our findings confirm the strong influence of wind speed and direction on cloud movement, providing valuable insights for localized weather forecasting. The robust performance of PatchTST in predicting meteorological conditions across regions such as Lhasa, Beijing, and Daihai, coupled with its low error rates, reinforces its potential for real-world applications. The IoT-based deployment of GB-EHRSDI allows for real-time localized weather forecasting, which can be particularly beneficial for agricultural planning and disaster prevention in weather-dependent industries.

Looking forward, the integration of IoT-based systems with cloud computing, satellite remote sensing technologies, and large-scale AI models will be a key focus. This will enable more comprehensive and accurate weather forecasting, enhancing decision-making and response times across various sectors. This system represents a significant step towards advancing environmental monitoring and meteorological prediction in both rural and urban settings.

References

1. Han, J., Kamber, M., Tung, A.: Spatial clustering methods in data mining: a survey. In: Geographic Data Mining and Knowledge Discovery, Research Monographs in GIS, Taylor and Francis, Abingdon (2001)
2. Chawla, S., Shekhar, S., Wu, W., Ozesmi, U.: Modeling spatial dependencies for mining geospatial data. In: Proceedings of the 2001 SIAM International Conference on Data Mining (2001)
3. Weare, B.C., Mokhov, I.I.: Evaluation of total cloudiness and its variability in the Atmospheric Model Intercomparison Project. J. Clim. **8**, 2224–2238 (1995)
4. Zhang, M.H., Cess, R.D., Xie, S.C.: Relationship between cloud radiative forcing and sea surface temperatures over the entire tropical oceans. J. Clim. **9**, 1374–1384 (1996)
5. Sun, B., Groisman, P.Y., Bradley, R.S., Keimig, F.T.: Temporal changes in the observed relationship between cloud cover and surface air temperature. J. Clim. **13**, 4341–4357 (2000)
6. Wielicki, B.A., et al.: Mission to planet earth: role of clouds and radiation in climate. Bull. Am. Meteor. Soc. **76**, 2125–2153 (1995)
7. Yao, M.-S., Del Genio, A.D.: Effects of cloud parameterization on the simulation of climate changes in the GISS GCM. J. Clim. **12**, 761–779 (1999)
8. Rajab, H., Cinkelr, T.: IoT based smart cities. In: 2018 International Symposium on Networks, Computers and Communications (ISNCC), pp. 1–4 (2018). https://doi.org/10.1109/ISNCC.2018.8530997
9. Nsabagwa, M., Mugume, I., Kasumba, R., Muhumu, J.: Condition monitoring and reporting framework for wireless network-based automatic weather stations. IEEE, pp. 1–8 (2018)
10. Parashar, A.: IoT based automated weather report generation and prediction using machine learning. In: 2019 2nd International Conference on Intelligent Communication and Computational Techniques (ICCT), pp. 339–344 (2019)
11. Liu, Z., et al.: Swin transformer: hierarchical vision transformer using shifted windows. In: Proceedings of the IEEE/CVF International Conference on Computer Vision (ICCV), pp. 10012–10022 (2021)
12. Nie, Y., Nguyen, N.H., Sinthong, P., Kalagnanam, J.: A time series is worth 64 words: long-term forecasting with transformers. arXiv:2211.14730 (2022)
13. Yu, R., Wang, B., Zhou, T.: Climate effects of the deep continental stratus clouds generated by the Tibetan Plateau. J. Clim. **17**, 2702–2713 (2004)
14. Chen, Q., Fan, J., Hagos, S.M., Gustafson, W.I., Berg, L.K.: Roles of wind shear at different vertical levels, Part I: cloud system organization and properties. J. Geophys. Res. Atmos. 120 (2015)

Dynamic Monitoring and Evaluation of Ecological Environment Quality in Urumqi Metropolitan Based on Google Earth Engine

Shaojie Bai[1,2], Abudukeyimu Abulizi[1,2](✉), Junxia Wang[1,2], and Le Yuan[1,2]

[1] College of Geography and Remote Sensing Science, Xinjiang University, Urumqi 830046, China
keyimabliz@xju.edu.cn

[2] Xinjiang Key Laboratory of Oasis Ecology, Xinjiang University, Urumqi 830046, China

Abstract. The Urumqi Metropolitan Area, situated in the arid northwestern region of China, faces a particularly fragile ecological environment. Monitoring and evaluating its ecological quality are crucial for sustainable regional development. This study focuses on the Urumqi Metropolitan Area, utilizing the Google Earth Engine (GEE) cloud platform to screen Landsat images from April to September in 2000, 2010, and 2020. A Remote Sensing Ecological Index (RSEI) is constructed to dynamically monitor ecological quality in the study area. Additionally, the Standard Deviational Ellipse (SDE) method is employed to analyze the spatial expansion characteristics of the ecological environment. The results reveal: (1) Temporally, the RSEI increased from 0.307 to 0.343 between 2000 and 2020, indicating an overall improvement in ecological quality. (2) Spatially, ecological quality exhibits significant differences, with higher values in the southern and central regions and lower values in the north. Monitoring results show that from 2000 to 2020, the area experiencing environmental deterioration (44.50%) was smaller than the area showing improvement (55.50%), suggesting an overall trend of environmental quality enhancement. (3) Based on changes in the standard deviation ellipse and centroids of different ecological classes, the overall ecologically better regions in the study area exhibit an east-west orientation, with SDE centroids shifting southeast. This study focuses on the spatio-temporal changes and spatial expansion characteristics of ecological environment quality in the Urumqi Metropolitan area located in the arid region, which is of significant practical and referential implications for the achievement of sustainable urban development.

Keywords: Remote sensing ecological index (RSEI) · Google earth engine (GEE) · Sustainable urban development

1 Introduction

Ecological environment quality refers to the degree of excellence of the ecological environment, reflecting its suitability for human survival and the sustainable development of society and economy. It is a complex system that affects both social and economic

G. Luo (Ed.): GFSISD 2024, SPPHY 417, pp. 57–76, 2025.
https://doi.org/10.1007/978-981-96-2490-4_5

sustainability [1]. The Urumqi Metropolitan Area is the region with the highest level of economic development, the fastest growth rate, and the strongest strength in Xinjiang. It benefits from well-matched resource combinations, including abundant natural and cultural tourism resources, specialized agricultural and pastoral resources, and rich deposits of oil, coal, and other minerals [2]. Given these exceptional development conditions and sound policy guidance, the Urumqi Metropolitan Area has experienced rapid growth. At the same time, real-time, rapid, and accurate monitoring of the ecological environment is crucial for protecting the ecology and advancing ecological civilization in the metropolitan area.

Remote sensing is characterized by its strong spatial resolution, stable revisit cycles, and ability to perform rapid, real-time monitoring over large areas. These features have led to its widespread application in the field of ecological environment monitoring [3–5], making it an effective tool for assessing regional ecological conditions. Xu Hanqiu [6] used remote sensing data, integrated multiple indicators that most intuitively reflect the ecological environment. They proposed a comprehensive indicator evaluation method based on Principal Component Analysis (PCA), known as the Remote Sensing Ecological Index (RSEI). This method not only provides a comprehensive reflection of the overall characteristics of an ecosystem but also mitigates the subjectivity in evaluation, enabling rapid monitoring and assessment of regional ecological environments. Google Earth Engine (GEE) is a cloud-based, planetary-scale geospatial analysis platform that has been frequently used worldwide in recent years. It exhibits prominent advantages in long-time-series remote sensing monitoring [7]. The GEE cloud processing platform also offers the benefit of covering multiple datasets, making it widely used by many scholars in fields such as land resources, agriculture, forestry, meteorology, and ecological environment to address various social and environmental issues [8]. Calculating and downloading the Remote Sensing Ecological Index (RSEI) based on the GEE cloud platform can save a significant amount of time and effort in data acquisition and preprocessing compared to traditional local solutions. Therefore, in recent years, large-scale environmental monitoring studies based on the GEE cloud platform have been continuously advancing. For instance, Chen et al. [9] used the GEE platform to develop an arid land oasis ecological index to assess the ecological quality of the oasis concentration region in the middle reaches of the Heihe River from 1985 to 2020. Yao et al. [10] utilized GEE to create a model for estimating aboveground biomass in grasslands and analyzed the spatiotemporal patterns and future trends of this biomass. Wang et al. [11] developed an RSEI covering the northern Anhui region using the GEE platform, quantitatively analyzed the characteristics of ecological spatiotemporal changes in this area, and explored the driving mechanisms of spatial differentiation in ecological quality. This study also provides additional evidence and references in this field.

The Urumqi Metropolitan Area is a key engine of regional economic development in Xinjiang and a strategic node for regional cooperation in Central Asia, playing a pivotal role in supporting high-quality development in Xinjiang and the country's westward opening-up [12]. At the Two Sessions of the Urumqi Municipal People's Congress and Political Consultative Conference held on January 12, 2021, it was clarified that during the 14th Five-Year Plan period, Urumqi will unswervingly focus on cultivating and developing the metropolitan area. This includes accelerating urban integration and enhancing its radiative and driving capabilities, with the Urumqi Metropolitan Area leading the effort. Cultivating the Urumqi Metropolitan Area represents a new mission for Urumqi in this historical phase, a strategic choice for improving urban service functions and overall competitiveness, and a major measure for implementing the national western development strategy and establishing the core area of the Silk Road Economic Belt. This goal is set for Urumqi's 14th Five-Year Plan and beyond, necessitating long-term persistence and continuous effort. Under the background of accelerated development of metropolitan areas, the pressure on ecosystems is further intensified [13]. It is necessary to monitor ecological changes in the region in a timely, rapid, and objective manner to provide references for building a beautiful ecological metropolitan area. However, in the current research on related fields, many scholars utilize remote sensing ecological indices to conduct environmental monitoring and evaluation of ecosystems at different scales such as national, provincial, county, and watershed levels [14–17], but there is still limited research specifically focused on metropolitan areas. The study area selected for this research is located in a key position for accelerated development in northwest China, making it highly necessary to conduct ecological environment monitoring research in this region. Based on this, the study focuses on the Urumqi Metropolitan Area, utilizing the GEE cloud platform to conduct a detailed monitoring and evaluation of ecological changes in the region over the past nearly 20 years. This will provide a clear understanding of the regional ecological environment and aims to offer scientific and effective data support and theoretical basis for addressing the conflicts between population, resources, and environment arising from rapid metropolitan development.

2 Materials and Methods

2.1 Study Area

The Urumqi Metropolitan (UM) covers a total area of approximately 176,000 square kilometers, accounting for 10.57% of the total area of Xinjiang. Based on the historical evolution, ecological carrying capacity, industrial division of labor, and spatial clustering analysis of the metropolitan area, it can be divided into four relatively independent spatial units: the Urumqi-Changji Core Area, the Western Region, the Eastern Region, and the Southern Region. Specifically, the Urumqi-Changji Core Area includes Urumqi City, Changji City, Fukang City, and Wujiaqu City; the Western Region includes Shihezi City, Manas County, and Hutubi County; the Eastern Region includes Qitai County, Jimusar County, and Mulei Kazakh Autonomous County; and the Southern Region includes Turpan City, Shanshan County, and Toksun County [18] (Fig. 1).

The Urumqi Metropolitan Area serves as a regional international commerce and trade center, financial center, and cultural exchange hub. It is also a forefront for bidirectional opening-up in western China, a base for energy resources and a representative of comprehensive utilization, a hub for new industries and export processing, as well as a tourist destination and distribution center. It represents an oasis ecological economy and a model for multi-ethnic social harmony in Xinjiang. In 2022, the Urumqi Metropolitan Area accounted for approximately one-third of Xinjiang's regional GDP and general public budget revenue. The value-added of industrial enterprises above designated size and the total retail sales of consumer goods accounted for about 40.0% of Xinjiang's total. The urbanization rate reached 88.8%, surpassing Xinjiang's average of 57.9%, with a population accounting for 22.53% of Xinjiang's total. The Urumqi Metropolitan Area demonstrates robust development trends and possesses potential advantages in economy, population, industrial structure, urbanization rate, fiscal revenue, and consumption potential. There is a need to enhance the core city's capacity, strengthen its radiation and driving capabilities towards surrounding areas, foster an integrated development trend within the metropolitan area, and develop it into a new engine leading Xinjiang's economic and social development [19]. Against the backdrop of high expectations and responsibilities placed on the Urumqi Metropolitan Area, ecological and environmental issues have also emerged.

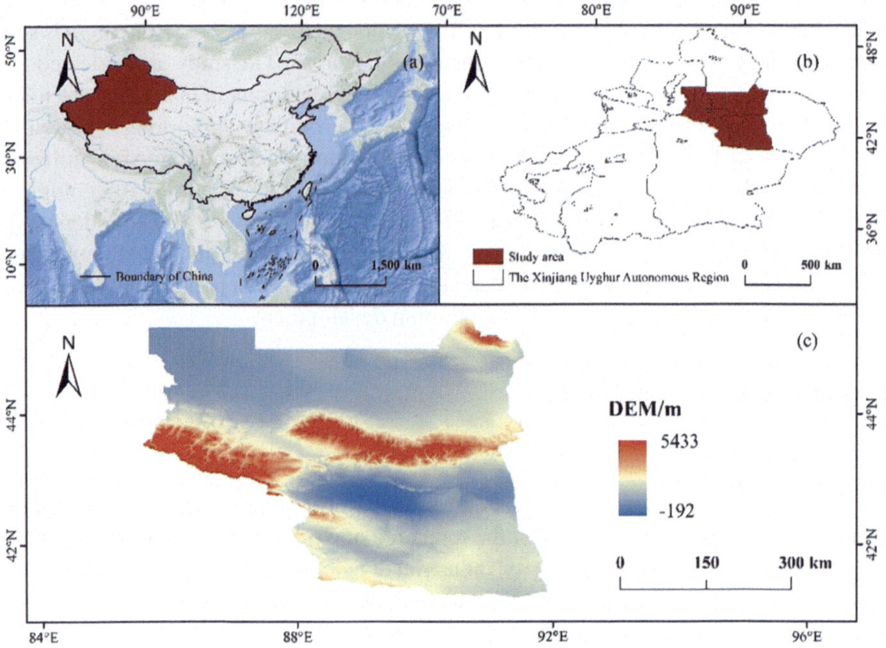

Fig. 1. Map of the study area: China (a), Xinjiang (b), Urumqi Metropolitan (c)

2.2 Data Sources and Pre-processing

This study focuses on the Urumqi Metropolitan Area as the research object. Based on the scope of the study area, Landsat images from April to September of the target year, processed to remove clouds, were generated on the GEE cloud platform. The Landsat data includes Surface Reflectance products from Landsat 5 (TM) and Landsat 8 (OLI), sourced from the United States Geological Survey (http://glovis.usgs.gov), with a spatial resolution of 30 m. The JRC Yearly Water Classification History data product was used for water body masking, followed by the application of Principal Component Analysis (PCA) to construct the Remote Sensing Ecological Index (RSEI).

2.3 Methods

Remote Sensing Ecological Index. In 2015, the Ministry of Ecology and Environment of China released the 'Technical Specification for Assessing Ecological Environment Conditions' (HJ192-2015). The Ecological Environment Index (EI) outlined in this specification includes five evaluation indicators: biological richness, vegetation coverage, water network density, land stress, and pollution load. While indicators such as biological richness, vegetation coverage, and water network density can be easily obtained through remote sensing, the latter two indicators are challenging to acquire directly. Additionally, the weighting of these five indicators is determined by humans, introducing a degree of subjective bias [7]. The Remote Sensing Ecological Index (RSEI), proposed by Xu Hanqiu [6], uses four indicators—greenness, wetness, dryness, and heat—to characterize ecological environment features. By utilizing Principal Component Analysis (PCA), RSEI can avoid the subjective bias introduced by human-determined weightings.

Greenness Index. The Normalized Difference Vegetation Index (NDVI) is a widely used indicator for characterizing the growth status of vegetation. It is closely related to leaf area, biomass, and vegetation cover, and is commonly applied in remote sensing monitoring of vegetation such as forests and grasslands [20]. Therefore, the greenness index can be represented by NDVI, and the formula is:

$$NDVI = (N - R)/(N + R) \tag{1}$$

In the formula, N represents the near-infrared band of Landsat TM data, and R represents the red band.

Wetness Index. The Tasseled Cap transformation can effectively reduce the dimensionality of remote sensing data, and the obtained wetness component can well reflect the humidity conditions of surface vegetation, water bodies, and soil, which is closely related to ecological and environmental changes such as soil degradation [21]. Therefore, the wetness index in this study is represented by the wetness component, and its expression varies for different sensor data. The extraction formulas for the wetness component based on TM and OLI data are respectively:

$$\begin{aligned} WET_{TM} = {} & 0.0315\rho_{blue} + 0.2021\rho_{green} + 0.3102\rho_{red} \\ & +0.1594\rho_{NIR} - 0.6806\rho_{SWIRI} - 0.6109\rho_{SWIR2} \end{aligned} \tag{2}$$

$$\begin{aligned} WET_{OLI} = {} & 0.1511\rho_{blue} + 0.1972\rho_{green} + 0.3283\rho_{red} \\ & 0.3407\rho_{NIR} - 0.7117\rho_{SWIRI} - 0.4559\rho_{SWIR2} \end{aligned} \tag{3}$$

where ρblue, ρgreen, ρred, ρ_{NIR}, ρ_{SWIRI}, ρ_{SWIR2} respectively represent the reflectance of the blue band, green band, red band, near-infrared band, shortwave infrared 1 band, and shortwave infrared 2 band corresponding to each image.

Dryness Index. In regional ecological environments, built-up land and bare soil are primary contributors to urban surface aridification. This study uses the average values of the Built-Up Index (IBI) and the Bare Soil Index (SI) to construct the Built-Up and Bare Soil Index (NDBSI) as an indicator of dryness. The calculation formula [22] is as follows:

$$NDBSI = (IBI + SI)/2 \tag{4}$$

$$IBI = \frac{\left\{\frac{2\rho_{SWIRI}}{\rho_{SWIRI}+\rho_{NIR}} - \left[\frac{\rho_{NIR}}{\rho_{NIR}+\rho_{red}} + \frac{\rho_{green}}{\rho_{green}+\rho_{SWIRI}}\right]\right\}}{\left\{\frac{2\rho_{SWIRI}}{\rho_{SWIRI}+\rho_{NIR}} + \left[\frac{\rho_{NIR}}{\rho_{NIR}+\rho_{red}} + \frac{\rho_{green}}{\rho_{green}+\rho_{SWIRI}}\right]\right\}} \quad (5)$$

$$SI = [(\rho_{SWIRI} + \rho_{red}) - (\rho_{blue} + \rho_{NIR})]/[(\rho_{SWIRI} + \rho_{red}) + (\rho_{blue} - \rho_{NIR})] \quad (6)$$

Heat Index. The heat index is represented by land surface temperature. In this study, the heat index is derived by selecting the LST_Day_1km band from the MOD11A2 land surface temperature product data and converting the temperature units [23].

$$LST = 0.02 \times \rho_{LST\_Day\_1km} - 273.15 \quad (7)$$

where $\rho_{LST\_Day\_1km}$ represents the LST_Day_1km band in the MOD11A2 dataset.

Due to the different dimensions of the four indicators, directly applying Principal Component Analysis (PCA) would result in imbalanced weights among the indicators. Therefore, before performing the PCA, it is necessary to normalize these indicators to unify their dimensions within the range of [0, 1]. The normalization formulas for each indicator are as follows:

$$NI_i = (I - I_{min})/(I_{max} - I_{min}) \quad (8)$$

where NI_i is the value of an indicator after regularization, and I_i is the value of the indicator in Quadrant *i*. I_{max} is the maximum value of the indicator; and I_{min} is the minimum value of the indicator.

After the four indicators are normalized, they can be used to calculate the first principal component PC1. To ensure that larger values represent better ecological conditions, we can further subtract PC1 from 1 to obtain the initial Remote Sensing Ecological Index (RSEI0):

$$RSEI0 = 1 - \{PC1 - [F(NDVI, WET, LST, NDBSI)]\} \quad (9)$$

To facilitate metrics and the comparisons of indicators, RSEI0 was regularized using the following equation:

$$RSEI = (RSEI0 - RSEI0_{min})/(RSEI0_{max} - RSEI0_{min}) \quad (10)$$

The RSEI is the constructed index, with values ranging between [0, 1]. A value closer to 1 indicates better ecological conditions, while a value closer to 0 signifies poorer ecological conditions. $RSEI0_{min}$ and $RSEI0_{max}$ represent the minimum and maximum values of RSEI0, respectively.

Spatial Autocorrelation Analysis. Moran's I index is a widely used method for measuring spatial autocorrelation. It can explore the clustering characteristics of spatial data and reveal spatial correlations through univariate spatial autocorrelation analysis, which is conducive to a more comprehensive understanding of the spatial relationships in regional ecological environment quality. The global Moran's I reveals the overall degree of spatial correlation of the RSEI in the Urumqi metropolitan area, while the Local Indicators of Spatial Association (LISA) index is used to reveal the local clustering patterns among regions.

The range of Moran's I index is from −1 to 1. An index closer to 1 indicates higher spatial correlation, meaning that the ecological environment quality in the Urumqi metropolitan area tends to be spatially clustered, while an index closer to −1 represents higher spatial heterogeneity, indicating that the ecological environment quality tends to be dispersed [24].

Standard Deviational Ellipse. The standard deviation ellipse is a spatial statistical method used to describe the spatial characteristics of a study area by calculating the features of the ellipse's major axis, minor axis, and centroid. The distance between the major and minor axes indicates the degree of spatial dispersion of regional features, while the centroid represents the relative location. This method can reveal the global geographic distribution characteristics of the study elements and is an important tool for spatial pattern statistical analysis [25].

3 Results

3.1 Results of PCA

The extracted indicators of greenness, wetness, dryness, and heat are standardized and combined, followed by Principal Component Analysis (PCA) to obtain the results. The purpose of PCA is to transform the original multiple indicators into a few principal components through linear transformation, which can explain most of the data's information. According to the results in Table 1, the contribution rates of the first principal component for the selected three years are 73.02%, 71.77%, and 74.86%, respectively. This indicates that the first principal component incorporates the majority of the information from the four indicators, making it effective in reflecting the comprehensive features of the remote sensing data. Consequently, we can use the first principal component to construct the Remote Sensing Ecological Index [26].

3.2 Ecological Quality Status

Spatial and Temporal Characteristics of Ecological Quality. The boxplot (Fig. 2) presents the statistical analysis results of the RSEI for the study area in 2000, 2010, and 2020. The boxplot shows the mean RSEI values [27], clearly reflecting the trend in environmental quality changes over time. The results indicate that the environmental quality in the Urumqi Metropolitan Area has generally improved. The mean RSEI was 0.307 in 2000, rising to 0.321 by 2010, an increase of 4.40%. From 2010 to 2020, the mean RSEI further increased to 0.343, a growth of 6.98%, suggesting that the pace of environmental quality improvement has accelerated over the past decade. However, despite the upward trend in the mean RSEI, the multi-year average value of 0.324 indicates that the overall environmental quality remains at a relatively poor level.

To further quantify and visualize the RSEI, the results from the three time periods are classified into five categories based on intervals of 0.2: Bad, Poor, Moderate, Good, and Excellent [28]. This classification method facilitates a more detailed analysis of the distribution of ecological quality. As shown in Table 2, the proportion of areas with

Table 1. Results of the PCA

Year	Indicators	PC1	PC2	PC3	PC4
2000	NDBSI	−0.258	0.375	−0.576	−0.679
	LST	−0.503	−0.839	−0.140	−0.154
	WET	0.330	−0.165	0.591	−0.717
	NDVI	0.756	−0.358	−0.547	−0.021
	Percentage Variance (%)	73.02	15.27	10.17	1.54
2010	NDBSI	−0.256	0.187	0.701	−0.639
	LST	−0.507	−0.830	−0.137	−0.190
	WET	0.321	0.070	−0.581	−0.745
	NDVI	0.758	−0.521	0.391	−0.027
	Percentage Variance (%)	71.77	17.38	9.67	1.18
2020	NDBSI	−0.147	−0.089	0.461	−0.870
	LST	−0.281	−0.672	−0.647	−0.226
	WET	0.412	0.541	−0.589	−0.437
	NDVI	0.854	−0.497	0.151	−0.014
	Percentage Variance (%)	74.86	17.27	7.69	0.18

Bad and Poor ecological environments has slightly decreased from 86.13% in 2000 to 85.39% in 2020. Although this proportion remains high, indicating that the Urumqi Metropolitan Area still predominantly consists of Bad and Poor areas, the decrease suggests an improvement in environmental quality. At the same time, the proportion of areas with Good and Excellent ecological environments increased from 7.75% in 2000 to 8.27% in 2020. Although the increase is modest, it indicates some progress in improving ecological quality. Overall, these data suggest that the ecological environment in the Urumqi Metropolitan Area has improved over time. The increase in areas with Good and Excellent environments, along with the decrease in Bad and Poor areas, reflects an overall optimization of the ecological environment, demonstrating the city's efforts in environmental protection and improvement.

Figure 3 provides a visual representation of the changes in the Remote Sensing Ecological Index (RSEI) in the Urumqi Metropolitan Area over the past 20 years. In the figure, the ecological levels are represented by red, orange, yellow, light blue, and dark blue, corresponding to Bad, Poor, Moderate, Good, and Excellent, respectively. Comparing the RSEI maps for the three periods, it is evident that from 2000 to 2020, most of the study area was predominantly in red and orange, indicating that the ecological environment was primarily in the Bad and Poor categories. Although the area of each ecological level did not change significantly, the distribution exhibits a clear spatial variation with higher values in the central region and lower values in the northern and southern regions. Specifically, areas with poorer environmental quality are located in large desert and unused areas in the south and north, while areas with better environmental

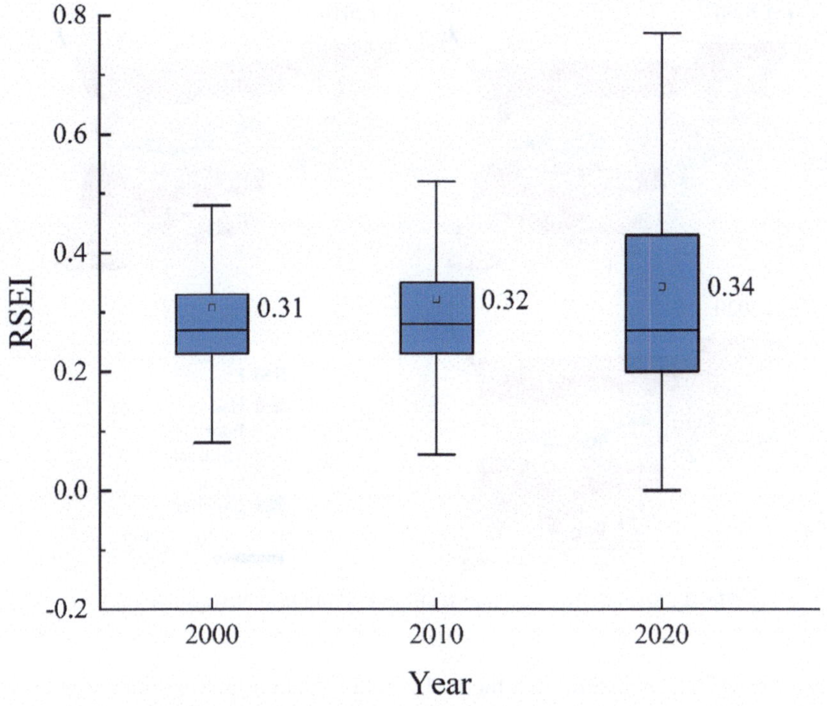

Fig. 2. Statistical distribution of RSEI in UM, 2000–2020.

Table 2. Statistics of the remote sensing ecological index (RSEI) level in UM, 2000–2020.

RSEI grade	2000		2010		2020	
	Area/10,000 km^2	%	Area/10,000 km^2	%	Area/10,000 km^2	%
Bad	6.62	38.21	6.05	34.77	6.57	37.58
Poor	8.3	47.92	8.64	49.64	8.36	47.81
Moderate	1.06	6.13	1.31	7.55	1.11	6.34
Good	0.64	3.72	0.74	4.22	0.68	3.91
Excellent	0.7	4.03	0.66	3.81	0.76	4.36

quality are concentrated in the central mountainous region, agricultural lands, and oasis zones.

Spatial Autocorrelation Analysis. Based on the univariate spatial autocorrelation model [29], the global Moran's I value of the Remote Sensing Ecological Index (RSEI) in the Urumqi metropolitan area for the years 2000, 2010, and 2020 were calculated using Geoda software to analyze the spatial distribution of RSEI. The global Moran's I value reflects the intensity and direction of spatial autocorrelation. As shown in Fig. 4, the global Moran's I value for RSEI in all years are greater than 0, and the p-values are

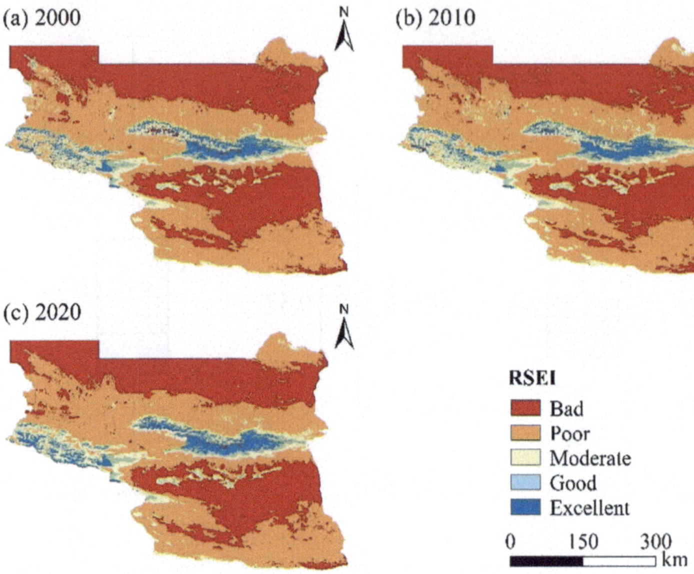

Fig. 3. Spatial distribution of RSEI in UM:2000(a),2010(b),2020(c)

all less than 0.001, indicating that the RSEI in the Urumqi metropolitan area exhibits significant spatial correlation in 2000, 2010, and 2020 [30]. This means that the spatial distribution of RSEI is not random and independent but shows a clear spatial clustering trend, with high and low value clusters. This spatial clustering phenomenon is mainly due to the relatively stable overall layout of regional land use types, which enhances the degree of spatial aggregation. For example, areas with better environmental quality are often concentrated in specific geographical areas such as mountainous regions or oases, while areas with poorer environmental quality are concentrated in deserts or unused land. From the perspective of evolutionary trends, the global Moran's I value increased from 0.920 in 2000 to 0.951 in 2020. This upward trend indicates that the spatial dependence of RSEI has strengthened over the years and exhibits strong spatial clustering characteristics. This change may reflect the long-term stability of regional land use policies and the spatial effects of environmental management measures.

Using Geoda software, LISA cluster maps of the Remote Sensing Ecological Index (RSEI) for the Urumqi Metropolitan Area for the years 2000, 2010, and 2020 were generated to analyze local spatial characteristics of RSEI distribution. In the maps (Fig. 5), red represents regions with high RSEI values, dark blue indicates regions with low RSEI values, pink shows areas where RSEI values are high relative to surrounding units, and light blue represents areas where RSEI values are low relative to surrounding units [31]. According to the LISA maps, the overall spatial clustering characteristics of RSEI in the Urumqi Metropolitan Area changed little from 2000 to 2020. More than 50% of the areas are not significant, with the spatial patterns mainly showing high-high and low-low clustering. The proportion of high-high clustering for the three periods was 14.71%, 15.55%, and 14.70%, respectively, predominantly located in the central mountainous

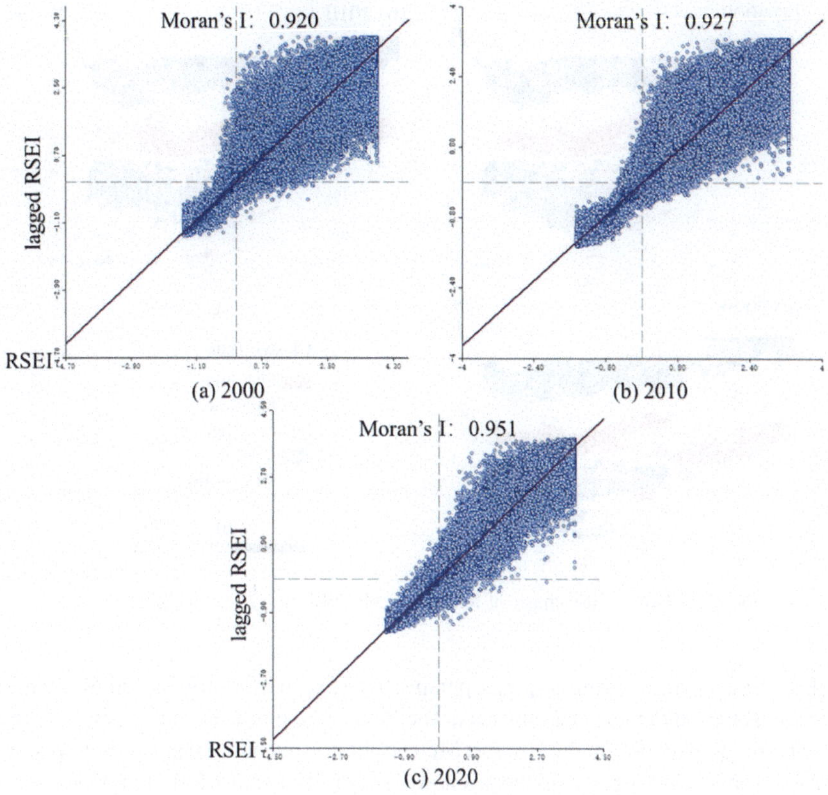

Fig. 4. Moran scatter plots of RSEI in UM: 2000 (a), 2010 (b), 2020 (c)

areas and urban center. Notably, the proportion of high-high clustering increased from 2000 to 2010 and decreased from 2010 to 2020, which may be related to the policy-driven urban planning and infrastructure development in the region in 2009. The proportion of low-low clustering was 30.38%, 30.80%, and 30.87%, respectively, mainly distributed in the southern and northern desert and unused land areas. This proportion shows a slight increasing trend and spatially spreads to the south. This trend may be due to efforts to promote urbanization in the metropolitan area, leveraging the radiating effect of Urumqi to the east, which has led to changes in land use patterns and ecological degradation, resulting in lower RSEI values. Additionally, the proportions of high-low and low-high clustering regions were both below 0.5%, with minimal range and little variation.

3.3 Analysis of Dynamic Changes in Ecological Quality

Spatial Distribution Changes. The standard deviational ellipse method is a typical approach used to describe the density and distribution range of spatial points [32], effectively revealing the movement direction and extent of ecological quality within a spatial range. This study applies the standard deviational ellipse method to analyze the changes in the direction and extent of overall ecological quality in the Urumqi Metropolitan Area

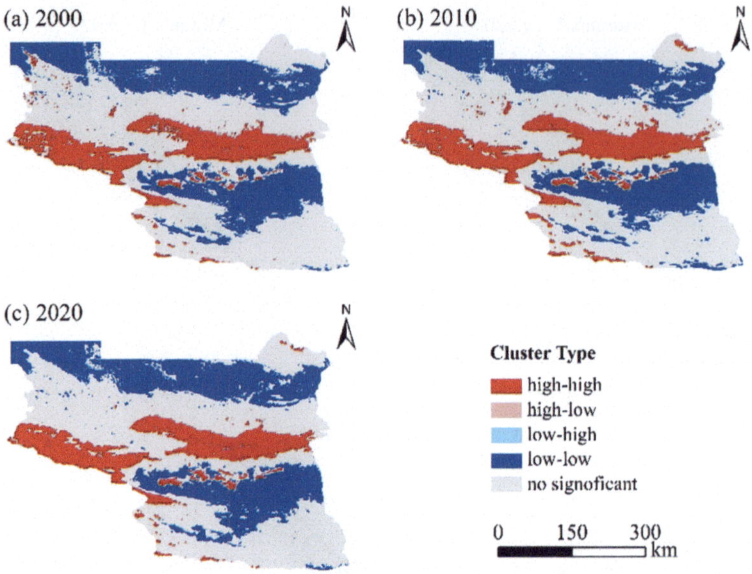

Fig. 5. LISA cluster maps of RSEI in UM: 2000 (a), 2010 (b), 2020 (c)

over three time periods. From a spatial perspective (Fig. 6), the orientation of the major axis of the standard deviational ellipses at the three time points is consistent, indicating that the ecological quality in the Urumqi Metropolitan Area has expanded east-west over the past 20 years. This aligns with the orientation of the Tianshan Mountains in the central region, reflecting the geographical consistency of ecological quality distribution. From a temporal perspective, the length of the major axis of the standard deviational ellipse in 2020 has increased compared to 2000 and 2010, suggesting that the future development direction of ecological quality in the Urumqi Metropolitan Area is primarily eastward, potentially related to regional development strategies. Regarding centroid distribution, the centroid of the standard deviational ellipses for areas with better ecological quality has significantly shifted southeast from 2000 to 2020. This movement is likely closely related to the ecological restoration and protection policies in the southeastern part of Urumqi, particularly in Dabancheng District, where the active implementation of wetland restoration policies has significantly improved the ecological quality in the area.

From the spatial shift characteristics of centroids for different ecological quality levels (Fig. 7), it can be observed that, overall, centroids for Bad and Poor ecological quality tend to shift eastward, while centroids for moderate and higher ecological quality shift westward. From 2000 to 2020, with the exception of the centroid for bad ecological quality, which generally moved west, centroids for better ecological quality levels all shifted eastward. This suggests that the primary direction of ecological improvement efforts in the Urumqi Metropolitan Area has been towards the east. Specifically, the centroids for bad and moderate ecological quality show significant movement and clear spatial shifts. The centroid for bad ecological quality moved from the west to the east and then shifted

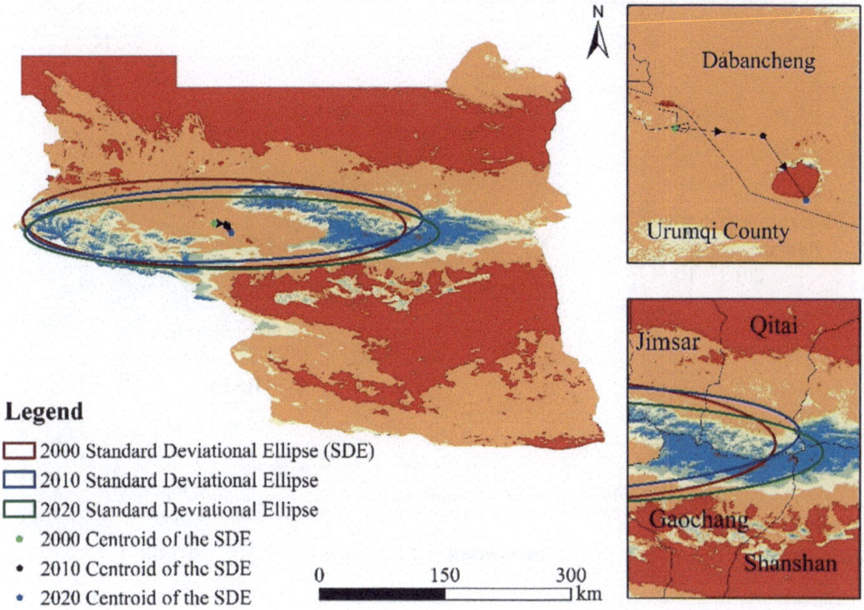

Fig. 6. Standard deviation spatial distribution of ecological quality of ellipse in UM.

southwest, while the centroid for moderate ecological quality moved from the west to the east and then shifted southeast. This indicates that changes in ecological quality and environmental conditions in these two levels are relatively frequent and pronounced in the Urumqi Metropolitan Area. In contrast, the centroids for good and excellent ecological quality exhibited smaller movements, remaining primarily in the area where Urumqi County borders Dabancheng District. This region shows higher stability in ecological changes, reflecting sustained and stable improvements in ecological management and environmental protection.

Difference Analysis. To conduct a thorough analysis of the spatiotemporal differences in ecological environment quality in the Urumqi metropolitan area over the past 20 years, this study utilizes the ENVI 5.3 platform to detect changes in the RSEI results. The results are presented in Fig. 8, where the degrees of change are categorized into six levels [33]: (0.4, 1.0], (0.1, 0.4], (0.0, 0.1], [−0.1, 0.0], [−0.4, −0.1), and [−1.0, −0.4), representing significant increase, notable increase, slight increase, slight decrease, notable decrease, and significant decrease, respectively, and distinguished by different colors. As can be observed from the Fig. 8, areas with improved ecological environments are primarily distributed in the central mountainous regions, indicating that ecological protection and restoration measures implemented in these areas have effectively enhanced environmental quality. Notably, ecological restoration projects in the central mountainous regions, such as vegetation restoration and water source protection, have significantly promoted local ecological environment improvement. In contrast, areas with deteriorated ecological environments are mainly concentrated in the rapidly expanding urban regions

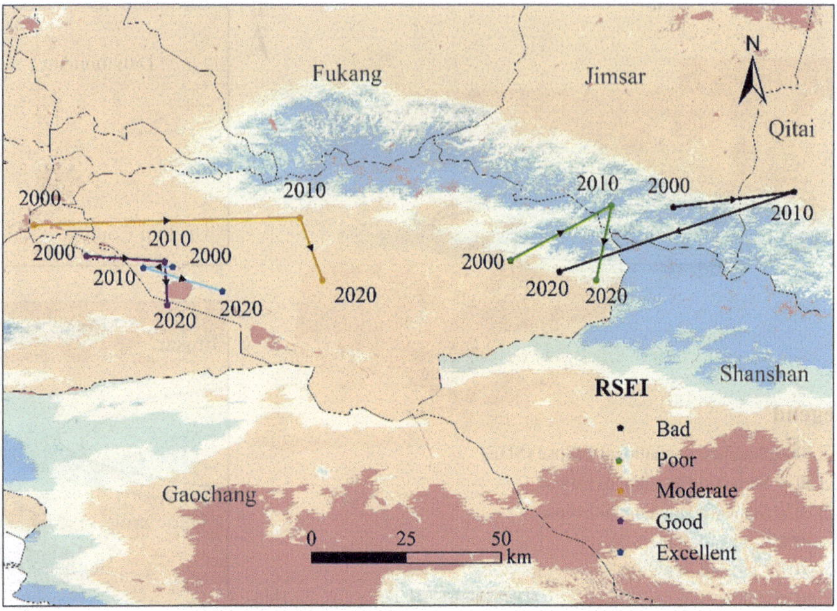

Fig. 7. Deviation diagram of RSEI's classified barycenter in UM.

surrounding the city, reflecting the negative impacts of urbanization on the environment. Urban sprawl has led to extensive land development and utilization, subsequently disrupting the original ecosystem and causing a notable decline in ecological quality.

Table 3 presents the RSEI change monitoring results for the Urumqi Metropolitan Area from 2000 to 2020. The data indicates that approximately 162,200 km^2, or 92.15% of the total area, experienced little to no change in ecological conditions during this period, suggesting that most regions maintained relative stability in their ecological environment. The area where ecological conditions significantly improved was 7,800 km^2, accounting for 4.41% of the total area, highlighting the notable improvements in environmental quality due to ecological governance and restoration measures. Conversely, the area where ecological quality significantly deteriorated was 6,100 km^2, or 3.45%, indicating that despite overall improvements, some regions experienced considerable ecological decline. Overall, the proportion of deteriorated ecological environment increased from 1.91% to 3.45%, while the proportion of improved ecological environment rose from 3.42% to 4.41%. This reflects a predominant trend of improvement in ecological conditions, though the rate of deterioration is also accelerating. Continued focus on ecological protection is essential to mitigate further decline and ensure long-term sustainability.

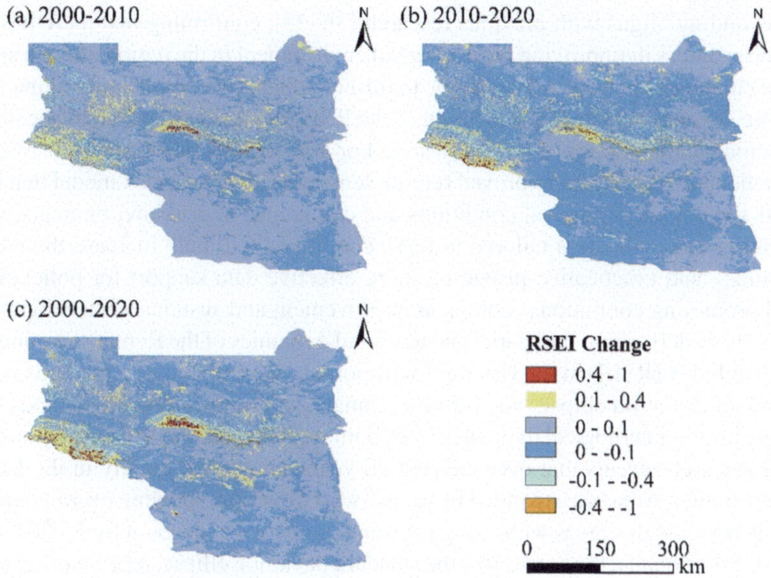

Fig. 8. Detection map of RSEI changes in UM, 2000–2020.

Table 3. RSEI detection in UM, 2000–2020.

Class	2000–2020			
	Level	Level Area/10,00 0 km^2	Class Area/10,000 km^2	Proportion/%
Improved	(0.4,1.0]	0.121	9.770	55.50
	(0.1,0.4]	0.655		
	(0.0,0.1]	8.994		
Degraded	[−0.1,0.0]	7.225	7.832	44.50
	[−0.4, −0.1)	0.565		
	[−1.0, −0.4)	0.042		

4 Discussion

4.1 Comparison with the Previous Literature

(1) This study uses remote sensing ecological indices to monitor and analyze the ecological environment quality of the Urumqi Metropolitan Area. The Remote Sensing Ecological Index (RSEI) is a stable, effective, and objective assessment tool that has been widely used in monitoring the ecological quality of cities of various sizes [15, 34, 35], offering accurate environmental quality evaluations. The results of this study indicate that the ecological environment in the Urumqi Metropolitan Area is generally showing a positive trend; however, it remains at a relatively poor level overall.

This finding aligns with previous research [36–38], confirming that there has been some progress in improving the ecological environment in the region, though significant challenges persist. Additionally, to further enhance the application of the RSEI, future development and optimization of the RSEI model could take into account the specific characteristics of the study area. For arid regions like Urumqi, it would be beneficial to develop an improved remote sensing ecological index model that better suits the unique ecological conditions and challenges of arid environments. Incorporating new indicators tailored to these conditions will help increase the model's accuracy and practicality, providing more effective data support for policymakers and promoting continuous ecological improvement and sustainable development.

(2) This study delves into the spatial and temporal dynamics of the Remote Sensing Ecological Index (RSEI) in the Urumqi Metropolitan Area, employing methods such as standard deviation ellipses and difference analysis. These combined methods effectively monitor ecological dynamics from both spatial and temporal dimensions [39]. The research reveals that over the past 20 years, ecological quality in the Urumqi Metropolitan Area has expanded in an east-west direction, showing overall improvement. Specifically, areas with rising environmental quality account for 55.50% of the total. Additionally, the centroid of the standard deviation ellipse, representing regions with better ecological quality, has shifted southeast. This movement aligns closely with regional policies that promote eastward development [40], indicating that policy implementation has positively influenced ecological quality improvement. Overall, the standard deviation ellipse analysis not only highlights the dynamic changes in ecological quality but also reflects the actual impact of policies and environmental factors, providing crucial insights for future ecological protection and environmental management strategies.

4.2 Policy Implications

The findings of this study provide crucial evidence for future ecological protection and environmental governance strategies and reflect the key role of policy adjustments and regional planning in improving the ecological environment. The RSEI of the Urumqi Metropolitan Area exhibits significant spatial clustering, indicating varying ecological conditions across different regions. This necessitates tailored measures based on regional characteristics to ensure continuous ecological improvement. Firstly, areas with good and excellent ecological environments are primarily located in the central mountainous regions, which accounted for 8.27% in 2020. These regions should further leverage their ecological resources while maintaining their ecological benefits, with a focus on policies that prioritize mountain forest maintenance and glacier protection. This not only affects the ecological balance of the area but also directly impacts water resource security [41]. Secondly, the regions with bad and poor ecological environments, which constituted 85.39% in 2020, require increased attention and policy support. Focus should be given to ecological restoration in natural ecosystems such as grasslands, wetlands, and deserts, and addressing the long-standing issue of water resource shortages. Accelerating water diversion projects could help alleviate these shortages. Additionally, the rapid development of the cultural tourism industry has positively impacted the regional economic structure but also presents challenges for ecological protection. Policymakers should dedicate efforts to addressing ecological concerns related to tourism, promote green tourism, and ensure a balance between economic development and ecological preservation.

Overall, ecological issues in the Urumqi Metropolitan Area require greater attention at the policy level. The government should continue to invest resources in ecological protection and restoration. At the same time, while promoting economic development in the eastern regions, it is essential to also consider ecological protection in other areas to achieve a more comprehensive improvement in environmental quality. This approach will provide a solid foundation for sustainable and high-quality development in the future.

5 Conclusion

This study conducted dynamic monitoring and evaluation of the ecological environment quality of the Urumqi metropolitan area, with specific analyses from two aspects: ecological quality status and dynamic changes. Based on the GEE cloud platform, the Remote Sensing Ecological Index (RSEI) was constructed and graded for evaluation. Furthermore, the standard deviation ellipse and difference analysis methods were used for further dynamic monitoring of the RSEI. The main research conclusions are as follows:

(1) From the perspective of ecological quality, the ecological environment quality of the Urumqi metropolitan area is relatively stable. The mean RSEI (Remote Sensing Ecological Index) values in 2000, 2010, and 2020 are 0.307, 0.321, and 0.343, respectively, showing an overall upward trend. The multi-year average RSEI value is 0.324, indicating that the average environmental quality of the Urumqi metropolitan area is at a relatively poor level. The proportion of areas with moderate or above ecological environment quality has increased from 13.88% in 2000 to 14.61% in 2020, suggesting an overall optimization of the ecological environment. The RSEI exhibits a spatial agglomeration pattern, with spatial dependence strengthening annually and demonstrating strong agglomeration characteristics, primarily manifesting as high-high agglomeration and low-low agglomeration.

(2) From the perspective of the direction of change, the ecological quality of the Urumqi metropolitan area has expanded in the east-west direction over the past 20 years. From 2000 to 2020, the center of the standard deviational ellipse with better ecological quality has moved in the southeast direction, indicating that the main direction of ecological governance in the Urumqi metropolitan area is toward the southeast. The movement amplitude of the center with good and excellent ecological environment is relatively small, concentrated at the junction of Urumqi County and Dabancheng District, suggesting that the ecological changes in this area are stable and improving.

(3) From the perspective of dynamic monitoring, the overall ecological environment of the Urumqi metropolitan area has improved from 2000 to 2020. Regions with improved environmental quality account for 55.50% of the total, while regions with deteriorating environmental quality account for 44.50%. The improvement in the ecological environment is mainly attributed to ecological restoration in the central mountainous areas, while the deteriorating areas are primarily the rapidly expanding

urban areas surrounding the city. Additionally, the area with a significant increase in RSEI is 0.121 million km^2, accounting for 0.69%, while the area with a significant decrease is 0.042 million km^2, accounting for 0.24%. However, the increase in the area of deteriorating ecological environment is faster than that of improving ecological environment, indicating that although the overall ecological environment shows improvement, attention must be paid to ecological environment restoration and protection to avoid accelerated deterioration.

Acknowledgment. This work was supported by the Third Xinjiang Scientific Expedition Program (Grant No. 2022xjkk1100), to which we are grateful.

References

1. Ge, Y., Jia, W., Zhao, H., Xiang, P.: A framework for urban resilience measurement and enhancement strategies: a case study in Qingdao, China. J. Environ. Manage. **367** (2024) https://doi.org/10.1016/j.jenvman.2024.122047
2. Luo, J., Luo, X., Gu, Z., Tang, M.: Integration of Xinjiang production and construction corps and local government: exploration on coordinated development of Urumqi metropolitan area. City Plan. Rev. **46**, 38–43 (2022)
3. Chen, G., Zhou, Y., Voogt, J.A., Stokes, E.C.: Remote sensing of diverse urban environments: from the single city to multiple cities. Remote Sens. Environ. **305** (2024). https://doi.org/10.1016/j.rse.2024.114108
4. Xu, Z., et al.: The high spatial resolution Drought Response Index (HiDRI): an integrated framework for monitoring vegetation drought with remote sensing, deep learning, and spatiotemporal fusion. Remote Sens. Environ. **312** (2024) https://doi.org/10.1016/j.rse.2024.114324
5. Zhi, J., et al.: Accurately mapping social functional zones of urban green spaces by integrating remote sensing images and crowd-sourced geospatial data. Int. J. Digit. Earth **17**, 1–26 (2024). https://doi.org/10.1080/17538947.2024.2376255
6. Xu, H.: A remote sensing index for assessment of regional ecological changes. China Environ. Sci. **33**, 889–897 (2013)
7. Zhu, X., Wang, H., Wu, Y., Wei, S., Yu, Y.: Research on the coupling coordination degree of urbanization and eco-environment of the urban agglomeration in the middle reaches of the Yangtze River based on google earth engine. Resour. Environ. Yangtze Basin **31**, 2706–2717 (2022)
8. Gorelick, N., et al.: Google earth engine: planetary-scale geospatial analysis for everyone. Remote Sens. Environ. **202**, 18–27 (2017). https://doi.org/10.1016/j.rse.2017.06.031
9. Chen, W., Zhao, R., Lu, H.: Response of ecological environment quality to land use transition based on dryland oasis ecological index (DOEI) in dryland: a case study of oasis concentration area in Middle Heihe River, China. Ecol. Indic. **165** (2024). https://doi.org/10.1016/j.ecolind.2024.112214
10. Yao, Y., Ren, H.: Estimation of grassland aboveground biomass in northern China based on topography-climate-remote sensing data. Ecol. Indic. **165** (2024). https://doi.org/10.1016/j.ecolind.2024.112230
11. Wang, X., Yao, X., Jiang, C., Duan, W.: Dynamic monitoring and analysis of factors influencing ecological environment quality in northern Anhui, China, based on the Google Earth Engine. Sci. Rep. **12** (2022). https://doi.org/10.1038/s41598-022-24413-0

12. Bai, C., Cang, C., Dong, Y.: Study on the coupling and coordinated development of industry-population-space in Urumqi metropolitan area. J. Xinjiang Normal Univ. Nat. Sci. Ed. **42**, 26–33+42 (2023). https://doi.org/10.14100/j.cnki.1008-9659.2023.03.003
13. Li, T., Chen, H., Ma, C.: China's urbanisation evolution and metropolitan area expansion, based on the Prolonged Artificial Nighttime-light Dataset (PANDA, 1984–2020). Int. J. Digit. Earth **17** (2024). https://doi.org/10.1080/17538947.2024.2347448
14. He, Y., et al.: Spatio-temporal changes in remote sensing-based ecological index in China since 2000. J. Nat. Resour. **36**, 1176–1185 (2021)
15. Liu, S., Zhang, F., Zhao, X., Huang, J., Gao, P.: Improvement of remote sensing ecological index in oasis in arid area. Sci. Survey. Mapp. **47**, 143–151, 203 (2022)
16. Wang, B., Fu, S., Hao, Z., Zhen, Z.: Ecological security pattern based on remote sensing ecological index and circuit theory in the Shanxi section of the Yellow River Basin. Ecol. Indic. **166** (2024). https://doi.org/10.1016/j.ecolind.2024.112382
17. Wu, A., Zhao, Y., Qin, Y., Liu, X., Shen, H.: Analysis of ecological environment quality and its driving factors in the Beijing-Tianjin-Hebei region of China. Sustainability **15** (2023). https://doi.org/10.3390/su15107898
18. Abuduwaili, A.: Urumqi Metropolitan level of Urbanization and Difference Research. Xinjiang University (2013)
19. Li, M., Fu, Q.: Study on fractal characters of the urban system of Urumqi Metropolitan Area. J. Geo-Inf. Sci. **16**, 242–248 (2014)
20. Wang, L., Jiao, L., Lai, F., Zhang, N.: Evaluation of ecological changes based on a remote sensing ecological index in a Manas Lake wetland, Xinjiang. Acta Ecol. Sin. **39**, 2963–2972 (2019)
21. Li, B., Ti, C., Yan, X.: Study of derivation of tasselled cap transformation for Landsat 8 OLI images. Sci. Survey. Mapp. **41**, 102–107 (2016)
22. Xu, H.: Assessment of ecological change in soil loss area using remote sensing technology. Trans. Chin. Soc. Agric. Eng. **29**, 91–97 (2013)
23. Hang, Z., Chen, J.: Study on the evolution of long-time series ecological environment quality in Pingxiang City based on GEE. Ecol. Environ. Monit. Three Gorges 1–20 (2024)
24. Jing, Y., et al.: Assessment of spatial and temporal variation of ecological environment quality in Ebinur Lake Wetland National Nature Reserve, Xinjiang, China. Ecol. Indic. **110** (2020). https://doi.org/10.1016/j.ecolind.2019.105874
25. Zhao, Z., et al. Spatiotemporal patterns and driving factors of ecological vulnerability on the Qinghai-Tibet Plateau based on the google earth engine. Remote Sens. **14** (2022). https://doi.org/10.3390/rs14205279.
26. Yu, T., et al.: Evolution of environmental quality and its response to human disturbances of the urban agglomeration in the northern slope of the Tianshan Mountains. Ecol. Indic. **153** (2023). https://doi.org/10.1016/j.ecolind.2023.110481
27. Ali, S., et al.: Spatio-temporal variations in trends of vegetation and drought changes in relation to climate variability from 1982 to 2019 based on remote sensing data from East Asia. J. Integr. Agric. **22**, 3193–3208 (2023). https://doi.org/10.1016/j.jia.2023.04.028
28. Yang, P., Zhou, A., Fu, K., Wang, Y., Luo, H.: Assessment and analysis on roadside ecological environment quality along typical sections of Qinghai-Tibet Highway based on remote sensing ecological indices. J. Arid Land Resour. Environ. **38**, 39–48 (2024)
29. Cao, N., et al.: Study on diversity and distribution characteristics of rodents and fleas in different landforms of desert landscape in Junggar Basin. Acta Ecol. Sin. **43**, 4270–4283 (2023)
30. Qiu, M., Zuo, Q., Wu, Q., Yang, Z., Zhang, J.: Water ecological security assessment and spatial autocorrelation analysis of prefectural regions involved in the Yellow River Basin. Sci. Rep. **12** (2022). https://doi.org/10.1038/s41598-022-07656-9

31. Wang, Y., Chen, H., Zhang, Y.: Spatial characteristics of coupling development of ecological protection and agricultural economy in China. Sustainability **15** (2023). https://doi.org/10.3390/su15119068
32. Zheng, X., et al.: Spatiotemporal evolution and influencing factors of provincial tourism ecological security in China. Ecol. Indic. **148** (2023). https://doi.org/10.1016/j.ecolind.2023.110114
33. Wang, J., Ding, J., Zhang, Z.: Change of ecological environment in Turpan and Hami cities based on remote sensing ecology index. Arid Land Geogr. **45**, 1591–1603 (2022)
34. Wang, H., Zhou, W., Guan, Y., Wang, J., Ma, R.: Monitoring the ecological restoration effect of land reclamation in open-pit coal mining areas: an exploration of a fusion method based on ZhuHai-1 and Landsat 8 data. Sci. Total Environ. **904** (2023). https://doi.org/10.1016/j.scitotenv.2023.166324
35. Wang, Z., Chen, T., Zhu, D., Jia, K., Plaza, A.: RSEIFE: a new remote sensing ecological index for simulating the land surface eco-environment. J. Environ. Manage. **326** (2023). https://doi.org/10.1016/j.jenvman.2022.116851
36. Kou, Y., et al.: Spatiotemporal patterns and coupling coordination analysis of multiscale social-economic-ecological effects in ecologically vulnerable areas based on multi-source data: a case study of the Tuha Region, Xinjiang Province. Land **13** (2024). https://doi.org/10.3390/land13030282
37. Tang, L., Kasimu, A., Ma, H., Eziz, M.: Monitoring multi-scale ecological change and its potential drivers in the economic zone of the Tianshan Mountains' northern slopes, Xinjiang, China. Int. J. Environ. Res. Public Health **20** (2023). https://doi.org/10.3390/ijerph20042844
38. Zhang, J., Zhou, Q., Cao, M., Liu, H.: Spatiotemporal change of eco-environmental quality in the Oasis City and its correlation with urbanization based on RSEI: a case study of Urumqi, China. Sustainability **14** (2022). https://doi.org/10.3390/su14159227
39. Sun, Y., Liu, S., Li, Y., Li, J.: Evaluation of ecological quality of eastern Yuxi City based on the improved remote sensing ecological index. Pratacult. Sci. **40**, 551–560 (2023)
40. Zhao, W., Wang, C.: Urumqi metropolis circle "production-living-ecological spaces" carbon metabolism analysis based on ecological network utility. Environ. Sci. 1–16 (2024). https://doi.org/10.13227/j.hjkx.202405272
41. Tamasiga, P., Mogomotsi, P.K., Onyeaka, H., Mogomotsi, G.E.J.: Amplifying climate resilience: the impact of social protection, social cohesion, and social capital on public support for climate change action. Sustainable Environ. **10**, 2361568 (2024). https://doi.org/10.1080/27658511.2024.2361568

Analysis of Trade-Offs and Synergies Between Urban Development and Ecosystem Services in the Northwestern Region Under SSP-RCP Scenarios

Juan Yang[1], Jianghua Zheng[1,2(✉)], ChuQiao Han[1], and Jiale Wu[1]

[1] College of Geography and Remote Sensing Science, Xinjiang University, Urumqi 830046, China
zheng.jianghua@xju.edu.cn

[2] Xinjiang Key Laboratory of Oasis Ecology, Xinjiang University, Urumqi 830046, China

Abstract. Understanding how climate change and human activities impact ecosystem services (ESs) is crucial for optimizing urban planning, land resource allocation, and ecological protection. This study simulated various land use scenarios in northwestern China using a coupled system dynamics model (SD) and a patch-generation land use simulation model (PLUS) under Shared Socioeconomic Pathways and Representative Concentration Pathways (SSP-RCP) scenarios. The In-VEST model was used to assess ecosystem service levels and trade-offs for 2030. We analyzed these using correlation coefficients and Geographically Weighted Regression (GWR) models. The results indicate: (1) The SSP126 scenario, which promotes ecological restoration, leads to a shift from cropland to grassland. The SSP245 scenario continues current development trends, optimizing unused land. The SSP585 scenario emphasizes economic growth, resulting in significant cropland expansion. (2) Ecosystem service levels for WY, CS, and SC are higher in the northwest and southeast but lower in the central region, with HQ showing the opposite pattern. The Comprehensive Ecosystem Service Index (CESI) under SSP126 is notably higher, making it preferable for long-term ecological development in the northwest. (3) Except for WY-HQ, most ecosystem services exhibit synergies, though trade-offs and synergies vary spatially. WY-HQ trade-offs are concentrated in the southeast, while SC-HQ synergies are strongest in the northwest. This research aims to mitigate the effects of human activities and climate change on ecosystem services, enhancing ecological security and sustainability in northwestern China and Central Asia's arid zone.

Keywords: CMIP6 · SD-PLUS · Ecosystem Services · Trade-offs/Synergies

1 Introduction

Ecosystem services (ESs) refer to the significant contributions of ecosystems to human well-being by meeting direct or indirect needs through the provision, regulation, support, and cultural services [1, 2]. However, uncontrolled urban expansion and unreasonable

G. Luo (Ed.): GFSISD 2024, SPPHY 417, pp. 77–94, 2025.
https://doi.org/10.1007/978-981-96-2490-4_6

human activities pose a severe risk of decline in global ESs [3]. A recent study reported that in 2017, approximately 785 million people globally lacked access to safe drinking water, and about 4.8 billion people faced threats to water security or biodiversity [4]. These hazards not only threaten regional and global ecological security but also pose severe challenges to achieving the Sustainable Development Goals for 2030 [5]. Therefore, assessing changes in ESs helps us understand their status, distribution, and evolutionary trends and supports policymakers in developing ESs management policies to ensure the sustainable development of human society [6].

Changes in land use have a profound impact on the structure and function of ecosystems. Existing studies primarily focus on the effects of historical land use changes on ESs, neglecting the potential consequences of future land use transitions. Given the uncertainty and complexity of future land use changes, scenario simulation and modeling are crucial for mitigating potential risks and enhancing ESs [7]. Scenario simulation assesses the impact of various future scenarios by setting different possible conditions and is widely used in urban planning and ecological environmental assessment [8]. The Coupled Model Intercomparison Project Phase 6 (CMIP6) recommends integrating socio-economic and climate scenarios (SSP-RCPs) in climate change research to help policymakers address ecological degradation caused by future land use changes [9]. The Patch-based Land Use Simulation (PLUS) model, which combines land expansion analysis and random patch seed models, has become an important tool for simulating future land use scenarios and can describe the evolution of land use types in detail [10].

Different ecological processes of ESs often form competitive or complementary relationships due to limited resources and stakeholders' preferences. These relationships manifest as trade-offs or synergies between services. Spatial trade-offs/synergies refer to the impact of changes in one service on the supply capacity of other services within the same or adjacent areas [11]. Various methods are currently used to analyze trade-offs and synergies between services, such as Spearman's rank correlation coefficient, Pearson's correlation coefficient, and binary spatial autocorrelation [12, 13]. However, these methods fail to reveal the spatial variability in service relationships. Therefore, we combine Spearman's rank correlation coefficient with Geographically Weighted Regression (GWR) to more comprehensively explore the spatial interactions of ESs.

Based on the above background, this study aims to: (1) Land use demand analysis based on the SD model: The SD model is used to forecast the land use demand in the Northwest Territories from 2020 to 2030. The changes in different development patterns of land use are quantified based on historical trends and future development needs. (2) Spatial simulation using the PLUS model: The spatial pattern of land use in the Northwest China region in 2030 was simulated using the PLUS model. The model integrates various factors affecting land use changes and provides a detailed spatial characterization of land configuration and distribution. (3) Evaluation of ESs based on the InVEST model: The InVEST model was used to evaluate the ESs in the Northwest China region in 2030. The model allowed us to assess the capacity of ecosystems to provide water yield (WY), soil conservation (SC), carbon storage (CS), and habitat quality (HQ) services. (4) Analysis of trade-offs and synergies among ESs: To explore the trade-offs and synergies among different ESs, we used Spearman's rank correlation analysis and the GWR model to explore the trade-offs/synergies among ESs at the global and spatial scales (Fig. 1).

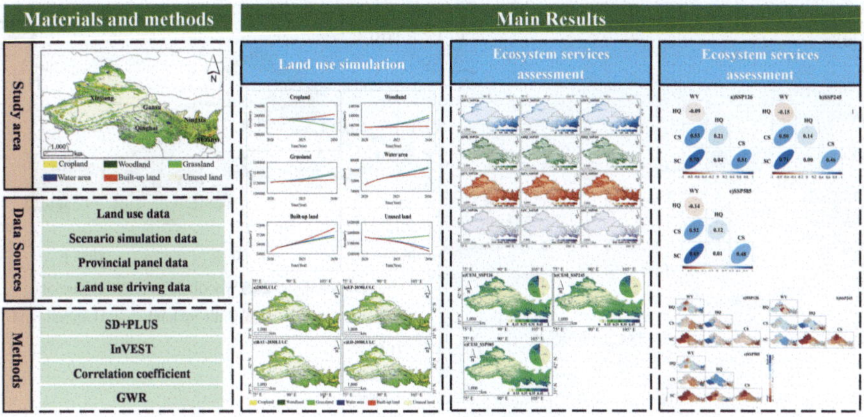

Fig. 1. Research framework.

2 Overview of the Study Area and Data Sources

2.1 Overview of the Study Area

The Northwest region (Fig. 2) is China's arid core area, encompassing Shaanxi, Gansu, Qinghai, Ningxia, and Xinjiang, with a total area of approximately 3 million km [2], accounting for 32% of China's land area [14]. Located between 73°–111°E and 31°–49°N, the region's terrain gradually rises from the northern lowlands to the southern plateau [15], with elevations ranging from −164 m to 8369 m. The climate is primarily temperate continental and alpine, with extremely low annual precipitation averaging less than 400 mm, and precipitation decreasing from southeast to northwest [16], showing very uneven distribution. Unused land, grasslands, and croplands dominate land cover types [17]. The region's ecological environment is fragile, with long-term drought and high evaporation leading to water resource shortages, severely constraining ecosystem and socio-economic development. Therefore, advancing ecological restoration and protection in the Northwest region is crucial for maintaining ecological balance and achieving sustainable socio-economic development.

2.2 Data Sources

The data required for this study include land use data, scenario simulation data, socio-economic panel data, land use driving factors data, and InVEST model data (Table 1). Among these, the land use data is reclassified according to the first-level classification standard, with land use types including: Arable land, forest land, grassland, water bodies, construction land, and unused land. Subsequent land use spatiotemporal simulations will use this standard. In the land use driving factors data, OSM data are calculated using the Euclidean distance method. All data are unified in projection coordinates and resampled to a 500 m resolution.

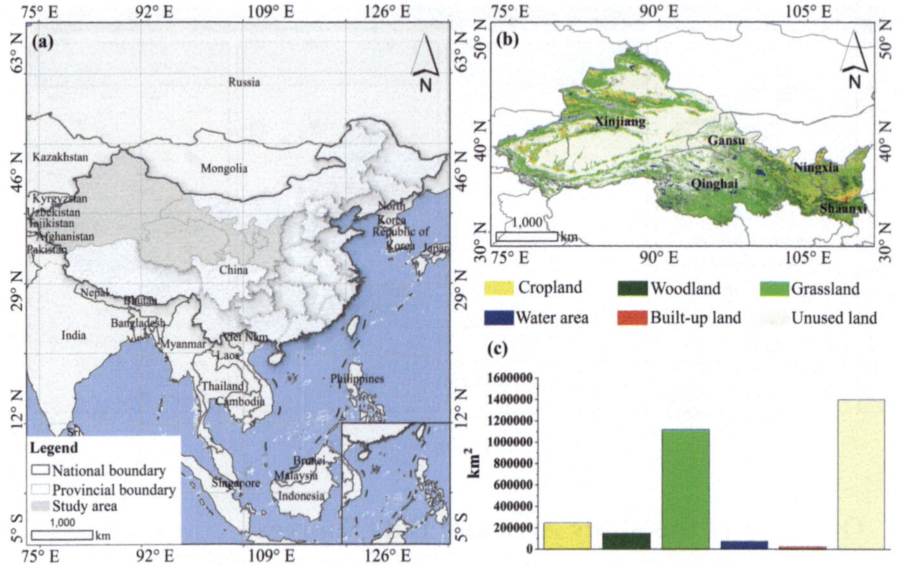

Fig. 2. Geographical location map of the research area. a. Geographical location of Northwest China, b. Spatial distribution of land use in Northwest China in 2020, c. Northwest China 2020 all types of land area.

3 Research Methodology

3.1 Scenario Setting

CMIP6 effectively integrates Shared Socioeconomic Pathways (SSPs) and Representative Concentration Pathways (RCPs) to generate various future scenarios for exploring potential development trends under different socioeconomic conditions and emission levels. These coupled scenarios provide macro-level predictions of human activities and their impacts [18]. We analyzed three SSP-RCP scenarios with different characteristics: (1) SSP1–2.6 emphasizes comprehensive management of ecosystems, including coordinated development of mountains, forests, water bodies, croplands, and deserts, with a focus on conserving natural resources. (2) SSP2–4.5 describes a scenario where socioeconomic and land-use trends are consistent with historical baselines, meaning economic and environmental conditions remain stable without significant changes. (3) SSP5–8.5 depicts a world with rapid fossil fuel use, highly uneven global economic development, and rapid population growth, assuming a lack of effective measures in climate policy and sustainable development, which indicates severe environmental pressures.

3.2 Land Use Demand Modelling

The Vensim software platform can be used to assess land use demands under various scenarios (Fig. 3). The system dynamics (SD) model consists of four core subsystems: population, economy, climate, and land use. The population subsystem focuses on population size and urbanization processes, where population growth increases living demands, thereby accelerating the development of various industries and increasing land demand. The economic subsystem affects fixed asset investment through changes in GDP, which in turn influences land demand. The climate subsystem analyzes the different impacts of temperature and precipitation on arable land, forest land, and grassland. The land use subsystem includes arable land, forest land, grassland, water areas, construction land, and unused land, with dynamic conversions and mutual constraints among these types.

Table 1. Data sources for this study.

Category	Data	Year	Source	Data resolution
Land use	Land use data	1990–2020	Data Center for Resources and Environmental Sciences, Chinese Academy of Sciences (http://www.resdc.cn)	30m
Landform	Elevation	2020		30m
	Slope			
	Aspect			
Soil	Soil type			1km
	Soil erosion intensity			1km
Atmosphere	Temperature		National Earth Science Data Center (https://www.geodata.cn/)	1km
	Precipitation			
Geographic data	Way		OpenStreetMap (https://www.openstreetmap.org/)	\
	Waters			
	Government			
	Railway			
Socio-economic data	Population		Data Center for Resources and Environmental Sciences, Chinese Academy of Sciences (http://www.resdc.cn)	1km
	GDP			
	Night light index			
	Statistical data	1990–2020		
Future data	Temperature	2021–230	NEX-GDDP-CMIP6(https://portal.nccs.nasa.gov/datashare/nexgddp_cmip6/) ACCESS-CE2 、BCC-CSM2-MR 、EC-Earch3 、GFDL-ESM4、 MIRO6	\

(*continued*)

Table 1. (*continued*)

Category	Data	Year	Source	Data resolution
	Precipitation			
	Population		Gridded datasets for population and economy under Shared Socioeconomic Pathways (https://doi.org/https://doi.org/10.57760/sciencedb.01683)	0.5° × 0.5°
	GDP			
	Urbanization rate			
	Evapotranspiration	2030	National Earth Science Data Center (https://www.geodata.cn/)	

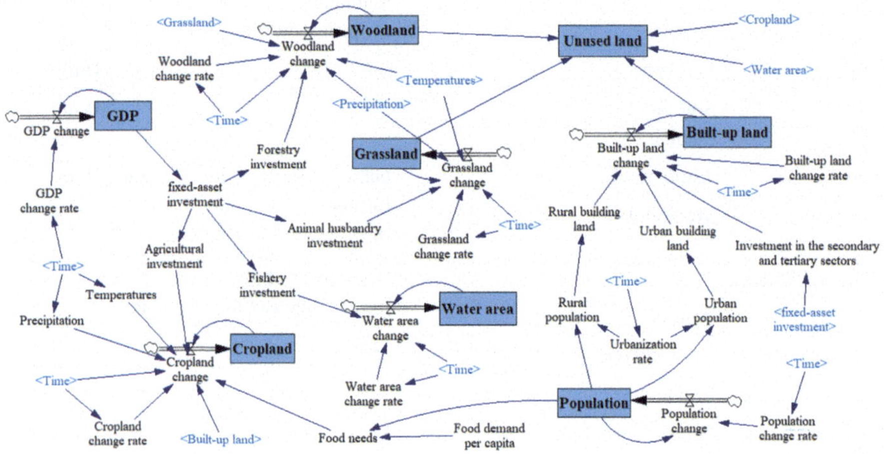

Fig. 3. Modelling of land-use demand forecasts.

3.3 Spatial Modelling of Land Use

This study utilizes the PLUS model to achieve spatial simulation of land use. First, in the LEAS module, the development potential of each land use type is obtained based on land use data from 2010 and 2020 and related driving factors (Fig. 4). Next, in the CARS module, neighborhood weight factors are determined by calculating the ratio of the expansion area of each land use type to the total expansion area, thereby realizing the spatial simulation of land use. The accuracy of the model is verified using the kappa coefficient and overall accuracy, with values ranging from 0 to 1; higher values indicate better simulation accuracy. A kappa coefficient greater than 0.75 is considered to indicate good simulation accuracy.

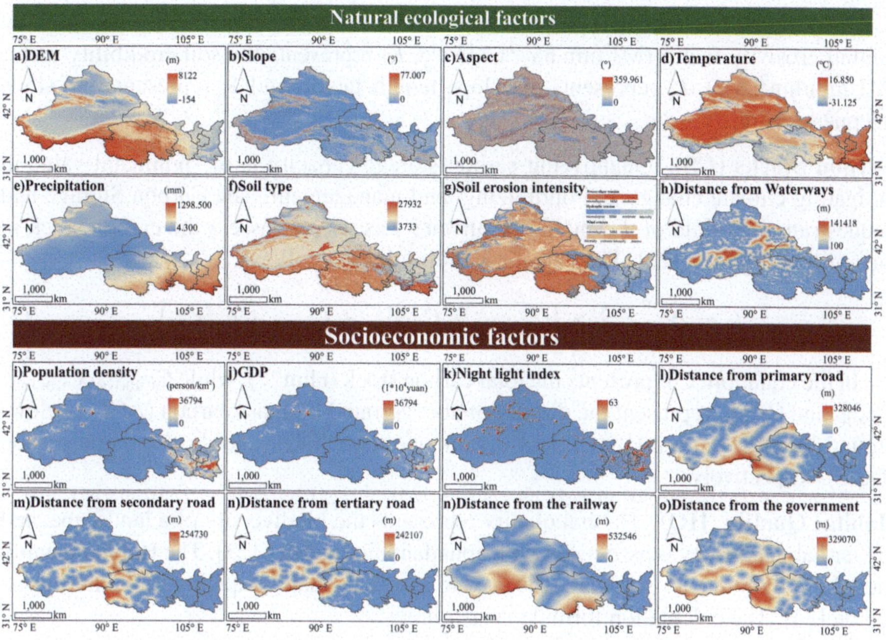

Fig. 4. Driving factors of land use in northwest China

3.4 Ecosystem Service Assessment

The InVEST model is a geographic model that utilizes land use data along with relevant biophysical, economic, and institutional variables specific to the study area [19]. This study employs the InVEST model, selecting WY, SC, CS and HQ to assess the ecosystem services generated under various scenarios in the Northwest region.

Water Yield (WY). The abundance of water resources plays a decisive role in safeguarding agricultural and ecological water security. The Water Yield module of the InVEST software was used to assess the water production function in the northwest region [20]. The calculation formula is as follows:

$$Y(i) = \left(1 - \frac{AET(i)}{P(i)}\right) \cdot P(i) \tag{1}$$

In the equation, $Y(i)$ represents the annual water yield (mm), $AET(i)$ represents the actual evapotranspiration (mm), and $P(i)$ represents the annual average precipitation (mm). During the modeling process, the Z parameter needs to be adjusted based on the average water yield modulus of the five provinces in the northwest region.

Soil Conservation (SC). Soil conservation plays a crucial role in preventing soil erosion and promoting regional sustainable development. The Sediment Delivery Ratio module of the InVEST software was used to assess the soil conservation function in the northwest region [21]. The calculation formula is as follows:

$$PKLS_i = R_i \times K_i \times L_i \times S_i \tag{2}$$

In the equation, $PKLS_i$ represents the potential soil erosion (t), R_i represents the rainfall erosivity factor (MJ mm·ha^{-1}h^{-1}a^{-1}), K_i represents the soil erodibility factor (MJ mm/hm^2 h a), L_i represents the slope length factor, and S_i represents the slope steepness factor.

Carbon Stocks (CS). Quantifying carbon storage capacity is of significant value in mitigating climate change and optimizing land management. The Carbon Storage and Sequestration module of the InVEST software was used to assess the carbon stock in the northwest region [22]. The calculation formula is as follows:

$$C = \sum\nolimits_{i=1}^{n} S_i \times \left(C_{i,above} + C_{i,below} + C_{i,soil} + C_{i,dead}\right) \tag{3}$$

In the equation, C represents the total carbon stock (t hm^{-2}), while $C_{i,above}$, $C_{i,below}$, $C_{i,soil}$, and $C_{i,dead}$ represent the density of aboveground biomass carbon (AGC), below-ground biomass carbon (BGC), soil organic carbon (SOC), and dead organic carbon (DOC), respectively.

Habitat Quality (HQ). Habitat quality represents the biodiversity of a landscape, and higher habitat quality is associated with abundant biodiversity [23]. The Habitat Quality module of the InVEST software was used to assess the habitat quality in the northwest region [24]. The calculation formula is as follows:

$$Q_{xj} = H_j \times \left[1 - \left(\frac{D_{xj}^z}{D_{xj}^z + k^z}\right)\right] \tag{4}$$

In the equation, Q_{xj} represents the habitat suitability index of grid cell x in land type j, H_j represents the habitat suitability of land type j, D_{xj}^z represents the habitat degradation level of grid cell x in land type j, k is the semi-saturation and coefficient, with a default value of 0.5 in the model. In this study, we consider cropland, construction land, and unused land as threat sources.

3.5 Trade-Offs and Synergies

Spearman Correlation Analysis. Using a 10 km × 10 km grid as the study unit, Spearman correlation is employed to identify the relationships between ecosystem services in the study area. If two ecosystem services are positively correlated, it is considered a synergistic relationship; if negatively correlated, it is deemed a trade-off relationship. The absolute value of the correlation coefficient can determine the strength of the relationship between two ecosystem services [11].

Geographically Weighted Regression Models. Based on traditional regression analysis methods, the GWR model has been further extended and refined. It allows for local rather than global parameter estimation by incorporating parameters that represent spatial autocorrelation and heterogeneity, thus reflecting the spatial variability in the contribution of samples to the regression equation and enhancing the reliability of the regression results [25]. The basic expression is as follows:

$$Y = \beta_0(u_i, v_i) + \sum\nolimits_k \beta_k(u_i, v_i)x_{ik} + \varepsilon_i \tag{5}$$

In the equation: $\beta_0(u_i,v_i)$ represents the intercept; $\sum_k \beta_k(u_i, v_i)$ denotes the regression coefficients for the variable x_{ik}; ε_i is the random error term.

4 Analysis of Results

4.1 Land Use Demand Modelling for 2020–2030

Comparing the simulation results for 1990, 2000, 2010, and 2020 with remote sensing monitoring results verifies whether the simulation accuracy meets the modeling requirements. The results show (Table 2) that the errors for each year and the average error are within 2%, indicating a satisfactory simulation performance suitable for predicting land use demands under different scenarios.

Table 2. SD model accuracy validation.

Type	1990	2000	2010	2020	Average
Cropland	0.000%	0.213%	0.504%	0.577%	0.323%
Woodland	0.000%	0.035%	0.192%	0.194%	0.105%
Grassland	0.000%	0.047%	0.119%	0.190%	0.089%
Water area	0.000%	0.153%	1.924%	2.164%	1.060%
Built-up land	0.000%	0.388%	5.115%	1.822%	1.831%
Unused land	0.007%	0.077%	0.362%	0.419%	0.216%

As shown in Fig. 5, there are significant differences in the development of land use types under the three scenarios. In all scenarios, forest land, grassland, water bodies, and built-up areas exhibit an upward trend. Specifically, forest land, grassland, and water bodies show similar patterns, with the highest levels in SSP126 and the lowest in SSP585. Conversely, cropland and built-up areas display different trends: These two land types significantly increase under the SSP585 scenario but decrease notably under the SSP126 scenario. Under the SSP245 scenario, cropland area shows no substantial change over the decade. Regarding unused land, there is no significant change in the SSP126 scenario, whereas SSP245 and SSP585 scenarios show considerable variations. This indicates that different scenarios have varying impacts on land use types, reflecting diverse outcomes of land management strategies and policy interventions under different predictive pathways [26].

4.2 Spatial Modelling of Land Use, 2020–2030

According to the 2023 land use spatial simulation results shown in Fig. 6, while the overall land use spatial patterns are similar across different scenarios, significant differences still exist at the local scale. These differences reflect the varying impacts of each

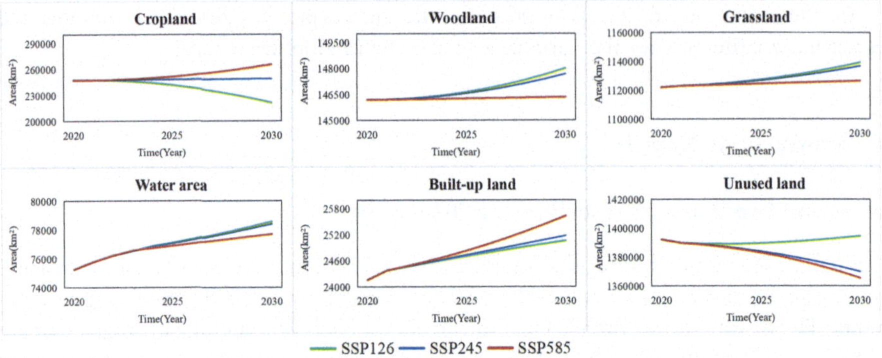

Fig. 5. Forecast results of land use demand in northwest China

scenario on land use patterns. The SSP126 scenario emphasizes ecological prioritization, leading to notable changes in land use. Specifically, the shift from arable land and unused land to ecological land is very apparent. In this scenario, the transition of arable land to grassland is particularly striking, with a total area of 17,480 square kilometers. This indicates that the SSP126 scenario has a significant effect on promoting ecological restoration and increasing ecological land [27]. The SSP245 scenario's land use pattern continues the current development trend, primarily characterized by further development of unused land. Specifically, the areas converted from unused land to arable land, grassland, and water bodies are 4,860 square kilometers, 17,698.5 square kilometers, and 5,464.75 square kilometers, respectively. This pattern reflects the continued utilization and expansion of existing resources in the SSP245 scenario. The SSP585 scenario emphasizes human activities and economic development, leading to a notable shift of forest and unused land to arable land. In this scenario, the area of forest converted to arable land reaches 5,061.5 square kilometers, while the area of unused land converted to arable land is 16,631.75 square kilometers. This indicates that the economic orientation of land use in the SSP585 scenario significantly drives agricultural expansion.

4.3 Results of Ecosystem Services Assessments

This study assessed the levels of ESs in the Northwest region under different scenarios for the year 2030 (see Fig. 7). The results reveal regional differences in various ecosystem services as follows: WY, CS, and SC are higher in the Northwest and Southeast regions, while they are relatively lower in the central region. This indicates that these services are concentrated more in the Northwest and Southeast regions, whereas the central region has lower service levels. In contrast, the distribution pattern of HQ is opposite to the three services, being higher in the central region and lower in the Northwest and Southeast regions.

Using the natural breaks method, various ESs were classified and compared across different scenarios. Specifically, WY performed best in the SSP585 scenario among the three scenarios. Particularly, the area of grades IV and V reached 318,671 square kilometers, reflecting the optimal state of hydrological services in this scenario. HQ in the

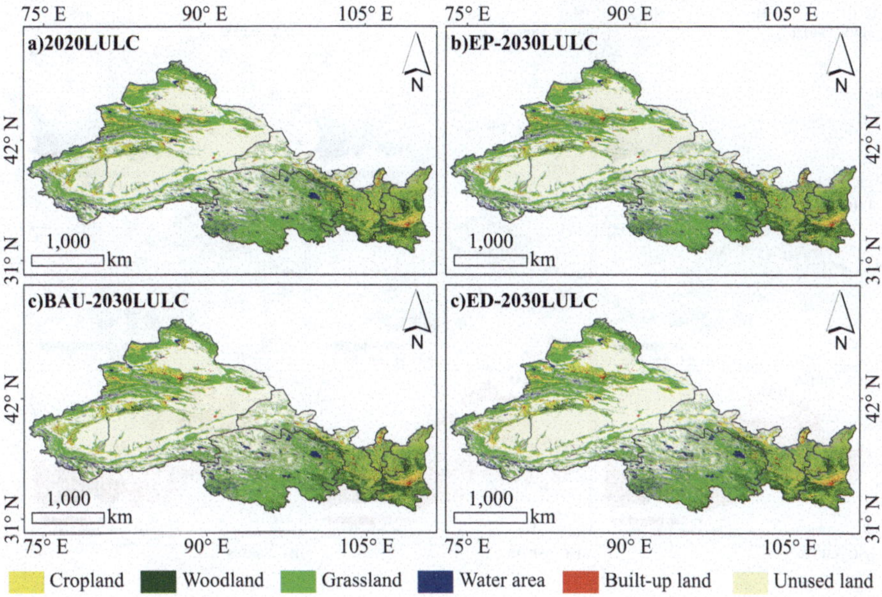

Fig. 6. Multi-scenario spatial simulation results of land use in the Middle East Northwest China

SSP126 scenario significantly outperformed the other two scenarios in habitat quality. Although HQ is grade II in most areas, the area of grade I in the SSP126 scenario decreased by 6.88% compared to the SSP585 scenario, indicating that the SSP126 scenario is more favorable for improving habitat quality. By 2030, CS in the Northwest region will reach 3192.50 Tg. Notably, in the SSP126 scenario, the proportion of carbon storage at grade IV reached 5.44%, which is significantly higher than the other two scenarios, showing better carbon storage protection in this scenario. SC in the Northwest region is mainly concentrated in grade I, with a proportion exceeding 75%. Only in the southeastern part of the study area is the grade slightly higher than I, indicating that in most regions, the SC level is relatively consistent, but the southeastern part has slightly higher reserves.

The four types of ESs were averaged using an equal-weight method to calculate the 2030 multi-scenario Comprehensive Ecosystem Service Index (CESI) for the Northwest region. The results show (Fig. 8) that the CESI in the Northwest region is predominantly at levels I and II, mainly distributed in the central part of the study area, with the primary land use type being unused land. This distribution indicates that the central region of the study area has high potential for ESs development. Among the three scenarios, the CESI for SSP126 is significantly higher than the other two scenarios, with levels IV and V accounting for 14.08%, covering an area of 423,167.25 km^2. This suggests that the SSP126 scenario provides the best conditions for ecosystem services in the Northwest region and is most suitable for promoting ecological sustainability in the area. In the SSP245 scenario, level I accounts for the smallest proportion, which means that according to the current development model, although there is some improvement in low-level ecological areas in the Northwest region, the overall ecological development

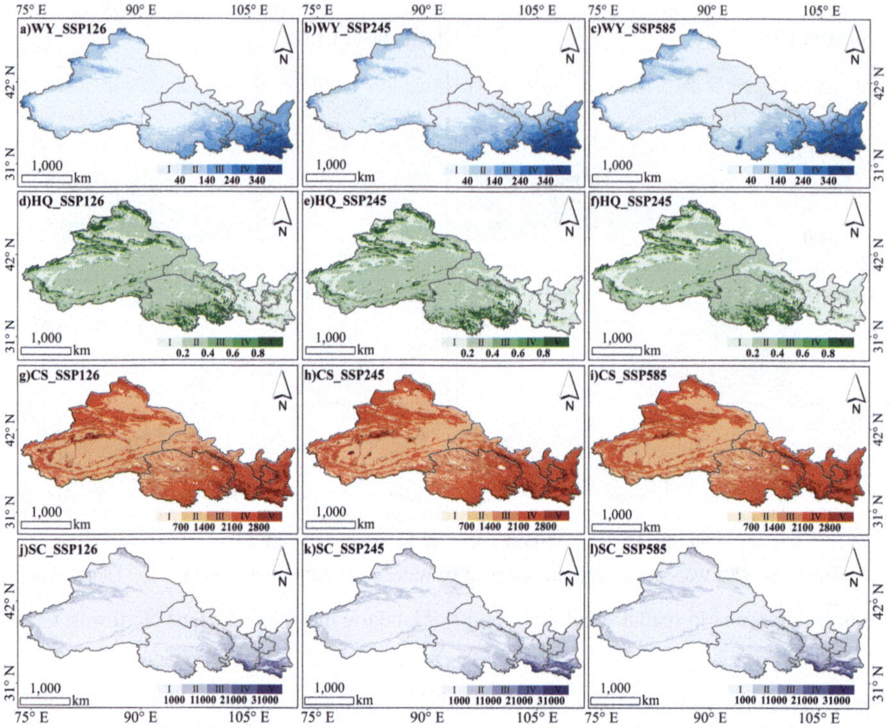

Fig. 7. Evaluation results of multi-scenario ESs in northwest China

level has not significantly increased. Even with some improvement measures, they are insufficient to elevate the entire region's ecosystem service level to a higher standard.

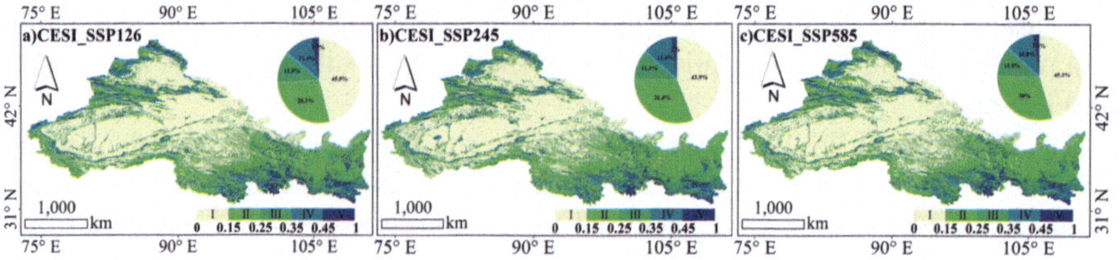

Fig. 8. Evaluation results of multi-scenario CESI in northwest China

4.4 Trade-Offs/synergies

According to the correlation analysis (Fig. 9), within the study area, other ESs generally exhibit synergies, except for WY-HQ, which shows a trade-off. Under the SSP126 scenario, the trade-off effect of WY-HQ is minimal, with a correlation coefficient of

−0.09. This indicates that under the SSP126 scenario, the mutual constraints of WY-HQ are the weakest, reflecting that ecological protection measures and rational land use in this scenario effectively mitigate the negative effects between these services and enhance the overall benefits of ecosystem services [28]. In contrast, under the SSP245 and SSP585 scenarios, the trade-off effect of WY-HQ is significantly higher than under SSP126, indicating stronger mutual constraints between ecosystem services in these two scenarios. SC-WY shows a significant synergistic relationship, especially under the SSP245 scenario (0.71), where this synergy is most pronounced. Although CS-HQ exhibits a weak similar effect, the synergistic effect between the two is significantly better in the SSP126 scenario compared to other scenarios, suggesting that the SSP126 scenario better achieves the synergy of CS-HQ, thereby enhancing the overall effectiveness of ecosystem services. Under the SSP585 scenario, the synergistic effects among most ecosystem services are the weakest. This suggests that in this scenario, the positive interactions between ecosystem services are significantly reduced, which may have a substantial negative impact on the overall function and service levels of the ecosystem.

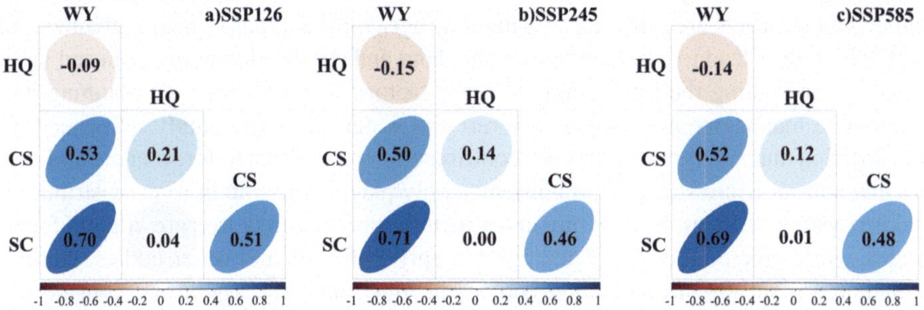

Fig. 9. Trade-offs/synergies among multi-scenario ESs in Northwest China in 2030

According to the results of the GWR analysis (Fig. 10), the trade-offs and synergies between different ESs show significant spatial distribution differences at the spatial scale. Specifically, the trade-off between WY and HQ is mainly concentrated in the southeastern part of the study area, and this trade-off becomes more pronounced as the geographic location extends eastward. This phenomenon may be related to the specific environmental conditions and land use patterns in that area [29]. Under the SSP126 scenario, the synergy between CS and HQ is significantly higher than in the other two scenarios. This is primarily because, in this scenario, the synergy between ESs in the central region of the study area is enhanced, whereas in other scenarios, especially in the SSP585 scenario, the trade-off between CS and HQ remains evident. The trade-off under the SSP585 scenario is particularly pronounced, indicating that negative interactions between ecosystem services are most significant in this scenario. Additionally, there is a strong synergy between SC and HQ, which is mainly due to the particularly strong synergy effects in the northwest part of the study area. The ecological conditions and environmental factors in this region may contribute to more pronounced positive interactions between SC and HQ, leading to the formation of strong synergies.

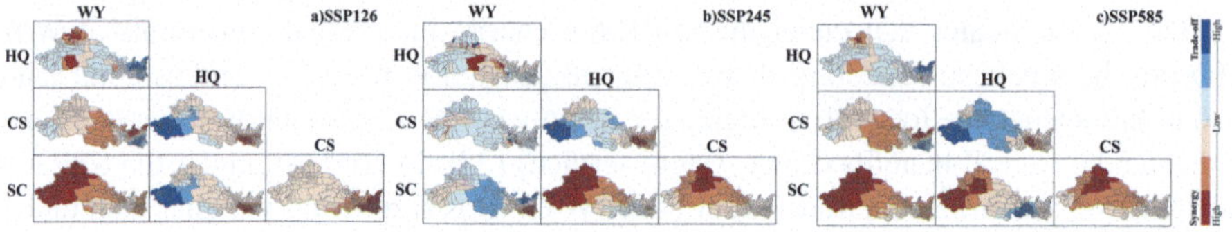

Fig. 10. Pairwise trade-offs/synergies among multi-scenario ESs in Northwest China in 2030

5 Discussion

5.1 Factors Affecting the Evolution of Land-Use Patterns

The sensitivity differences of various land types to driving factors reveal their dynamic response mechanisms in environmental changes (Fig. 11). Arable land and unused land are significantly affected by population density and GDP, indicating that economic development and population growth directly drive the increased demand for these land types, promoting their development and transformation. For example, high population density and economic levels are often accompanied by agricultural expansion and urbanization [30], which may lead to a reduction in arable land and the development of unused land. Thus, understanding the interactions of these factors is crucial for land planning and sustainable management. Changes in forest and water areas are mainly influenced by precipitation and soil types. Soil types contribute up to 26.328% to forest areas, indicating that soil moisture storage and nutrient supply play a core role in forest distribution and ecosystem services. Soil texture and structure directly affect tree growth and forest health, while precipitation regulates water supply to forests and water bodies, thereby influencing plant growth and water ecological functions [31]. Therefore, considering soil characteristics and precipitation changes helps in predicting and managing forest and water area changes. Grassland changes are mainly driven by temperature and GDP. Temperature changes affect the growing season and growth rate of grasslands, while GDP changes reflect the demand for grassland utilization driven by economic activities. For example, economic growth may increase grazing pressure on grasslands or lead to their conversion for agricultural use [32]. Therefore, climate change and economic development trends significantly impact grassland management. The expansion of construction land is primarily influenced by precipitation and nighttime light index. Precipitation affects land developability, while the nighttime light index reflects the extent of urbanization. The nighttime light index, with a high contribution of 22.532%, indicates that urban light pollution and human activities are key factors driving the expansion of construction land [33]. This expansion not only alters land functions but may also put pressure on surrounding ecosystems. Therefore, closely monitoring urbanization processes and light pollution changes is crucial for developing land management strategies.

5.2 Strengths and Limitations of This Study

In this study, we employed various SSP-RCP scenarios provided by CMIP6 to simulate the spatiotemporal changes in land use and its multiple ESs using the SD, PLUS, and

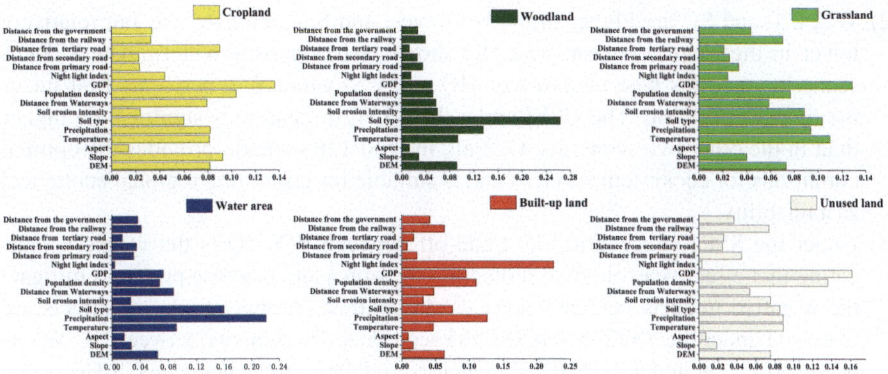

Fig. 11. Class driving factor contribution in northwest China

InVEST models. Specifically, we used annual data on temperature, precipitation, population, GDP, and urbanization rates from multiple SSP-RCP scenarios to set the simulation parameters for the SD model, enabling robust predictions of land use changes. The coupling of the SD and PLUS models allows for a comprehensive consideration of land use demand and spatial distribution patterns from both top-down and bottom-up perspectives, thereby providing more accurate predictions of the impacts of climate change and socio-economic development on land use trends.

This study focuses on three representative economic development and climate change scenarios (SSP126, SSP245, and SSP585) [34] and does not address specific ecosystem service policies [35]. Additionally, although the InVEST model is a commonly used tool for assessing ecosystem services, its parameter settings involve a degree of subjectivity and lack objective standards [36]. Future research is recommended to consider more SSP-RCP scenarios and integrate field research with predictive models to improve the accuracy of ecosystem service assessments.

6 Conclusion

This study is based on the scenario parameters provided by CMIP6 and uses the coupled SD-PLUS model to simulate the multi-scenario land use spatial patterns in the Northwest region for 2030. Subsequently, the InVEST model is employed to simulate the development levels of various ESs under multiple scenarios and explore the trade-offs and synergies between these ESs. Our research results indicate significant heterogeneity in land use and ESs among the three scenarios. We have drawn the following conclusions:

(1) The SSP126 scenario promotes ecological restoration, significantly reducing cropland and increasing ecological land such as grasslands. The SSP245 scenario continues the existing development trend, converting unused land into cropland, grasslands, and water bodies. The SSP585 scenario emphasizes economic growth, leading to the conversion of forest and unused land into cropland and promoting agricultural expansion.

(2) WY, CS, and SC are higher in the Northwest and Southeast regions but relatively lower in the Central region, with HQ showing the opposite pattern. The SSP126 scenario performs best in terms of HQ and CS, while WY performs best under the SSP585 scenario. The CESI under the SSP126 scenario is significantly higher than in the other two scenarios. Overall, the SSP126 scenario provides the optimal conditions for ecosystem services and is suitable for promoting regional ecological sustainability.

(3) Under the SSP126 scenario, the trade-off between WY-HQ is the weakest, indicating that effective ecological protection measures and land use practices mitigate the negative impacts between services. In contrast, stronger trade-off effects are observed under the SSP245 and SSP585 scenarios. The synergy between SC-WY is most significant under the SSP245 scenario, and the synergy between CS-HQ under the SSP126 scenario is superior to other scenarios, suggesting that SSP126 better enhances the overall effectiveness of ecosystem services. Conversely, the synergy under the SSP585 scenario is the weakest, which may negatively impact ecosystem functions. Additionally, trade-offs and synergies exhibit notable spatial distribution differences, such as the trade-off between WY-HQ being concentrated in the Southeast, while the strong synergy between SC-HQ is mainly concentrated in the Northwest.

References

1. Chang, B., et al.: Analysis of trade-off and synergy of ecosystem services and driving forces in urban agglomerations in Northern China. Ecological Indicators **165**, 112210 (2024). https://doi.org/10.1016/j.ecolind.2024.112210
2. Hua, Y., Yan, D., Liu, X.: Assessing synergies and trade-offs between ecosystem services in highly urbanized area under different scenarios of future land use change. Environmental and Sustainability Indicators **22**, 100350 (2024). https://doi.org/10.1016/j.indic.2024.100350
3. Wang, K., Gao, J., Liu, C., Zhang, Y., Wang, C.: Understanding the effects of socio-ecological factors on trade-offs and synergies among ecosystem services to support urban sustainable management: A case study of Beijing. China. Sustainable Cities and Society **100**, 105024 (2024). https://doi.org/10.1016/j.scs.2023.105024
4. Ma, X., et al.: Assessing the relative contributions, combined effects and multiscale uncertainty of future land use and climate change on water-related ecosystem services in Southwest China using a novel integrated modelling framework. Sustainable Cities and Society **106**, 105400 (2024). https://doi.org/10.1016/j.scs.2024.105400
5. Qiu, J., Yu, D., Huang, T.: Influential paths of ecosystem services on human well-being in the context of the sustainable development goals. Science of The Total Environment **852**, 158443 (2022). https://doi.org/10.1016/j.scitotenv.2022.158443
6. Zheng, S., et al.: Linking cultural ecosystem service and urban ecological-space planning for a sustainable city: Case study of the core areas of Beijing under the context of urban relieving and renewal. Sustainable Cities and Society **89**, 104292 (2023). https://doi.org/10.1016/j.scs.2022.104292
7. Li, Y., et al.: The role of land use change in affecting ecosystem services and the ecological security pattern of the Hexi Regions, Northwest China. Science of The Total Environment **855**, 158940 (2023). https://doi.org/10.1016/j.scitotenv.2022.158940

8. Liao, W., et al.: Projections of land use changes under the plant functional type classification in different SSP-RCP scenarios in China. Sci Bull (Beijing) **65**, 1935–1947 (2020). https://doi.org/10.1016/j.scib.2020.07.014
9. Wang, W., Yu, H., Tong, X., Jia, Q.: Estimating terrestrial ecosystem carbon storage change in the YREB caused by land-use change under SSP-RCPs scenarios. Journal of Cleaner Production **469**, 143205 (2024). https://doi.org/10.1016/j.jclepro.2024.143205
10. Liang, X., et al.: Understanding the drivers of sustainable land expansion using a patch-generating land use simulation (PLUS) model: A case study in Wuhan, China. Computers, Environment and Urban Systems **85**, 101569 (2021). https://doi.org/10.1016/j.compenvurbsys.2020.101569
11. Huang, Y., Wu, J.: Spatial and temporal driving mechanisms of ecosystem service trade-off/synergy in national key urban agglomerations: A case study of the Yangtze River Delta urban agglomeration in China. Ecological Indicators **154**, 110800 (2023). https://doi.org/10.1016/j.ecolind.2023.110800
12. Yang, Y., et al.: Spatiotemporal variation of essential ecosystem services and their trade-off/synergy along with rapid urbanization in the Lower Pearl River Basin. China. Ecological Indicators **133**, 108439 (2021). https://doi.org/10.1016/j.ecolind.2021.108439
13. Qiao, X., et al.: Temporal variation and spatial scale dependency of the trade-offs and synergies among multiple ecosystem services in the Taihu Lake Basin of China. Sci. Total Environ. **651**, 218–229 (2019). https://doi.org/10.1016/j.scitotenv.2018.09.135
14. Zhang, L., et al.: Spatio-temporal characteristics and driving mechanism of land degradation sensitivity in Northwest China. Science of The Total Environment **918**, 170403 (2024). https://doi.org/10.1016/j.scitotenv.2024.170403
15. Chen, N., et al.: Contribution of multi-objective land use optimization to carbon neutrality: A case study of Northwest China. Ecological Indicators **157**, 111219 (2023). https://doi.org/10.1016/j.ecolind.2023.111219
16. Su, J., Fan, L., Yuan, Z., Wang, Z., Wang, Z.: Quantifying the drought sensitivity of grassland under different climate zones in Northwest China. Science of The Total Environment **910**, 168688 (2024). https://doi.org/10.1016/j.scitotenv.2023.168688
17. Xu, X., et al.: Persistence of increasing vegetation gross primary production under the interactions of climate change and land use changes in Northwest China. Science of The Total Environment **834**, 155086 (2022). https://doi.org/10.1016/j.scitotenv.2022.155086
18. Cai, G., et al.: Predicting the ecosystem service values and constructing ecological security patterns in future changing land use patterns. Ecological Indicators **154**, 110787 (2023). https://doi.org/10.1016/j.ecolind.2023.110787
19. Wenbo, X., Hengzhou, X., Xiaoyan, L., Hua, Q., Ziyao, W.: Ecosystem services response to future land use/cover change (LUCC) under multiple scenarios: A case study of the Beijing-Tianjin-Hebei (BTH) region. China. Technological Forecasting and Social Change **205**, 123525 (2024). https://doi.org/10.1016/j.techfore.2024.123525
20. Liu, M., et al.: Spatiotemporal changes in future water yield and the driving factors under the carbon neutrality target in Qinghai. Ecological Indicators **158**, 111310 (2024). https://doi.org/10.1016/j.ecolind.2023.111310
21. Xu, J., Chen, J., Liu, Y., Fan, F.: Identification of the geographical factors influencing the relationships between ecosystem services in the Belt and Road region from 2010 to 2030. Journal of Cleaner Production **275**, 124153 (2020). https://doi.org/10.1016/j.jclepro.2020.124153
22. Guo, W., et al.: A new assessment framework to forecast land use and carbon storage under different SSP-RCP scenarios in China. Science of The Total Environment **912**, 169088 (2024). https://doi.org/10.1016/j.scitotenv.2023.169088

23. Sharps, K., et al.: Comparing strengths and weaknesses of three ecosystem services modelling tools in a diverse UK river catchment. Sci. Total Environ. **584–585**, 118–130 (2017). https://doi.org/10.1016/j.scitotenv.2016.12.160
24. Wang, B., Oguchi, T., Liang, X.: Evaluating future habitat quality responding to land use change under different city compaction scenarios in Southern China. Cities **140**, 104410 (2023). https://doi.org/10.1016/j.cities.2023.104410
25. Zhang, Q., Shen, J., Sun, F.: Spatiotemporal differentiation of coupling coordination degree between economic development and water environment and its influencing factors using GWR in China's province. Ecological Modelling **462**, 109794 (2021). https://doi.org/10.1016/j.ecolmodel.2021.109794
26. Chen, Q., Ning, Y.: Projecting LUCC dynamics and ecosystem services in an emerging urban agglomeration under SSP-RCP scenarios and their management implications. Science of The Total Environment **949**, 175100 (2024). https://doi.org/10.1016/j.scitotenv.2024.175100
27. Wang, Z., et al.: Dynamic simulation of land use change and assessment of carbon storage based on climate change scenarios at the city level: A case study of Bortala. China. Ecological Indicators **134**, 108499 (2022). https://doi.org/10.1016/j.ecolind.2021.108499
28. Zhang, Y., Yu, P., Tian, Y., Chen, H., Chen, Y.: Exploring the impact of integrated spatial function zones on land use dynamics and ecosystem services tradeoffs based on a future land use simulation (FLUS) model. Ecological Indicators **150**, 110246 (2023). https://doi.org/10.1016/j.ecolind.2023.110246
29. Gong, J., et al.: Tradeoffs/synergies of multiple ecosystem services based on land use simulation in a mountain-basin area, western China. Ecol. Ind. **99**, 283–293 (2019). https://doi.org/10.1016/j.ecolind.2018.12.027
30. Hou, D., Meng, F., Prishchepov, A.V.: How is urbanization shaping agricultural land-use? Unraveling the nexus between farmland abandonment and urbanization in China. Landscape and Urban Planning **214**, 104170 (2021). https://doi.org/10.1016/j.landurbplan.2021.104170
31. Liyun, W., Weibin, Y., Zhirong, J., Shihong, X., Dongjin, H.: Ecosystem health assessment of Dongshan Island based on its ability to provide ecological services that regulate heavy rainfall. Ecol. Ind. **84**, 393–403 (2018). https://doi.org/10.1016/j.ecolind.2017.09.006
32. Dong, H., et al.: Enhancing sustainability: Exploring the evolution and dynamics of the grassland social-ecological system in Inner Mongolia over the past 260a. Environmental Impact Assessment Review **106**, 107537 (2024). https://doi.org/10.1016/j.eiar.2024.107537
33. Lu, Y., Gong, M., Lu, L., Wang, Y., Wang, Y.: Urban polycentrism and total-factor energy efficiency: An analysis based on the night light data. Technological Forecasting and Social Change **198**, 122984 (2024). https://doi.org/10.1016/j.techfore.2023.122984
34. Gomes, E., et al.: Future land-use changes and its impacts on terrestrial ecosystem services: A review. Science of The Total Environment **781**, 146716 (2021). https://doi.org/10.1016/j.scitotenv.2021.146716
35. Ai, X., et al.: Climate and land use changes impact the trajectories of ecosystem service bundles in an urban agglomeration: Intricate interaction trends and driver identification under SSP-RCP scenarios. Science of The Total Environment **944**, 173828 (2024). https://doi.org/10.1016/j.scitotenv.2024.173828
36. Vallet, A., et al.: Relationships Between Ecosystem Services: Comparing Methods for Assessing Tradeoffs and Synergies. Ecol. Econ. **150**, 96–106 (2018). https://doi.org/10.1016/j.ecolecon.2018.04.002

Advancing Decision Support Through Knowledge Services in the Qinling Region's Natural Resource Analysis and Evaluation

Xiaoyang Zhang[1,2], Sijia Wang[3,4](✉), Bingliang Cui[1,2], Panding Zhi[1,2], and Jiacheng Xu[5]

[1] Guangzhou Alpha Software Information Technology Co., Ltd., Guangzhou 510630, China
[2] Key Laboratory of Earth Surface System and Human-Earth Relations, Ministry of Natural Resources of China, Beijing 518055, China
[3] School of Geosciences and Info-Physics, Central South University, Changsha 410083, China
wsjgps@163.com
[4] Key Laboratory of Metallogenic Prediction of Nonferrous Metals and Geological Environment Monitoring Ministry, Ministry of Education, Beijing 410083, China
[5] College of Geoscience and Surveying Engineering, China University of Mining and Technology, Beijing 100083, China

Abstract. This study introduces a pioneering knowledge service framework designed to bridge the gap between the abundance of information and the scarcity of actionable knowledge in the natural resource survey and monitoring sector in the Qinling region. The framework is designed to significantly improve the utility of survey and monitoring data in decision-making processes through natural resource analysis and assessment knowledge services. The paper first establishes a comprehensive knowledge system that intricately integrates various dimensions, such as geographic characteristics, resource conditions, human activities, and spatial control, to provide a comprehensive view of the characteristics, conditions, and evolutionary trends of natural resources. The study then develops a scenario-oriented knowledge hypergraph model to support decision-making in natural resource management. This model uses hypergraph theory to highlight the complex connections and synergies among knowledge elements within different scenarios. The study concludes with the development of a knowledge service system specifically curated for management applications, introducing a suite of services including knowledge visualization, thematic knowledge, knowledge search functionalities, and an interactive knowledge Q&A. These features are designed to precisely meet the users' needs and provide customized knowledge recommendations. The results of the study validate the viability of the proposed knowledge service framework and highlight its potential for practical implementation. In conclusion, this research not only extends the theoretical underpinnings of knowledge services, but also presents a case study that illustrates the application of knowledge services in strengthening decision support for ecological balance and sustainable development within the region.

Keywords: Natural resource analysis and evaluation · Natural resource survey and monitoring · Knowledge system · Scenario-oriented knowledge hypergraph model · Knowledge services

G. Luo (Ed.): GFSISD 2024, SPPHY 417, pp. 95–111, 2025.
https://doi.org/10.1007/978-981-96-2490-4_7

1 Introduction

One of the main objectives of natural resources surveying and monitoring is to systematically collect and analyze data, identify current conditions, trends, and underlying issues of natural resources, and provide the basis for integrated conservation and restoration of landscapes, including mountains, waters, forests, farmlands, lakes, and grasslands, and support the scientific planning and management of territorial spaces. [1]. The scope of this work goes beyond data collection and compilation. It also includes in-depth analysis and understanding of both static and dynamic data to ensure a scientific and accurate basis for resource management and decision making [2]. In recent years, the Qinling region has successfully used survey and monitoring data to develop a comprehensive set of analytical assessments. These assessments detail the types, distribution, quantity, quality, and fluctuating trends of natural resources, thereby providing a basic framework for their effective management and conservation [3]. Nevertheless, the challenge of further harnessing the potential of these analytical findings to furnish knowledge services conducive to decision-making support remains a formidable endeavor.

Knowledge services for the analysis and evaluation of natural resources within the Qinling region encompass a myriad of facets, including the effective structuring and stewardship of survey and monitoring data, the intensification of analytical depth, and the refinement of service methodologies. Presently, the majority of research is concentrated on the technical frameworks and indicator systems for surveys and monitoring, exemplified by the routine monitoring schemes and integrated survey and monitoring paradigms proposed by Guo Haijing et al. [4, 5]. While these studies offer a methodological foundation at the technical level, they exhibit a paucity of focus on the systematic knowledge structuring of analytical outcomes. Concurrently, certain studies, such as those by Zhao Kan [6], focus solely on the extraction and analysis of individual data elements, neglecting an in-depth examination of the survey and monitoring results from a business perspective. This omission hinders decision-makers from extracting comprehensive operational support and insights from the data. Moreover, research conducted by He Lihua et al. [7], despite its emphasis on analysis and assessment, primarily targets the optimization of the indicator system rather than the enhancement of knowledge service capabilities. This approach somewhat curtails the holistic and profound excavation of knowledge. Furthermore, the prevailing models of knowledge service [8], including large-screen presentations and interactive 2D/3D maps, while facilitating the display of information, restrict the capacity for multimodal knowledge responses. Consequently, the systematic organization and mining of knowledge from natural resource surveys and monitoring, coupled with the provision of a continuum from data to information to knowledge services, is pivotal to the empowerment of research within the domain of survey and monitoring data [9].

Significant strides have been achieved in the exploration of constructing a service architecture that leverages the spatial and semantic interrelations among entities, with a core focus on spatiotemporal data analytics, intelligent analysis, and knowledge services. This research trajectory not only epitomizes the vanguard scientific challenges within the domain but also stands as a pivotal conduit for augmenting the intrinsic value and service efficacy of spatiotemporal big data [10]. The ongoing evolution of knowledge graph technologies has engendered novel analytical perspectives within the realms of Earth sciences, atmospheric pollution regulation, and typhoon disaster mitigation [11–14]. These knowledge graphs, characterized by their nuanced semantic network topologies, adeptly elucidate the spatiotemporal interconnections and profound semantic linkages among geographical entities, encompassing pivotal factors such as geographical conditions, resource valuations, and evolutionary trajectories [15]. Moreover, the advent of knowledge-steered spatiotemporal narrative visualization research, which amalgamates multimodal information streams, has not only diversified the modalities of knowledge representation but also substantially demystified the intricacies of data interpretation [16]. The trajectory of the field is increasingly oriented towards the cultivation of spatiotemporal situational awareness and cognitive analytic capabilities, thereby offering context-sensitive services that cater to a spectrum of operational requirements [17]. The GeoGraphVis solution exemplifies this trend by dissecting expert-driven disaster response workflows, translating them into machine-interpretable decisional frameworks, and employing knowledge graphs to navigate the exploration of knowledge [18, 19]. It has further introduced scenario-driven visualization tactics that, through interactive and heuristic analytical methods, facilitate a more profound comprehension of disaster impacts and the crafting of humanitarian assistance strategies. Despite the novel insights and instruments furnished by these investigations, a dedicated knowledge service framework tailored to the systematic analysis and evaluation of natural resources within the Qinling region remains nascent, thereby constraining the efficacy of decision-making support in this locale.

Consequently, addressing the systematic modeling of knowledge, the profound interlinking of knowledge elements, and the scenario-oriented representation of knowledge to efficiently deliver knowledge services for the management of natural resources in the Qinling region (specifically referring to the Shaanxi segment of the Qinling area), this paper introduces a knowledge service framework for the analysis and evaluation of natural resources in the Qinling. This framework is anchored by three pivotal stages: "construction of an integrated knowledge system, expression of knowledge within applied scenarios, and development of a knowledge service system", which elucidates the intrinsic relationships and evolutionary laws within the spatiotemporal data of natural resource survey and monitoring analyses, with the objective of bolstering the decision-making support capabilities for natural resource management in the Qinling region.

This paper provides an overview of the general framework in the second section. The third section offers an in-depth exposition of the integrated knowledge system. The fourth section investigates scenario-oriented knowledge modeling and representation. The fifth section showcases the development of the knowledge service system. A comprehensive synthesis of conclusions will be encapsulated in the sixth section.

2 General Framework

To fully unearth the profound value embedded within the analytical and evaluative outcomes of the natural resources in the Qinling region, this research has meticulously structured a knowledge system around four pivotal themes, geographic characteristics, resource conditions, human activities, and spatial control, through structured modeling of "themes, connotations, and knowledge points", along with associative processing of "hierarchical, geographical, and event-based" associations, thereby the paper providing a comprehensive overview of the attributes, conditions, and evolutionary trends of natural resources. Subsequently, a scenario-oriented knowledge hypergraph model is proposed, leverages advanced relational capacities to synthesize complex, interdisciplinary knowledge points. It presents a rich tapestry of "graphical, textual, numerical, and linguistic" multiple spatiotemporal information, thereby enhancing the comprehension of the intricate relationships and mutual influences among these knowledge points. Ultimately, this paper has developed a knowledge service system for the analysis and evaluation of natural resources in the Qinling, grounded in the aforementioned knowledge system and hypergraph model. The system is equipped with features such as knowledge visualization, thematic knowledge modules, knowledge search capabilities, and knowledge Q&A interfaces, designed to precisely address and anticipate the profound needs of users, offering tailored knowledge recommendations and bolstering the decision-making support capabilities (Fig. 1).

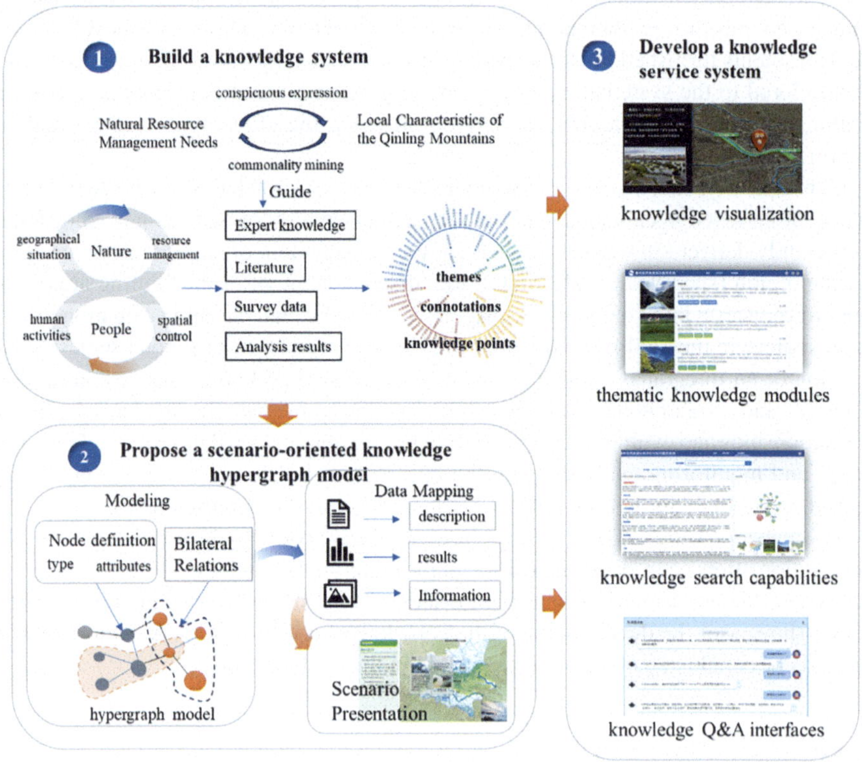

Fig. 1. General framework for knowledge service of the Qinling natural resources analysis and evaluation.

3 Building the Knowledge System for the Qinling Natural Resources Analysis and Evaluation

The knowledge system encapsulates a distillation of expert knowledge and empirical insights, culminating from iterative brainstorming and knowledge refinement processes. Its structured and graphically mapped form facilitates the clear and organized presentation of intricate subject matter [20, 21]. In this study, an in-depth analysis of the survey and monitoring requirements for the natural resources of the Qinling has been conducted. By conducting exhaustive business surveys and accumulating expert knowledge, a comprehensive organization of the analytical and evaluative outcomes pertaining to the natural resources of the Qinling has been achieved. Employing structured modeling and associative processing techniques, the study systematically integrates a multitude of dimensions, including geographic characteristics, resource conditions, human activities, and spatial control, into a cohesive knowledge framework. Utilizing knowledge graph technology, a comprehensive knowledge system for the analysis and evaluation of the Qinling's natural resources has been constructed. This system is designed to augment the systematic comprehension of various natural resource elements in the Qinling, such as mountains, waters, forests, farmlands, lakes, and grasslands, in relation to their resourcefulness, asset value, and capital potential. Moreover, it seeks to deepen the understanding of the complex interrelationships and dynamic interactions among these elements, thereby laying a robust intellectual foundation for the scientific stewardship and efficacious decision-making regarding these resources, as depicted in Fig. 2.

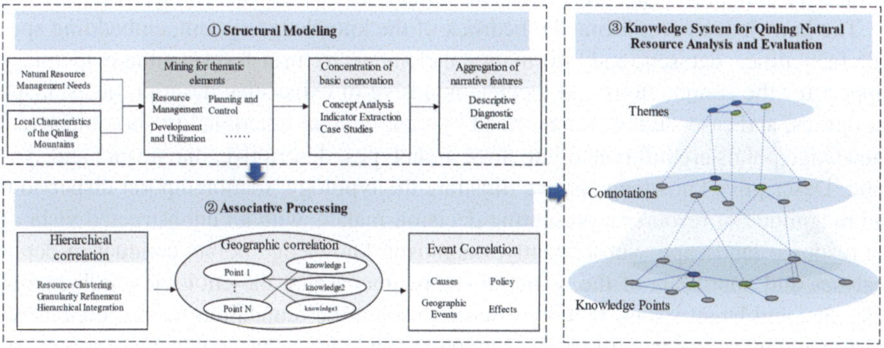

Fig. 2. The methodology for building the knowledge system for the Qinling Natural Resource analysis and evaluation

3.1 Structural Modeling

Structural modeling is foundational to the development of the analytical and evaluative knowledge system for the natural resources of the Qinling region. This process entails the systematic organization of fundamental components including concepts, knowledge points, and entities, along with their interconnections, into a structured and logically coherent knowledge graph [22]. This study meticulously aligns with the practical business requirements and geographic specificities of the Qinling area, employing a top-down approach to methodically distill and synthesize the pertinent knowledge for the analysis and evaluation of its natural resources. To elaborate, key insights are culled from an extensive array of literature, surveillance data, and outcome assessments of metric analyses, ensuring that the resulting knowledge graph is both logically exhaustive and pragmatically viable. Subsequently, by enriching and refining the associated concepts, a tiered knowledge architecture is established, which incrementally transitions from overarching themes to nuanced knowledge points, encapsulating a comprehensive three-tiered structure of "themes—intricacies—knowledge points".

The theme constitutes the apex of the knowledge system, providing a macroscopic viewpoint and categorical criteria for the assessment of natural resources. Encompassing four primary themes—"geographic characteristics, resource conditions, human activities, and spatial control"—it orchestrates and thoroughly evaluates the comprehensive framework, asset valuation, human-environment interactions, and regulatory mechanisms of the Qinling's natural resources, as delineated in Table 1.

The connotation serves as the mezzanine level of the knowledge system, delving into the intrinsic qualities and attributes of each theme, transmuting abstract concepts into tangible and operational analytical parameters. Via targeted core concept analysis, extraction of pivotal indicators from operational documentation, and case study examination, this stratum discerns the salient aspects of each theme, thereby distilling the essential connotations for natural resource assessment. For example, within the theme "Anthropogenic Activities", connotations encompass "Exploitation and Utilization", "Protective Outcomes", and "Resource Sustainability".

The knowledge point forms the bedrock of the knowledge system, embedding specific factualities, datasets, and informational elements to furnish substantial evidentiary support for the connotations and themes, inclusive of exhaustive descriptions, statistical figures, and case studies. Categorized by their content functionality and objectives, knowledge points are bifurcated into three archetypes: descriptive, diagnostic, and synoptic. Descriptive knowledge points illustrate the typology, spatiotemporal dispersion, and magnitude of resources, presenting decision-makers with an unobstructed vista of the resource landscape. Subsequently, diagnostic knowledge points conduct in-depth analyses and appraisals of the resource status, uncovering the etiologies, influencing elements, and latent issues, thereby offering a scientific foundation for resource stewardship and intervention. Ultimately, synoptic knowledge points amalgamate and distill the former two types of knowledge points from a more elevated standpoint, surpassing the scope of isolated knowledge points to integrate various interconnected knowledge points, culminating in a holistic comprehension of a specific domain or issue.

This tiered and interlinked structural design facilitates the downward propagation of knowledge and the upward accumulation of insights. Decision-makers can swiftly ascertain the macroscopic orientation from the thematic stratum, comprehend the nuanced circumstances through the connotative stratum, and descend to the knowledge point stratum to unearth pertinent data and analytical findings, thereby laying a solid intellectual groundwork for decision-making support and resource management.

Table 1. Themes of the Qinling Natural Resource Analysis and Evaluation

Theme	Definition
Geographic Characteristics	This theme emphasizes the overall pattern and characteristics of natural resource elements and human geographical elements. It primarily includes the spatial distribution pattern of natural resources, regional differences, genesis mechanisms, and inter-element interaction relationships, as well as the economic, cultural, and social aspects of human geography
Resource Conditions	This theme focuses on the quantitative statistical analysis of the stock, changes, and quality of natural resources, as well as the accounting of the asset value of each natural resource element, including their physical value and ecological service value
Human Activities	This theme is concerned with the multifaceted and bidirectional relationships between "people" and "land". It mainly includes the development and utilization of various natural resource elements, analysis of the sustainable use of resources in the Qinling region, and the protection and restoration of resource elements by humans
Spatial Control	This theme focuses on the planning, implementation, and supervision of the use and protection of natural resources. It mainly includes bottom-line control, focusing on the three zones and three lines in the Qinling region and the regulation of the use of nature reserves; and the overall ecological pattern, introducing the planning background, content, and overall requirements of the Qinling region

3.2 Associative Processing

This study achieves an organic integration between"themes - connotations - knowledge points" through hierarchical association, geographical association, and event association.

Hierarchical association clarifies the different levels within the knowledge system for the Qinling natural resources analysis and evaluation, establishes logical connections between elements at different levels, ensures the hierarchical nature and information transfer of the knowledge system, and makes the logical chain from macro themes to micro knowledge points clearly visible. For instance, by utilizing knowledge graph clustering, resources in the Qinling area are effectively categorized by type, such as arable land resources and forest resources, optimizing the graph and streamlining information. Moreover, by conducting a more granular analysis of the distribution, status, and trends of natural resources, the logical connections between macro themes and micro knowledge points are strengthened, presenting a clear hierarchical structure in the vertical dimension of the knowledge system (Table 2).

Geographical association focuses on the spatiotemporal aspect of the knowledge for the Qinling natural resources analysis and evaluation, closely integrating content with geographical locations to accurately reflect the spatiotemporal distribution, status, and trends of natural resources. This paper employs GIS and spatial data mining techniques for multidimensional and multi-scale analysis of natural resources. For example, through GIS, forest cover data and knowledge points of water resource distribution can be associated with specific topographical maps of the Qinling Mountains, thereby intuitively displaying the spatial distribution of forest resources and geographical differences in ecosystem services. Additionally, geographical association assesses the impact of human activities on natural resources by analyzing the spatial interdependencies and interactions, such as the impact of urban sprawl on surrounding arable land.

Event association leverages the event attributes of the knowledge for the analysis and evaluation of natural resources in the Qinling, linking specific events to various aspects of natural resource management, and providing a dynamic perspective on the causal impact of events on natural resources. By weaving isolated events into meaningful sequences, a deeper understanding and prediction of natural resource trends are achieved. For example, through causal chain analysis, how the geographical environment affects the distribution and quality of forest resources and how forest conservation leads to enhanced biodiversity can be understood. Furthermore, by integrating event forecasting models, the impacts of different management strategies and changes in external conditions on natural resources are simulated.

3.3 Knowledge System for Qinling Natural Resource Analysis and Evaluation

This paper establishes a knowledge system for the Qinling natural resource analysis and evaluation through structured modeling and associative processing, as illustrated in Fig. 3. A directed graph is utilized to represent the framework, detailing the inter-level linkages and the trajectory of knowledge flow. The formal representation is expressed as:

$$0 = (T_j, I_j, K_j | REL | PRO), j = 1,2,3 \ldots, n \tag{1}$$

0 symbolizes the knowledge system;T_j denotes the theme, structured around four pivotal topics shedding light on the natural resources' spatial-temporal distribution dynamics, trend patterns, and etiological mechanisms;I_j signifies the connotations, mapping the specific traits and attributes under each theme, K_j encapsulates knowledge points, including descriptive, diagnostic, and summarizing forms of knowledge;REL denotes the relationships, which incorporate hierarchical, event, and geographic ties between themes, connotations, and knowledge points, such as the spatial distribution of resources, regional disparities, causality, and interplays among elements; PRO represents the properties of objects, such as the attributes of survey and monitoring outcomes, and the characteristics of economic, humanistic, and cultural facets of human geography.

By integrating multi-level and multi-theme knowledge, Knowledge System for the Qinling Natural Resource Analysis and Evaluation not only provides a solid knowledge base for the management of natural resources, but also offers robust decision support for achieving regional ecological balance and sustainable development. These outcomes provide significant theoretical support and practical guidance for the subsequent research on scenario-oriented knowledge modeling and presentation, as well as the development of knowledge service systems in this paper.

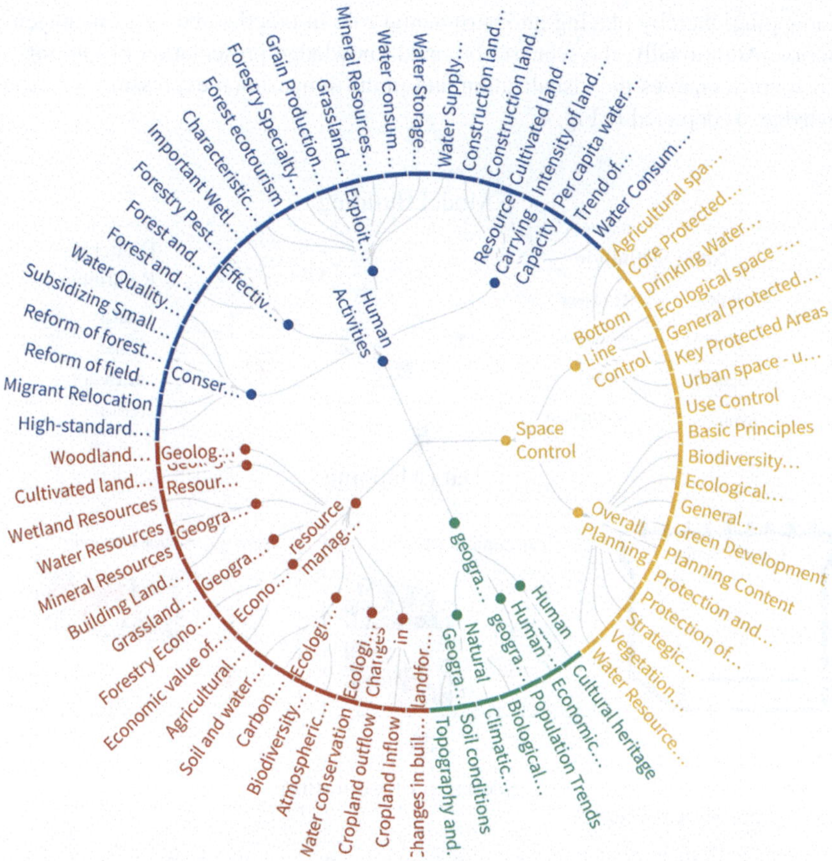

Fig. 3. "Themes-Connotations-Knowledge Points" Knowledge System Structure

4 Scenario-Oriented Knowledge Modeling and Presentation of Qinling Natural Resources

To enhance the multi-dimensional cognition and understanding of the Qinling natural resources analysis and evaluation, the implementation of scenario-oriented knowledge modeling and presentation is imperative [23]. A hypergraph, within the realm of graph theory, is a mathematical construct that permits a hyperedge to link more than two vertices, thereby facilitating the presentation of intricate relationships, particularly those of a higher order, among set elements. This attribute bestows hypergraphs with a distinctive capability for managing interactions within complex, multi-element systems [24]. In this regard, the present study leverages hypergraphs for scenario-oriented knowledge modeling, synthesizing diverse hierarchical levels, categories of events, and multi-scale temporal and spatial data. This synthesis provides robust model support for the scenario-oriented analysis and interpretation of the intricate spatiotemporal dynamics of natural resources. Thereafter, the model facilitates the translation of abstract knowledge points into tangible, actionable scenario knowledge through multidimensional spatiotemporal data mapping, thereby playing an instrumental role in practical business management scenarios. Additionally, the scenario-oriented knowledge presentation of Qinling natural resources enables the visualization and multi-dimensional expression of scenario knowledge, as depicted in Fig. 4.

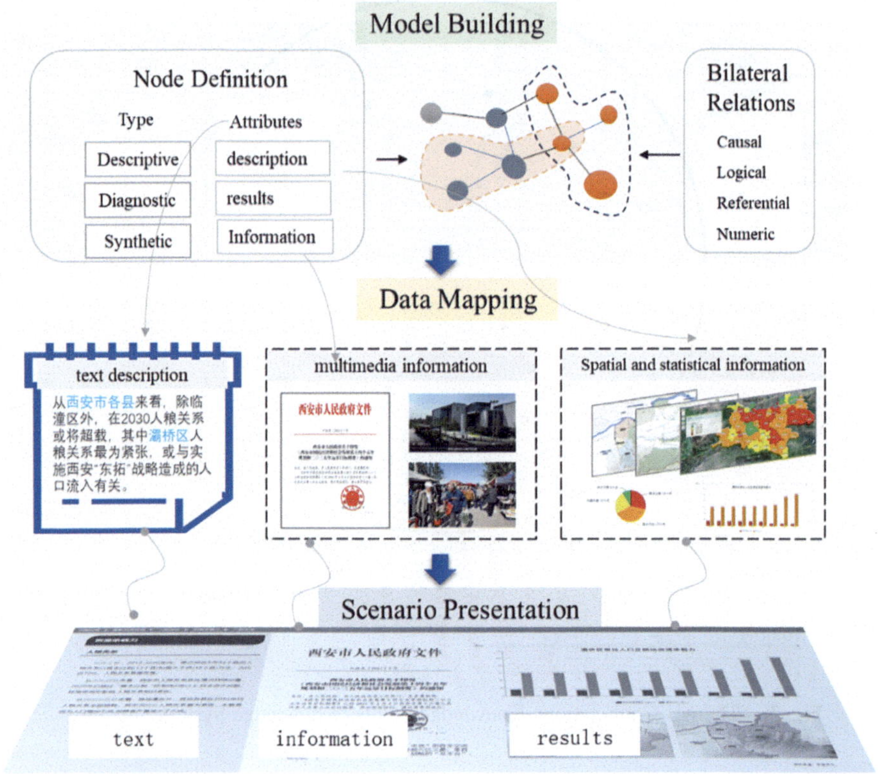

Fig. 4. The methodology for scenario-oriented knowledge modeling and presentation of Qinling natural resources

4.1 Scenario-Oriented Knowledge Modeling

In the construction of the scenario-oriented knowledge model for the natural resources, this paper begins by meticulously defining and distinguishing knowledge points. The knowledge points are stratified into three categories: descriptive, which visually outlines the current conditions, geographical distribution, and evolutionary trends of natural resources; diagnostic, conducting an in-depth examination of potential issues within resource management, encompassing data analysis and historical contrast; and summarizing, delivering an exhaustive synthesis of specific topics by amalgamating contextual background, significance, and an overall assessment. Each knowledge points is endowed with comprehensive textual narratives and a spectrum of multivariate spatiotemporal data, as demonstrated in Table 1, thereby presenting the knowledge points in a multifaceted manner through an integrated approach of textual descriptions, numerical data, and multimedia content. Subsequently, by establishing the edge structure within the hypergraph model, we interlink two or more knowledge points, articulating their intricate interrelationships through the edges' relational attributes—inclusive of causal, logical, mathematical, and referential connections. These relational attributes not only unveil the causal linkages among phenomena and stipulate the prerequisites and sequence for the validation of knowledge points but also quantify the interconnections between knowledge points with precision, concurrently highlighting the interdependencies and mutual reinforcement that exists among them.

Table 2. Attributes of Multivariate Spatiotemporal Data

No	Multidimensional Spatiotemporal Data	Description	Presentation Form
1	Graphical Charts	To intuitively display the spatial distribution and trends in quantity changes of data	Distribution maps, Trend graphs
2	Video and Audio	To provide multimedia records of field investigations and expert interviews	Video links, Audio recordings
3	Web Links	To link to policy documents or academic research, offering background information	Link 1, Link 2,...
4	Database Links	To access detailed datasets and records, supporting in-depth analysis	Database name, Dataset ID
5	Temporal Tags	To indicate the timeliness of information and historical changes	$t1,2,\ldots$
6	Spatial Coordinates	To locate the geographical position of nodes, reflecting the geographical distribution of resources	$(x1,1),(x2,y2),\ldots$
7	Extended Materials	To provide in-depth analysis and discussion, enhancing comprehensive understanding of knowledge points	References, Research reports

4.2 Multivariate Spatiotemporal Data Mapping

The mapping of multivariate spatiotemporal data is instrumental in translating the abstract knowledge points within the model into tangible, operational scenario knowledge. This is presented to users in an intuitive and multidimensional format, designed to augment the expressiveness and flexibility of knowledge representation. Furthermore, it lays the groundwork for knowledge expression applications across diverse scenarios. This study employs a data-enrichment mapping technique, aligning a plethora of multivariate spatiotemporal data—including textual narratives, statistical diagrams, geographic coordinates, and multimedia content—with the nodes and edges within the model to manifest a multifaceted knowledge display. A case in point is the node "River Distribution in the Qinling", where textual data offers specifics such as river names, lengths, and basin areas, complemented by maps and charts linked to databases that vividly portray the rivers' geographic positions, flow directions, and spatiotemporal dynamics of river flow and water quality. Multimedia links provide access to ecological records of the rivers, while web links direct users to pertinent hydrological research and conservation policies. Subsequently, this paper constructs various edges to articulate the intricate interconnections between the "River Distribution in the Qinling" node and other knowledge points. For example, a hyperedge connects with the "Water Resource Quantity" node to elucidate the influence of water resource volume on river distribution. Concurrently, another edge is linked to the "Climate Change" node to underscore the significance of climate change on the distribution and seasonal patterns of rivers.

4.3 Scenario-Oriented Knowledge Presentation for Qinling Natural Resource

This paper enhances the visualization of information by presenting visual hierarchies and meticulously arranging scenarios, adopting a scenario-oriented approach to provide users with a clear and intuitive pathway for understanding. Leveraging the scenario model and multivariate spatiotemporal data mapping previously constructed, we have developed visual tools and interfaces that graphically represent intricate knowledge, data, and relationships, facilitating a deeper understanding and analysis of information.

The presentation of scenarios incorporates critical elements such as time, space, and process. Time and space establish the contextual backdrop, while processes illustrate the specific content and dynamic evolution of the scenario. For instance, examining the human-grainline relationship in the Qinling region, the scenario is delineated into three sub-scenarios. Each sub-scenario, through distinct analytical perspectives and visual transformations, presents an overview, trends, and case analyses of specific areas, as depicted in the Fig. 5. To elaborate, the narrative commences with a macroscopic view of the Qinling area, utilizing text, cartography, and data visualization to holistically portray the human-grainline relationship, outlining the current status across various administrative divisions, such as demand surpluses or supply-demand imbalances. This aims to capture the fundamental structure of the human-grainline relationship in the region. The scenario then transitions to an analysis of the temporal trends, employing interactive charts to vividly demonstrate recent shifts, uncovering the temporal impact on these relationships and the disparities among regions. Lastly, the focus narrows to particular administrative regions, like Xi'an city and its environs, where comparative analysis delves into the specific causes of strained human-grainline relationships, highlighting the underlying factors driving change, such as urbanization and policy alterations. This spatiotemporal narrative technique not only visually communicates the complexity of the human-grainline relationship in the Qinling area but also augments the capacity to interpret and comprehend spatial information, thereby enhancing the depth of understanding and analysis of the region's human-grainline dynamics.

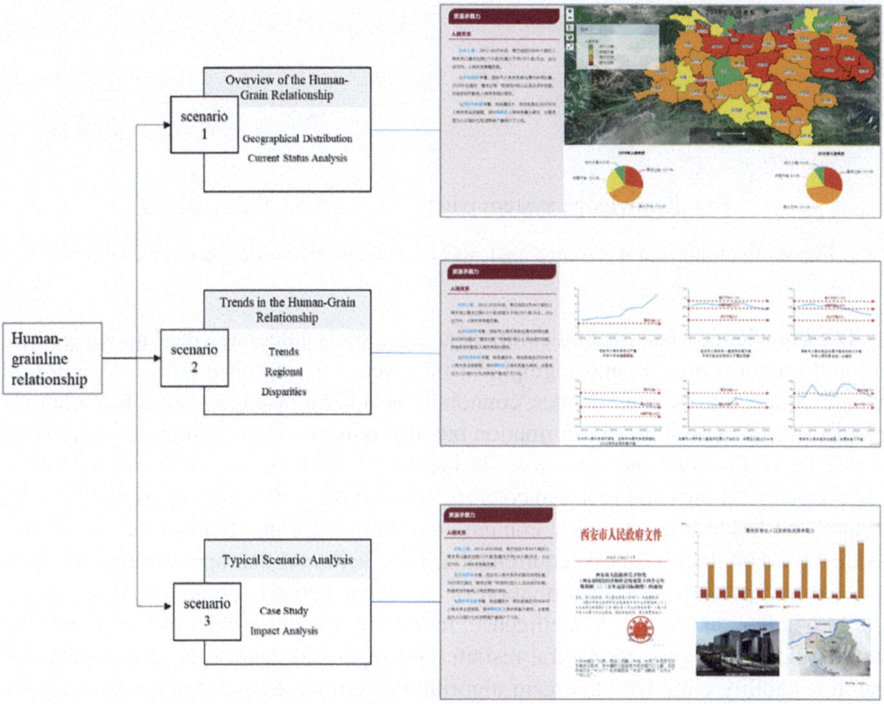

Fig. 5. Depiction of the human-grain relationship in scenario-oriented knowledge representation

5 Development of the Qinling Natural Resource Analysis and Evaluation Knowledge Service System

The development of the Qinling Natural Resource Analysis and Evaluation Knowledge Service system represents the final phase of the knowledge service initiative. It is designed to prominently feature the analytical and evaluative knowledge system and the scenario-oriented knowledge models of the Qinling's natural resources (Fig. 6).

5.1 Development of the Knowledge Service System

The system, underpinned by the Qinling Integrated Monitoring Infrastructure Cloud Platform and a three-dimensional spatiotemporal database, is designed to emphasize four core functionalities: knowledge visualization, thematic knowledge, knowledge search, and knowledge query response.

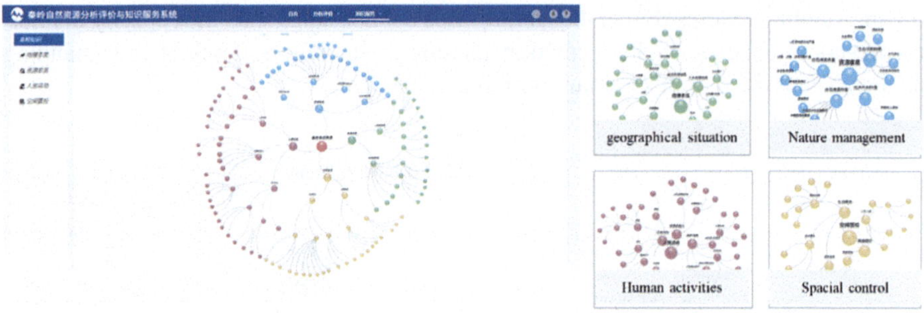

(a) Knowledge system page (b) theme pages

Fig. 6. Qinling natural resource analysis and evaluation knowledge service system

The knowledge visualization component seamlessly integrates the knowledge system and scenario models into a graphical interface, offering an intuitive depiction of the interconnections among themes, connotations, and knowledge points. This not only enhances the accessibility of information but also bolsters users' comprehension of the knowledge architecture and scenarios. The thematic knowledge feature is tailored to provide adaptable knowledge scenario construction and services, enabling users to swiftly retrieve and link pertinent topics, content, and knowledge points from the knowledge system based on their business requirements. This facilitates the rapid creation of cross-tiered scenarios and the transformation of abundant natural resource spatiotemporal data into actionable business intelligence, markedly enhancing the efficiency and efficacy in addressing intricate natural resource management challenges. The knowledge search capability employs intelligent algorithms from the knowledge graph to enable rapid and precise identification of the desired knowledge points and associated information. Through straightforward query operations, users can swiftly access a wealth of knowledge resources, including texts, diagrams, and multimedia, significantly augmenting the efficiency of information retrieval. The knowledge query response function leverages advanced language models and knowledge graph technologies to accurately discern users' intent and deliver descriptive, diagnostic, and summarizing insights on the analysis and evaluation of Qinling's natural resources. This interactive service not only greatly enriches the user experience in natural resource management but also extends the system's utility to non-specialist users, democratizing access to specialized knowledge.

5.2 Application Effects and Prospects for Promotion

The application effects of the Qinling Natural Resource Analysis and Evaluation Knowledge Service are manifested at multiple levels, significantly enhancing the system's practicality and interactivity, as well as markedly strengthening users' cognitive and handling capabilities regarding the complexity of natural resource management. At the knowledge level, the knowledge service system integrates knowledge graphs and hypergraph theories to deeply analyze and comprehensively evaluate natural resources, transforming them into deeply mined knowledge. This enables decision-makers to grasp the distribution, trends, and potential risks of natural resources from a macro perspective, providing intellectual guidance for formulating reasonable resource management and protection strategies. At the technical level, the system utilizes GIS technology and visualization knowledge expression techniques to present complex natural resource spatiotemporal data in intuitive two-dimensional and three-dimensional knowledge maps, greatly improving the readability and usability of spatiotemporal data. This multidimensional expression and interactive approach not only allows non-expert users to quickly understand complex natural resource information but also provides professional users with tools for in-depth analysis and exploration. At the service level, the knowledge service system offers diversified services, meeting the needs of different users in various scenarios, optimizing user experience, and improving decision-making efficiency. With the continuous improvement and optimization of these functions, the knowledge service system is expected to become an important tool for natural resource management and protection in the Qinling region, providing solid data support and decision-making basis for sustainable development in the area.

Currently, the Qinling Natural Resource Analysis and Evaluation Knowledge Service system has been applied in the Shaanxi Natural Resources Management Department, providing robust support for the sustainable management and ecological protection of natural resources in the Qinling area and beyond. However, as natural resource survey and monitoring data involve various formats and multiple business needs, the natural resource analysis and evaluation knowledge service program of this study requires localization processing and knowledge representation based on regional characteristics. Further validation of its universality and adaptability is needed during promotion to meet diverse management objectives.

6 Conclusion

This paper introduces a comprehensive knowledge service framework aimed at enhancing the utilization of survey and monitoring data pertaining to natural resources in the Qinling region. The framework encompasses the construction of an integrated knowledge system, the scenario-oriented knowledge modeling and presentation, and the development of a knowledge service system. This measure transitions the analysis and evaluation of natural resources from singular data and information services to a comprehensive service centered on knowledge. It enriches the theoretical underpinnings of the knowledge service domain and offers a tool for the management of natural resource knowledge to the Qinling region and its administrative departments. Future research endeavors could delve into mechanisms for knowledge renewal and user feedback to augment the iterative and optimization capabilities of the knowledge service strategy.

Funding. This research was funded by the National Natural Science Foundation of China Grant number 2022YFB390420501.

References

1. Chen, W., et al.: Building natural resources surveying and monitoring technological system: Direction and research agenda. Acta Geographica Sinica **77**(05), 1041–1055 (2022)

2. Jun, C., Hao, W., Wanzeng, L., et al.: Technical connotation and research agenda of natural resources spatio-temporal information. Acta Geodaetica et Cartographica Sinica **51**(7), 1130–1140 (2022)
3. Zenglin, H., Lei, F.: The application of the natural resource remote sensing monitoring system in the natural resource monitoring of the Qinling Mountains. Satellite Application **01**, 28–32 (2024)
4. Haijing, G., Yuanjun, Z., Nana, Y.: Exploration and practice of normalized monitoring mode of natural resources in Guangdong Province. Journal of Spatio-temporal Information **30**(4), 631–641 (2023)
5. Yao, Y., Yaolong, Z., Xiaoding, L., et al.: The pattern of natural resources survey and monitoring from the perspective of management. J. Nat. Resour. **38**(03), 808–821 (2023)
6. Kan, Z., Yun, S., Minjie, N.: Research on the temporal and spatial changes of vegetation coverage in Qinling mountains based on Google Earth Engine platform. Bulletin of Surveying and Mapping **05**, 49–55 (2022)
7. Lihua, H., et al.: Improvement and Application of Analysis and Evaluation System of Natural Resource Survey Monitoring Data. Geospatial Information **22**(02), 39–43 (2024)
8. Mingwei, H., Mengmeng, L., Chao, C., et al.: Application of 3D realistic geospatial scene in natural resources monitoring. Bulletin of Surveying and Mapping **10**, 150–153 (2024)
9. Jun, C., et al.: Basic issues and development directions of high-quality development empowered by spatio-temporal information. Journal of Spatio-temporal Information **30**(01), 1–11 (2023)
10. Wenju, Yun, Chunze, Li., Chao, Zhang, et al.: Assumption for the Integrated Investigation and Monitoring Technical System Construction of the Cultivated Land Quantity-quality-ecology. Geomatics World **29**(05), 14–19 (2022). +27
11. Lu, F., Zhu, Y.Q., Zhang, X.Y.: Spatiotemporal knowledge graph: Advances and perspectives. Journal of Geo-information Science **25**(6), 1091–1105 (2023)
12. Yipeng, W., Xueying, Z., Yulong, D., et al.: Knowledge graph representation of typhoon disaster events based on spatiotemporal process. Journal of Geo-information Science **25**(6), 1228–1239 (2023)
13. Xiaoshuang, W., et al.: Logic graph construction of air pollution law enforcement event considering the spatiotemporal features. Geography and Geo-information Science **38**(3), 1–8 (2022)
14. Dandan, Z., Chenchao, X., Dandan, W., et al.: Study on the technical framework of carbon sink monitoring for ecological protection and restoration projects based on spatial-temporal-spectral data. Spacecraft Recovery & Remote Sensing **43**(06), 119–128 (2022)
15. Yunqiang, Z., Xiaoliang, D., Jie, Y., et al.: One-stop Sharing and Service System for Geoscience Knowledge Graph. Geological Journal of China Universities **29**(03), 325–336 (2023)
16. Zhu, J., Lai, J., Xie, Y., et al.: Knowledge-guided Spatiotemporal Narrative 3D Visualization Method for the Bridge Construction Process.Geomatics and Information Science of Wuhan University, pp. 1–15. (2024)
17. Li, S., Zhu, X., Zhilin, L., et al.: From geographic information service to geographic knowledge service: research issues and development roadmap. Acta Geodaetica et Cartographica Sinica **50**(9), 1194–1202 (2021)
18. Li, W., Sizhe, W., Chen, X., et al.: Geographvis: a knowledge graph and geovisualization empowered cyberinfrastructure to support disaster response and humanitarian aid. ISPRS International Journal of Geo-Information **12**(3), 112 (2023). https://doi.org/10.3390/ijgi12030112
19. Wenwen, L., Wang Sizhe, W., Sheng, et al.: Performance benchmark on semantic web repositories for spatially explicit knowledge graph applications. Comput. Environ. Urban Syst. **98**, 101884 (2022)

20. Jiaxin, D., Shaohua, W., Xunyue, Y., et al.: GIS-KG: building a large-scale hierarchical knowledge graph for geographic information science. Int. J. Geogr. Inf. Sci. **36**(5), 873–897 (2022)
21. Yuying, Y., Xuesheng, Z., Huiyuan, L., et al.: Building a knowledge graph for wetlands based on landcover data. Journal of Geo-information Science **25**(6), 1240–1251 (2023)
22. Baskarada, S., Koronios, A.: Data, information, knowledge, wisdom (DIKW): A semiotic theoretical and empirical exploration of the hierarchy and its quality dimension. Australasian Journal of Information Systems **18**(1), 748 (2013)
23. Wu, J.H., Wang, Y., Wang, S.J., Wang, P., Chen, W.Y., Cui, J.Y.: Knowledge model of comprehensive evaluation of urban physical examination: A case study of Guangzhou. Journal of Spatio-temporal Information **31**(03), 455–468 (2024)
24. Zeng, L., Yang, J.R., Huang, G., et al.: Survey on Hypergraph Application Methods: Research Problem, Progress, and Challenges. Journal of Computer Applications, http://118.89.95.112:8085/kcms/detail/51.1307.TP.20240124.1837.006.html, last accessed 2024/7/21

Characterization and Integrated Risk Evaluation of Earthquake Disaster in China Mainland

Yahui Chen, Xiaoyue Gao(✉), Yihao Duan, and Keifeng Wang

China Earthquake Networks Center, Beijing 100045, China
{chenyahui,xygao}@seis.ac.cn

Abstract. China is an earthquake-prone country with a history of significant seismic disaster losses. It is crucial to assess the extent of potential losses in the event of an earthquake, with a view to enhancing emergency rescue and preparedness measures. This paper collates data on earthquake losses in mainland China between 1991 and 2022, and provides an account of the spatial and temporal distribution characteristics of the disasters. Mathematical statistical methods were employed to analyse the temporal trends of each data set, and the results of the seismic kernel density estimation were found to be highly consistent with the seismic zone data. Furthermore, the weights of five disaster evaluation indexes were calculated using the entropy weight theory, including fatalities, injuries, direct economic losses, affected population and affected area. A fuzzy comprehensive evaluation model was employed to categorise the 32-year losses into five grades. With the exception of 2008, 2013 was classified as an extra-large grade, 2010 and 2014 were assigned a large grade, and 1995, 1996, 2003, 2012 and 2022 were classified as moderate. The middle-western region of China has experienced a significant number of severe and catastrophic earthquakes, while the eastern part of the country has a relatively low frequency of seismic activity and a high population density, with occasional large earthquakes resulting in significant damage and loss of life. In general, the number of fatalities and direct economic losses caused by earthquakes has been observed to decrease at a slow rate. This can be attributed to two key factors: the enhanced anti-seismic capacity of buildings and the heightened awareness of the risks posed by earthquakes..

Keywords: Seismic hazard · Entropy weight theory · Fuzzy comprehensive evaluation model · Loss grade

1 Introduction

China is located between the Pacific Rim seismic zone and the Asia-Europe seismic zone, and is one of the most important seismic zones in the global continental region. The seismic activity in China is high in frequency, intensity and distribution with serious disaster losses, which occupies a large proportion of many natural disasters. The earthquake situation in China is serious and complex. There were 95 destructive earthquakes

G. Luo (Ed.): GFSISD 2024, SPPHY 417, pp. 112–128, 2025.
https://doi.org/10.1007/978-981-96-2490-4_8

(DEs) of magnitude 5 or greater in China mainland from 1949 to 1981, causing a total of 272,576 deaths and affecting an area of nearly 300,000 km^2 accounting for about 3.2% of the area of China mainland [1].

It is important to make a scientific and accurate comprehensive evaluation of the disaster losses in these earthquakes through summarizing the massive data including earthquake damage information and characteristics. This paper collects and organizes historical earthquake data in China mainland for 32 years from 1991 to 2022, and uses entropy weight method and fuzzy comprehensive evaluation model to obtain the relationship between the degree of earthquake damage and disaster loss evaluation indicators for each year to achieve comprehensive evaluation of earthquake disaster loss. It provides a basis for earthquake emergency preparedness and also provides decision support for post-earthquake emergency rescue. And it is of great significance to improve capacity of earthquake disaster prevention and mitigation in China.

2 Data and Methods

2.1 Data Source

China attaches great importance to earthquake disaster prevention and mitigation as well as post-earthquake on-site emergency work due to the severe earthquake situation. The China Earthquake Administration (CEA) organized the earthquake on-site disaster damage investigation and assessment for the first time after Datong-Yanggao 6.1 on the Richter magnitude scale (M_S) in 1989, and achieved scientific and reliable assessment results. The CEA formally incorporated disaster damage assessment into the earthquake on-site emergency work since then, and standardized the assessment work in a unified manner [2]. Since 1991 the CEA has collected more complete information on earthquake disasters and published "Compilation of earthquake loss evaluation" every five years [3–6]. This paper focuses on DEs that occurred in China mainland from 1991 to 2022 (no data for Hong Kong, Macau, and Taiwan), and some earthquakes in neighboring countries that affected China. Paper obtains earthquake catalogs, fatalities and injuries, direct economic losses, affected population and affected area data in the compilations as well as every DE's Report from 2011 to 2022 as the base data. These information were obtained by CEA from the unified standard reliable damage investigation and assessment. The national natural disaster data from 1991 to 2022 used in the article for comparative analysis are from the annual statistical report on the development of civil affairs published by the Ministry of Civil Affairs of the People's Republic of China [7]. For comparability, the consumer price index (CPI) was applied to calculate the direct economic losses as 2022 comparable prices [8].These data vary greatly between years, despite the massive records in on-site investigation and statistics, which needs an objective evaluation model to compare the disaster levels in different years the on the same scale revealing.

2.2 Evaluation Method

In recent years, the disaster classification has been used in different fields to evaluate the damage level, but there is no uniform classification standard and evaluation method.

Ma firstly proposed the concept of calamity degree and classified natural disasters losses into 5 grades based on the fatalities and losses of social property as the grading standard [9]. Since then, some scholars have improved the calamity degree model to be better [10]. Feng calculated the seismic disaster index by the factors of fatalities, injuries and economic losses, and divided the index into 12 grades to describe the disaster degree [11]. Yang used gray correlation analysis method to calculate the indicator correlations based on the function transformation of the 4 factors, and obtained the earthquake disaster grades [12].

The following evaluation modeling methods are commonly used: gray correlation theory, fuzzy comprehensive evaluation model, entropy weight theory, and analytic hierarchy process, etc. Li studied the earthquake disaster loss characteristics of 31 provinces in China from 1993 to 2016. The trends of direct economic losses, casualties and affected population are analyzed by constructing two relative earthquake loss indexes (losses/GDP and losses/E) and employing gray correlation theory. The loss grade was obtained based on the degree of correlation for each year [13]. Ge chose modified-fuzzy comprehensive evaluation method to evaluate and rank the risk level and loss level of regional flood disaster [14]. Song presented comprehensive weight composed of importance and improved entropy weight, and combined fuzzy weighting method to assess risk level of a reservoir after earthquake [15].

This paper selected the fatalities and losses, injuries, affected population, direct economic losses, and the affected area as disaster loss evaluation indexes in the statistics of earthquake disaster information from 1991 to 2022, considering that the damage of earthquakes to the natural and social environment of human existence is usually measured by intensity and economic loss [14]. Comparing the research methods, the entropy weight method is calculated by data without subjective assignments to make the results more accurate, and the fuzzy comprehensive evaluation model is evaluated by numerical quantification to obtain more scientific and reasonable evaluation results. Therefore, this paper adopts the entropy weight theory to determine the weights of each index, and provides the criteria of index rank. Then the fuzzy comprehensive evaluation model is used to rank and evaluate the earthquake disaster losses from 1991 to 2022, and objectively analyze the earthquake disaster losses and their impacts on the society and economy in China mainland in the past 32 years.

Entropy weight method. Among various methods to determine the weights of evaluation indexes, the entropy weighting method is entirely from data without considering the actual context of the problem, so as to guarantee the objectivity and accuracy of the results. The entropy weighting method determines the ranking of the index among all indexes depending on the amount of information contained within each index. The greater the information entropy is, the smaller the amount of existing information contains, which improves the scientificity of index weighting rather than subjective weighting. The specific calculation steps are as follows [16, 17]:

Step 1 Normalize earthquake loss evaluation indexes from the original data matrix. Assume that the original matrix $X = (x_{ij})_{m \times n}$, $i = (1, 2, \ldots, m)$, $j = (1, 2, \ldots, n)$, is composed of m loss indexes and n samples, and x_{ij} represents the value of sample j in the loss index i. In order to enable all indexes to be compared and calculated under a standardized scale, we normalize the index as:

$$zij = \frac{xij - \min ixij}{\max ixij - \min ixij} \tag{1}$$

This is the formula for the positive indexes, all the data involved in this paper are positive. Z_{ij} is the result of normalization, i = (1, 2,…,n), j = (1, 2,…,m). $\text{Max}_i x_{ij}$ and $\min_i x_{ij}$ are the maximum value and minimum value of the sample *j* in index *i*. According to normalize the *m* loss indexes of *n* samples, we can constitute the initial characteristic matrix Z as:

$$Z = \begin{bmatrix} z11 & z12 & \cdots & z1m \\ z21 & z22 & \cdots & z2m \\ \vdots & \vdots & \ddots & \vdots \\ zn1 & zn2 & \cdots & znm \end{bmatrix} \tag{2}$$

Step 2 Calculate the entropy of indexes. The formulas are shown as:

$$ei = -\frac{1}{\ln m} \sum_{i=1}^{m} PijlnPij \tag{3}$$

$$Pij = zij \Big/ \sum_{j=1}^{n} zij \tag{4}$$

where, e_i is the entropy value of index *i*, P_{ij} is the relative weight of the sample *j* in the index *i*, which is used in the entropy calculation.

Step 3 Calculate the entropy weight of indexes. Coefficient d_i represents the usefulness of index *i* in evaluation. The larger its value, the greater the effect on the evaluation results. The entropy weight of each index is obtained by normalizing the coefficient d_i with the following formula. The entropy weight w_i is used to measure the importance of index *i* in the evaluation. The higher the value of the entropy weight, the greater the contribution of the index to the overall evaluation.

$$di = 1 - ei \tag{5}$$

$$wi = di \Big/ \sum_{i=1}^{m} di \tag{6}$$

Fuzzy comprehensive evaluation model. The fuzzy comprehensive evaluation model is based on a fuzzy mathematical method to quantify fuzzy concepts and calculate various indexes quantitatively. Then the earthquake loss evaluation is carried out by weighted summation, which greatly avoids subjective judgments in the process of calculation and improves the accuracy of evaluation results. The fuzzy comprehensive evaluation model is calculated in the following steps [18, 19]:

Step 1 Determine the factor set. $U = \{u_1, u_2, u_3, u_4, u_5\}$, u_1, u_2, u_3, u_4 and u_5 are the disaster loss evaluation indexes of fatalities and losses, injuries, affected population, direct economic losses, and affected area. Evaluating the annual earthquake disaster damage level through five indexes of the factor set.

Step 2 Create the judging set. The judging set describes the evaluation results of the earthquake disaster damage level,which are divided into 5 grades: extra-large, large, moderate, general and lighter, corresponding judging set as $V = \{v_1, v_2, v_3, v_4, v_5\}$. The rating levels v_j are fuzzy concepts.

Step 3 Establish fuzzy comprehensive judgment matrices. For a certain evaluation index u_i, we want to get the membership degree of each sample corresponding to rating levels v_j, and use the concept of membership degree to determine whether the earthquake loss in a given year is large, lighter, or some other grades. The degree of membership is the degree to which a sample belongs to a fuzzy set, and is usually determined by the membership function. In this paper, a fuzzy set is assigned as the index ranking standards for rating levels v_j. The commonly used trapezoidal distributed membership function is used to calculate the membership degree of the samples. The membership functions are expressed as

$$R1(ui) = \begin{cases} 1 & , \quad ui \geq ui1 \\ \left|\frac{ui-ui2}{ui2-ui1}\right| & , \quad ui2 < ui < ui1 \\ 0 & , \quad ui \leq ui2 \end{cases} \tag{7}$$

$$Rj(ui) = \begin{cases} \left|\frac{ui-ui,j-1}{uij-ui,j-1}\right| & , \quad uij < ui < ui,j-1 \\ 0 & , \quad ui \leq ui,j+1, ui \geq ui,j-1 \\ \left|\frac{ui-ui,j+1}{uij-ui,j+1}\right| & , \quad ui,j+1 < ui < uij \end{cases} \tag{8}$$

$$R5(ui) = \begin{cases} 1 & , \quad ui \leq ui5 \\ \left|\frac{ui-ui4}{ui5-ui4}\right| & , \quad ui5 < ui < ui4 \quad (\mathrm{i} = 1, 2, ..., 5; \quad \mathrm{j} = 1, 2, ..., 5) \\ 0 & , \quad ui \leq ui4 \end{cases} \tag{9}$$

where u_i is the value of the sample corresponding to the index *i*, $R_j(u_i)$ is the membership degree of the study object u_i to the rating level v_j. The value is between 0 and 1. U_{ij} is the rank standard of index *i* corresponding to the rating level v_j.

The ranking standard for each index in the membership function are referenced as follows (with partial reference to [20]).

According to the membership function and the standard of index ranking, the judgment matrix R_{1991} in 1991 could be obtained, and similarly, R_{1992} ~ R_{2022}.

$$R1991 = \begin{bmatrix} 0 & 0 & 0 & 0 & 1 \\ 0 & 0 & 0.54 & 0.946 & 0 \\ 0 & 0 & 0 & 0.178 & 0.822 \\ 0 & 0 & 0 & 0 & 1 \\ 0 & 0 & 0 & 0.381 & 0.619 \end{bmatrix} \tag{10}$$

Step 4 Determine Index weights. We apply the entropy weighting method to assign weights to the evaluation indexes and obtain the weight vector $A =$ (0.35,0.20,0.22,0.13,0.10).

Table 1. Grading standards for evaluation index.

Indexes Rating Levels	v_1 Extra-Large	v_2 Large	v_3 Moderate	v_4 General	v_5 Lighter
u_1 Fatalities and Disappearances	3600	600	120	50	5
u_2 Injuries	15000	5000	1500	500	200
u_3 Direct Economic Losses (100 million RMB)	1000	300	100	30	10
u_4 Affected Population (ten thousand people)	1200	600	300	100	30
u_5 Affected Area(km^2)	160000	80000	40000	20000	10000

Step 5 Comprehensive evaluation. The comprehensive membership degree of earthquake disaster losses for each year is calculated with the matrix multiplication method.

$$B = A \cdot R = \{\ b1,\ b2\ ,\ b3,\ b4\ ,\ b5\} \tag{11}$$

where any value b_j in B is the membership degree of the evaluation index in the rating level v_j. The rating level corresponding to its maximum value is the final result of the fuzzy comprehensive evaluation.

3 Data Processing and Analysis

During the period from 1991 to 2022, there were 741 earthquakes of magnitude 5.0 (inclusive) or above on the China mainland. There were 363 disaster events caused by earthquakes in China mainland and neighboring areas, involving 417 earthquakes, of which 3 had their epicenters in the offshore and 12 in neighboring countries. These DEs caused 73,898 people deaths, 18,330 missing and 476,100 injured, more than 160 million people and about 1,679,000 square kilometers area affected. The cumulative direct economic losses could reach 1,626 billion RMB at comparable prices of 2022, approximately equivalent to 1.34% of the China GDP in 2022. Among these DEs, there were 65 DEs with magnitude less than M_S5.0, 254 DEs of M_S5.0-M_S5.9, 84 DEs of M_S6.0-M_S6.9, 11 DEs of M_S7.0-M_S7.9, and 3 DEs of M_S8.0 or greater. A total of 234 DEs caused casualties, accounting for 56.1% of all. Based on the disaster loss statistics from 1991 to 2022, Fig. 1 shows the magnitude classification and spatial distribution of DEs in China mainland over the past 32 years.

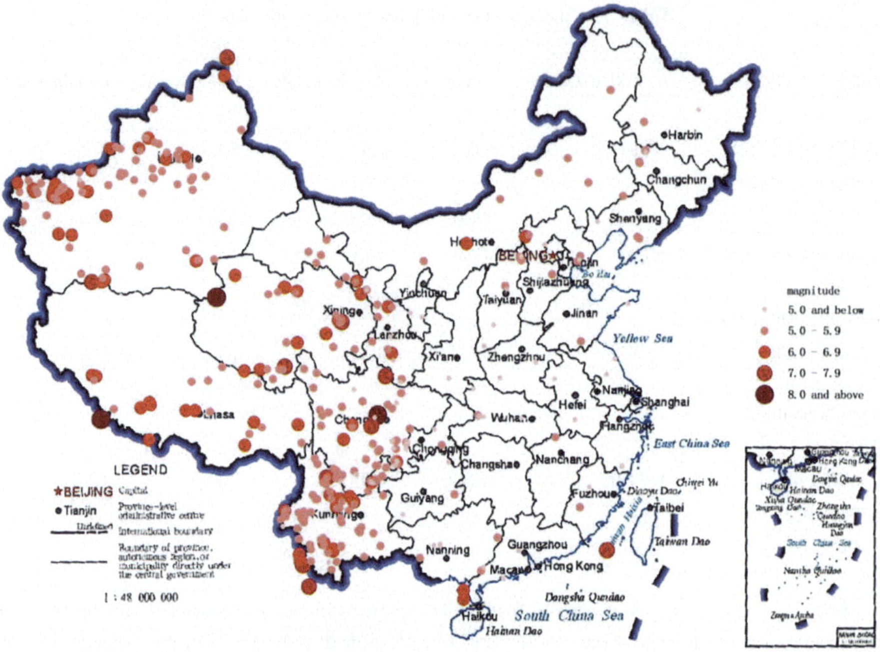

Fig. 1. Distribution of DEs in China mainland from 1991 to 2022.

3.1 Characteristics of the Temporal Distribution of DEs in China Mainland from 1991 to 2022

According to the statistical report on the development of civil affairs from 1991 to 2022, the number of deaths and missing people was obtained, as well as the direct economic losses in natural disasters nationwide in the past 32 years, from which the percentage of earthquake disaster losses in the national natural disaster losses was calculated (Fig. 2). The data show that the number of earthquake deaths accounted for 98.1% of the total number of deaths in natural disasters due to the Wenchuan M_S8.0 in 2008, and the direct economic losses accounted for 73.1% of the total economic losses. The Wenchuan M_S8.0 was the most costly natural disaster of the year with an epicenter intensity of XI degrees, and the earthquake caused 69,227 people killed, 17,923 people missing, and more than 375,000 people injured, with the direct economic losses of more than 852.3 billion RMB (2008 prices). In addition, the top 5 years in terms of the number of dead and missing were 2008, 2014, 2010, 2022 and 2003, and the top 5 years in terms of direct economic losses were 2008, 2013, 2014, 2022 and 2015. Generally, in addition to 2008, the earthquake disaster losses in 2014 and 2022 were also more serious.

With regards to the frequency of DEs, the average number is about 13 over the 32-year period, with 13 years above the average (Fig. 3). The number of DEs generally maintained a stable level with a slight decreasing trend. The frequency of 1995 was the maximum with 23 DEs, while that of 2007 was the minimum with 3 DEs. The frequency of 2003, 2008 and 2019 is significantly higher than the previous year with 22, 18 and 18 disaster earthquakes, respectively.

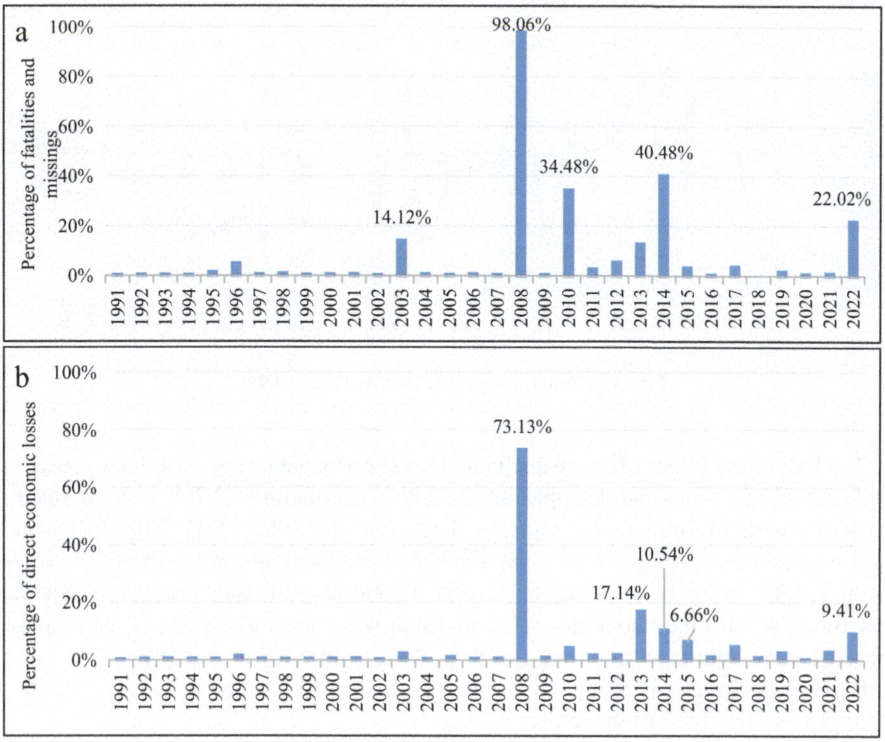

Fig. 2. The percentage of the number of annual fatality/missing people and seismic direct economic losses in the total losses of national natural disaster.

The number of earthquakes with magnitude less than 5.0 accounted for 15.6%, these earthquakes have smaller magnitude but caused casualties and losses. Usually these areas are not hit by earthquakes frequently, and should be taken seriously once earthquakes arrive. The number of DEs of magnitude 5.0 to 6.9 accounts for 81.1% of the total number, which happened the most and also affected the economic development to some extent. The number of earthquakes above magnitude 7.0 accounts for only 3.3%, but the fatalities and injuries account for 98.3% and 88.2%, respectively. The larger the magnitude, the greater the energy released and the more unpredictable the disaster damage.

The statistics of DE damage in China mainland for the period 1991–2022 are presented by year and show the inter-annual variation in the number of fatalities and injuries, direct economic losses, and the affected population (Fig. 4). The number of deaths and missing people due to earthquakes during the 32-year period was 92,121, an average of 2,879 per year, and the general trend remained essentially stable with an insignificant slight increase (Fig. 4a). The number of injuries caused by DEs was about 5.2 times greater than fatalities and disappearances. The most serious of these was 2008, when the number of injured was 4.3 times as high as that of fatalities and disappearances, and in which almost all the losses were caused by the Wenchuan $M_S8.0$ earthquake.

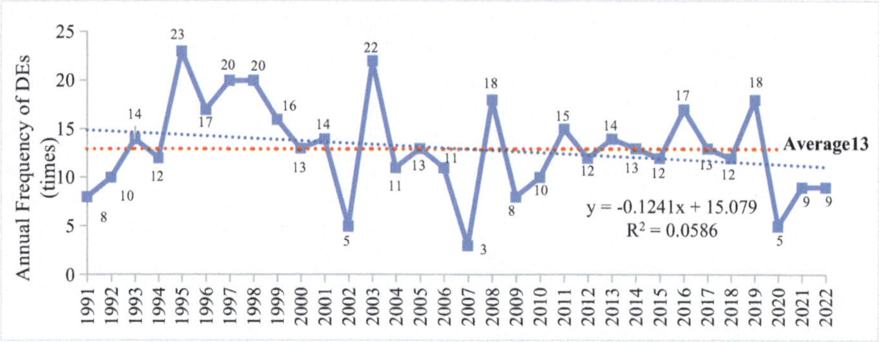

Fig. 3. Annual frequency distribution of DEs.

Excluding the effect of the Wenchuan M_S8.0 earthquake (Fig. 4b), DEs caused a total of 4,970 fatalities and disappearances, with a maximum of 2,705 in 2010, due to the M_S7.1 earthquake in Yushu, Qinghai. The values for 1996, 2003, 2010, 2013, and 2014 exceeded the average of 155, the number of dead and missing in these five years accounted for 86.7% of the sum of 32 years. In addition, the total number of injuries was 100,234, with the maximum value in 1996, when the Lijiang M_S7.0 earthquake caused 17,057 injuries, accounting for 95% of the total for that year. The seven years 1995, 1996, 1998, 2003, 2010, 2013, and 2014 had above average values, accounting for approximately 84% of the total.

Figure 4c shows the direct economic damage caused by the DEs, with the maximum value of 1129 billion RMB in 2008. Excluding the effect of Wenchuan MS8.0 earthquake, the value of direct economic losses increased slowly before 2013, reaching a peak in 2013 (RMB 127.7 billion, 2022 comparable price), about 35 times larger than the average value (RMB 36.0 billion). The M_S7.0 earthquake in Lushan, Sichuan caused losses of about RMB 85.4 billion (2022 comparable prices), about 67% of the total losses in 2013, which is the most serious earthquake in 32 years in terms of direct economic loss except for the Wenchuan M_S8.0 earthquake. In disaster area of Lushan, the simple houses accounted for 66.7% destroyed in the earthquake, most of which were not protected by seismic fortification measures. Even the brick-concrete houses, only a few of them had seismic fortification measures, and most of them were old masonry houses and rural self-built houses. Therefore, the damage of houses is extremely serious and difficult to repair with great economic loss [21].

3.2 Spatial Distribution Characteristics of Disaster Earthquakes in China Mainland from 1991 to 2022

In terms of spatial distribution, there are 27 out of 31 provinces (excluding Hong Kong, Macao and Taiwan) with earthquake disasters from 1991 to 2022 (Fig. 5a). Yunnan and Xinjiang have the highest number, reaching more than 90 times in cumulative data for the past 32 years, which is followed by Sichuan, Qinghai, Tibet, and Gansu, with a relatively higher frequency of earthquakes exceeding the average of 13. The total number of DEs in these 6 provinces was 327 accounting for 81.3% of all. Following the mapping method

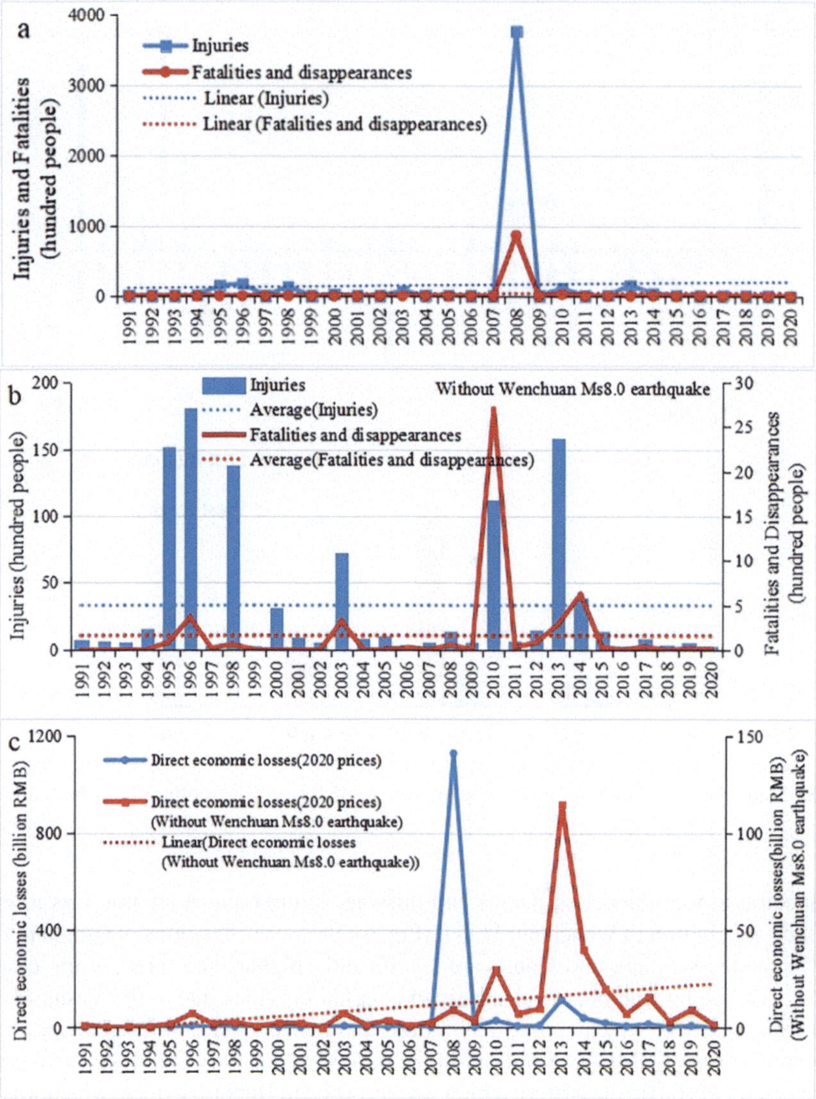

Fig. 4. Statistics on the number of fatalities/disappearances, injuries and direct economic losses (caused by DEs) in China over the period of 1991–2022.

of the China seismic intensity zoning map, the eastern and western regions were divided by 105°E as the boundary, and the number of DEs of different magnitudes in the two regions were counted separately [22]. Figure 5b shows that there are relatively more earthquakes of magnitude less than $M_S5.0$ in the eastern China, and there are frequent moderate to strong earthquakes in the western region.

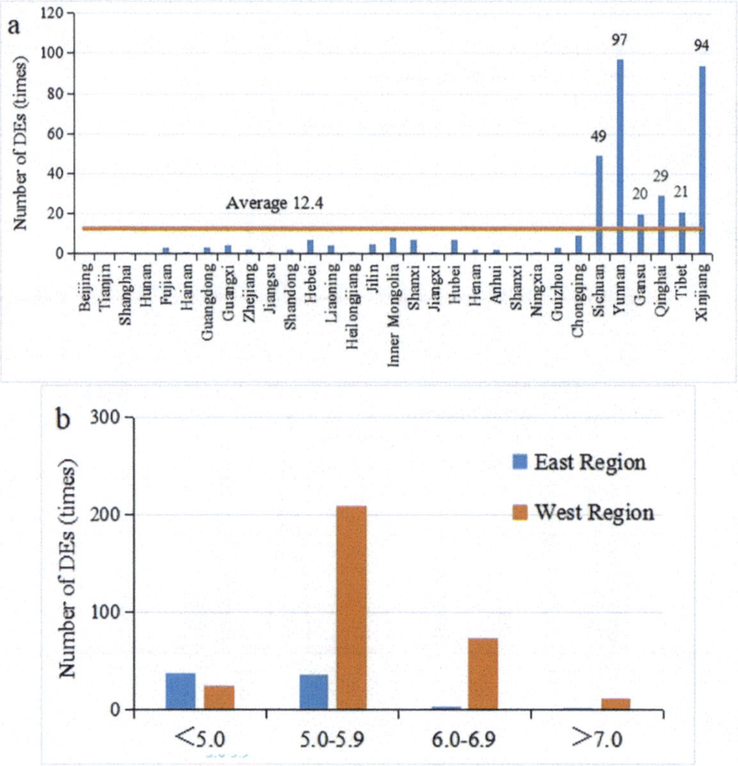

Fig. 5. Statistics on the number of DEs by provinces, eastern and western regions from 1991 to 2022.

Specific to the number of deaths and missing, injured and economic loss in each province, in addition to Wenchuan M_S8.0 (Fig. 6), the western regions of Qinghai, Yunnan, Xinjiang, Sichuan, and Gansu are significantly higher than those in the eastern regions. The highest number of deaths is in Qinghai province due to the Yushu M_S7.1 earthquake in 2010, which caused 2,698 dead and missing, 11,000 injured, and direct economic losses of RMB 22.8 billion. In the past 32 years, only one destructive earthquake of over M_S6.0 in the eastern region was recorded in Hebei province, which was the Zhangbei M_S6.2 earthquake in 1998. The affected area was densely populated, involving 5 towns in Zhangbei and Shangyi counties, resulting in 49 fatalities, 11,439 injuries and more than 400,000 houses damaged to different degrees. Back in 1976, a M_S7.8 earthquake hit Tangshan, Hebei, located 360 km from that earthquake, and the fatalities was as high as 242,769. When the magnitude of the earthquake increases by one level, the energy released will increase by 30 times. If an earthquake of M_S7.0 or higher is located in the eastern region of China, the disaster damage may increase by much more than 30 times. Generally, the frequency of earthquakes in the eastern is low, with occasional small earthquakes may causing moderate disasters and moderate earthquakes may causing major disasters.

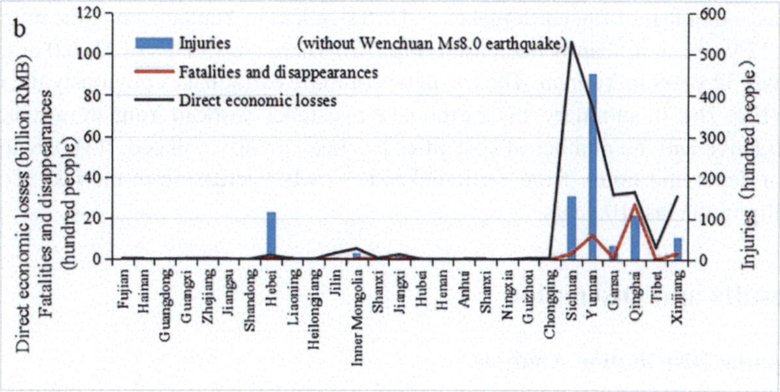

Fig. 6. Number of dead, missing, injured and direct economic losses in 32 years of DEs by province.

As an example, in Yunnan province, which has a high number of DEs, there were 103 DEs in this province from 1991 to 2022, with 74.8% of them below $M_S6.0$. In the first 20 years from 1991 to 2009, there was an insignificant upward trend in the number

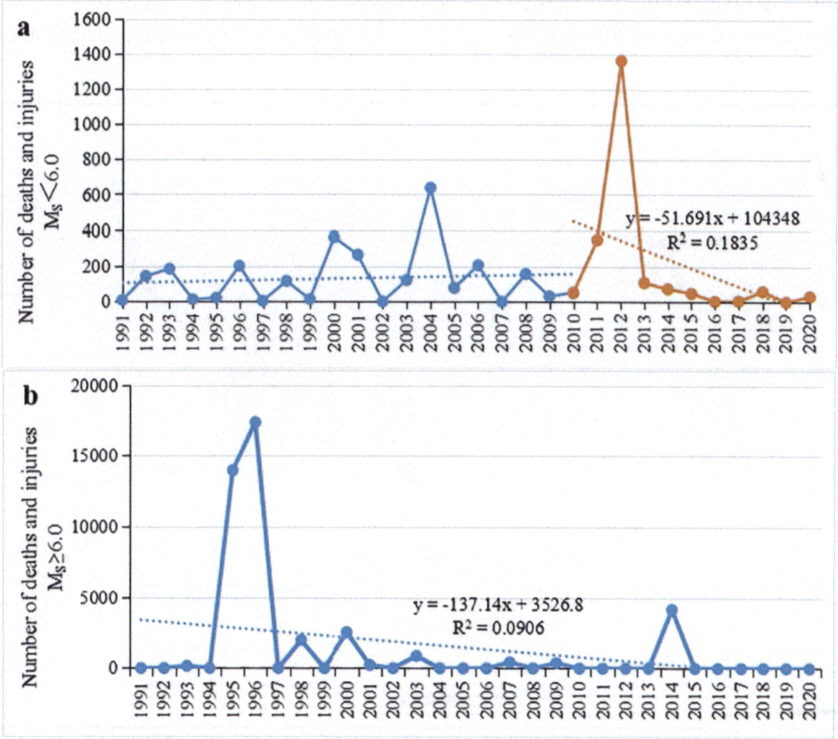

Fig. 7. Casualties of DEs in Yunnan province from 1991 to 2022.

of deaths and injuries from earthquakes of M_S6.0 or less in Yunnan, while the number in the last 12 years significantly decreased (Fig. 7a). There are 15 DEs of M_S6.0 or greater in the past 32 years in Yunnan. The casualties of these earthquakes obviously decreased yearly (Fig. 7b). In summary, the earthquake resistance work in Yunnan province has been effective and the number of casualties has been greatly reduced. It shows that the threat of death and injury from earthquakes is slowly decreasing in the last 32 years, especially in the last 10 years.

4 Results and Discussion

4.1 Spatial Distribution Analysis

Actually, the earthquake spatial distribution in different provinces is limited to administrative division lack of natural distribution characteristics which could be obtained through the kernel density analysis tool in ArcMap. The point density of DEs from 1991 to 2022 can be obtained by calculation and shown on the map as the earthquake distribution range. The kernel density analysis is a widely used surface density calculation method, which is mainly applied to the empirical analysis of aggregation class. It calculates and estimates the data aggregation through the sample data, and measures the change of the factors density through the established distance decay function, so as to explore the distribution and change characteristics of hot spots in spatial regions [23]. The database of all earthquakes above M_S4.0 in China mainland is preprocessed, and

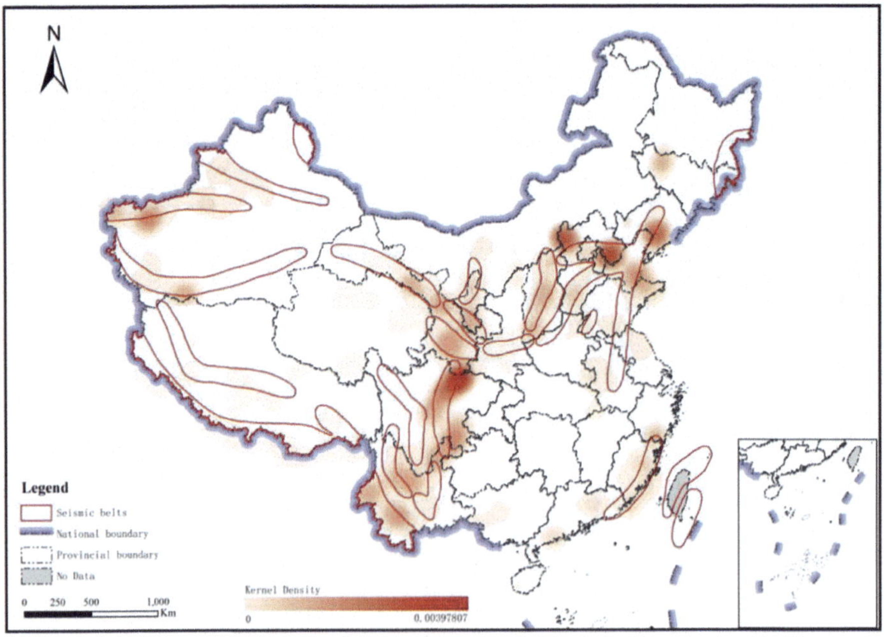

Fig. 8. The result of kernel density analysis.

the population data are overlaid with earthquake epicenters to remove earthquakes that fall in unpopulated areas. Considering the distribution difference between the eastern and western China, the kernel density analysis was based on the earthquakes $M_S4.0$ or greater in the eastern and $M_S5.5$ or greater in the western China (Fig. 8).

The kernel density analysis reveals that in western China, earthquakes are mainly distributed in Xinjiang, Tibet, Gansu, Yunnan, Sichuan provinces, especially in Sichuan, Yunnan and western Xinjiang with high frequency and density. While in the eastern region, there are relatively more earthquakes in some coastal areas such as Hebei, Liaoning and Shandong provinces. In addition, by overlaying and analyzing the data with the seismic belts, it could be found that the earthquake distribution range coincides well with the belts. The greatest density of kernel density analysis is 0.00397807 in the south-north seismic zone, followed by the north China and Tianshan seismic belts, which indicates the 3 areas with the highest frequency of earthquakes in China.

4.2 Comprehensive Evaluation on Each year

The weights of five indexes were calculated through entropy weight method to differ the earthquake disaster losses in each year and to evaluate the disaster severity. The comprehensive membership degrees of rating levels were obtained from the fuzzy comprehensive evaluation model according to the grading standards for evaluation index (Table 1). And then the disaster losses of each year were classified into five grades (Table 2).

The evaluation results revealed that, except for 2008, China mainland suffered extra-large grade earthquake disaster losses in 2013. The losses in 2010 and 2014 were of the large grade, and that of 1995, 1996, 2003, 2012 and 2022 were of moderate grade. The earthquakes of these 9 years resulted in 91,806 deaths, which exceeded 50% of the total number of deaths from natural disasters and accounted for 99.7% of the total number of deaths and missing from earthquakes over the past 32 years. The number of injuries and direct economic losses from earthquakes in these 9 years accounted for 94.4% and 93.1% of the total, respectively. The 8 years of 1998, 2000, 2001, 2005, 2009, 2011, 2015, and 2017 had relatively light disaster losses and were classified as general grade. The comprehensive membership degrees of the other 15 years has a higher probability of being lighter grade thus judging them as lighter grade.

The characteristics of earthquakes in China can be summarized as follows: strong earthquakes are frequent in the west, with many mountainous region sand poor geological conditions, resulting in more serious damage. While the eastern part has a low frequency of earthquakes and a high population density, with occasional large earthquakes causing disasters. The earthquake damage is a complex result of the combined effect of destructive energy released by the earthquake, surface structures and population distribution. On the one hand, it is necessary to strengthen the seismic protection of buildings and structures improving their seismic resistance; on the other hand, it is necessary to enhance the awareness of earthquake disaster prevention and mitigation of the public improving the post-earthquake risk avoidance.

These are the basic reasons for the difference earthquake disaster losses in different regions ans years, which provides a better decision direction for pre-earthquake emergency rescue preparation and post-earthquake emergency services.

Table 2. Result of Fuzzy Comprehensive Evaluation.

Year	Membership Degree of Rating Levels					Grade
	v_1 Extra-Large	v_2 Large	v_3 Moderate	v_4 General	v_5 Lighter	
1991	0	0	0.012	0.3	0.693	Lighter
1992	0	0	0.014	0.326	0.659	Lighter
1993	0	0	0.055	0.249	0.696	Lighter
1994	0.133	0	0.191	0.104	0.573	Lighter
1995	0.218	0.009	0.368	0.327	0.078	Moderate
1996	0.218	0.255	0.475	0.053	0	Moderate
1997	0	0	0.007	0.418	0.575	Lighter
1998	0.188	0.03	0.084	0.698	0	General
1999	0	0	0.017	0.12	0.863	Lighter
2000	0	0.092	0.172	0.371	0.365	General
2001	0	0	0.11	0.534	0.356	General
2002	0	0	0	0.116	0.884	Lighter
2003	0.047	0.407	0.49	0.057	0	Moderate
2004	0	0	0.045	0.477	0.478	Lighter
2005	0	0	0.191	0.492	0.317	General
2006	0	0	0	0.257	0.743	Lighter
2007	0	0	0.004	0.45	0.546	Lighter
2008	0.079	0.023	0.485	0.413	0	Moderate (exclusive of Wenchuan *M*s8.0)
2009	0	0	0.079	0.506	0.415	General
2010	0.384	0.386	0.052	0.149	0.029	Large
2011	0	0.034	0.271	0.556	0.138	General
2012	0	0.078	0.637	0.285	0	Moderate
2013	0.424	0.258	0.318	0	0	Extra-Large
2014	0.143	0.674	0.182	0	0	Large
2015	0	0.203	0.279	0.387	0.131	General
2016	0	0.042	0.206	0.111	0.641	Lighter
2017	0	0.109	0.225	0.433	0.233	General
2018	0	0	0.02	0.2	0.78	Lighter
2019	0	0.004	0.203	0.355	0.439	Lighter

(*continued*)

Table 2. (*continued*)

Year	Membership Degree of Rating Levels					Grade
	v_1 Extra-Large	v_2 Large	v_3 Moderate	v_4 General	v_5 Lighter	
2020	0	0	0	0.13	0.87	Lighter
2021	0	0.101	0.203	0.096	0.6	Lighter
2022	0	0.165	0.487	0.337	0.011	Moderate

References

1. Chen, Z., et al.: China earthquake yearbook 1990. Seismological Press, Beijing (1992)
2. Song, L.: Exploring earthquake site investigation and assessment work promoting standardization and standardization of earthquake damage assessment in China. City and Disaster Reduction **06**, 8–12 (2017)
3. CEA and NBS (China Earthquake Administration and National Bureau of Statistics): Compilation of earthquake loss evaluation (1990–1995). Seismological Press, Beijing (1996)
4. DEMPCEA (Department of Earthquake Monitoring and Prediction of China Earthquake Administration): Compilation of earthquake loss evaluation (1996–2000). Seismological Press, Beijing (2001)
5. DEERCEA (Department of Earthquake Emergency Rescue of China Earthquake Administration): Compilation of earthquake loss evaluation (2001–2005). Seismological Press, Beijing (2010)
6. DEERCEA (Department of Earthquake Emergency Rescue of China Earthquake Administration): Compilation of earthquake loss evaluation (2006–2010). Seismological Press, Beijing (2015)
7. MCAPRC (Ministry of Civil Affairs of the People's Republic of China): Statistical report on the development of civil affairs, 1991–2020. https://www.mca.gov.cn/n156/n2679/index.html, last accessed 2023/10/31
8. NBS (National Bureau of Statistics of China): China statistical yearbook. China Statistics Press, Beijing (2021)
9. Ma, Z. Li, M.: Natural disaster evaluation, disaster magnitude and strategy. China Natural Disaster Reduction Research. Collection of National Natural Disaster Reduction Research. Chinese Science and Technology Press, Beijing (1990)
10. Yu, Q.: The limitation and improvement of the method of disaster situation grade measurement. Journal of Natural Disasters **2**(2), 8–11 (1993)
11. Feng, Z., Hu, Z., He, J.: Loss evaluation of earthquake disaster and classification of the disaster grade. Journal of Catastrophology **9**(1), 13–16 (1994)
12. Yang, S.: Analysis and comparison of seismic disaster situations and damage grade division by using grey system theory. Northwestern Seismological Journal **19**(2), 49–57 (1997)
13. Li, X., Li, Z., Yang, J., Liu, Y., Fu, B., Qi, W., Fan, X.: Spatiotemporal characteristics of earthquake disaster losses in China from 1993 to 2016. Natural Hazards **94**, 843–865 (2018)
14. Ge, P.: The study of fuzzy evaluation model of regional flood disaster. Nanjing University of Information Science and Technology, Nanjing (2013). Dissertation
15. Song, Y., Jin, C., Zhang, X., Li, J.: Risk assessment of reservoir by fuzzy comprehensive evaluation method based on information entropy. Appl. Mech. Mater. **438–439**, 1612–1618 (2013)

16. Zhang, F., Deng, S., Du, H., Lv, J., He, S.: Grading of the historical earthquake disasters in the counties of Yunnan on the basis of the entropy weight method. J. Seismol. Res. **41**(2), 319–327 (2018)
17. Liu, J., Su, J., Wang, W., Ma, D.: Information entropy method for evaluating regional earthquake relative disaster-carrying capability. Appl. Mech. Mater. **1799**, 2070–2073 (2012)
18. Zhang, M., Jin, Y., Wang, Q.: Loss assessment of farmland from typhoon disasters based on fuzzy comprehensive evaluation in Zhejiang province. Hubei Agricultural Sciences **53**(20), 5002–5006 (2014)
19. Zhang, H., Li, W., Miao, P., Sun, B., Kong, F.: Risk grade assessment of sudden water pollution based on analytic hierarchy process and fuzzy comprehensive evaluation. Environ. Sci. Pollut. Res. **27**(1), 469–481 (2020)
20. General Office of the State Council of the People's Republic of China: National Earthquake Emergency Plan. http://www.gov.cn/yjgl/2012-09/21/content_2230337.htm, last accessed 2012/09/21
21. Liang, H., Shen, Y., Mao, L., Xiao, B.: Comparison of building damage characteristics of the Lushan MS7.0 and Jiuzhaigou MS7.0 earthquakes. South China Journal of Seismology **39**(3), 14–22 (2019)
22. Wang, S., Yu, Y., Gao, A., Yan, X.: Development of attenuation relations for ground motion in China. Earthquake Research in China **16**(2), 99–106 (2000)
23. Chen, X., Peng, F., Cai, Y., Hang, C., Gao, Y., Wang, B.: Spatial distribution characteristics and influencing factors of Hubei national wetland park based on ArcGIS kernel density analysis. Journal of Green Science and Technology **23**(2), 1–3 (2021)

Analysis of Spatial Patterns and Urban Road Network Correlations of Tourism Elements: A Case Study of Xiamen

Ying Yuan[1,2], Lian Hong Ye[1,2(✉)], Hui Li[1,2], Zong Cai Huang[1,2], Chao Chen[1,2], Xiao Hui Xia[1,2], and Qian Ye Huang[1,2]

[1] College of Computer & Information Engineering, Xiamen University of Technology, Xiamen 361024, China
905649266@qq.com

[2] Institute of Spatial Information Technology, Xiamen University of Technology, Xiamen 361024, China

Abstract. Examining the spatial distribution of tourism elements and their relationship with the transportation network structure is crucial for optimizing tourism efficiency and enhancing regional competitiveness. This study employs kernel density estimation, synergistic location quotient, and space syntax to analyze the spatial patterns of key tourism elements—restaurants, accommodations, transportation, and attractions—in relation to the urban road network. The analysis is validated through an empirical case study in Xiamen City.The results are as follows: ① In terms of spatial patterns, the core density of tourism elements is predominantly concentrated within the island area, exhibiting relatively weak spatial linkages between different elements at the city-wide level. However, on a more localized scale, the spatial association between restaurants and other tourism elements is the most pronounced. ② In terms of the association with the urban road network, restaurants and transportation elements demonstrate moderate to strong positive correlations with road network integration, local choice, and street density, but only a weak positive correlation with global choice. In contrast, accommodations and attractions show weak positive correlations with road network integration, local choice, and street density, with no significant correlation with global choice. Xiamen City should focus on enhancing the transportation service capacity of its scenic areas. Furthermore, linking tourism elements within strongly associated spaces could foster the development of tourism and leisure districts. This study provides both theoretical insights and practical recommendations for promoting the sustainable development of the tourism industry.

Keywords: tourism elements · spatial pattern · urban road network · space syntax · association analysis

1 Introduction

Accelerating the integration of transportation and tourism is a crucial initiative for promoting high-quality development in the tourism industry [1]. Continuously enhancing transportation services can inject strong momentum into the tourism sector, driving

G. Luo (Ed.): GFSISD 2024, SPPHY 417, pp. 129–147, 2025.
https://doi.org/10.1007/978-981-96-2490-4_9

it toward higher standards of excellence. Optimizing transportation connections and improving service infrastructure are key strategies for elevating the level of tourism services and enhancing the attractiveness of tourist destinations, which are essential for fostering the prosperity of the tourism industry.

Tourism elements are the fundamental components of tourism activities and mainly include six aspects: "Food, Accommodation, Transportation, Sightseeing, Shopping, and Entertainment". These aspects represent the various needs of tourists, such as restaurants, lodging, travel, sightseeing, shopping, and entertainment [2]. These six elements not only meet the basic needs of tourists but also cover the entire activity process of the tourism industry chain [3]. Understanding the spatial distribution characteristics and patterns of these elements is crucial for grasping the current state and future direction of tourism development, as well as for rationally planning its scale. This has significant practical implications [4]. In research on tourism space, scholars both domestically and internationally have explored distribution characteristics, evolutionary patterns, and influencing factors in areas such as scenic spots [5, 6], specific landscapes [7, 8], accommodation industry [9, 10], and transportation facilities [11, 12]. For example, Zhang Wei [6] investigated the spatial pattern of 5A and 4A tourist resources in China and found that they exhibit a concentrated distribution and are somewhat coupled with urban and economic belts. Agusti [7] examined spatial distribution differences in Uruguay's tourism landscapes by collecting media materials from relevant institutions. Long Fei [9] and colleagues discovered that the spatial distribution of homestays in the Yangtze River Delta follows a multi-core distribution pattern, with a significant positive impact from the number of A-level scenic spots and population density. Transportation, as a key link connecting urban clusters and various points within a region, is considered a crucial factor influencing the spatial pattern of tourism and has received widespread attention. Relevant studies have addressed the development mechanisms [13], models [14, 15], frameworks [16], strategies [17, 18], and planning [19, 20] of transportation-tourism integration. Zhao Lili [14] summarized the current status and characteristics of transportation-tourism integration in China and categorized it into four models: element enhancement model, resource-based model, interactive development model, and product innovation model. Lin Rongjie [16] proposed a big data application framework based on transportation-tourism integration by analyzing the characteristics and issues of transportation tourism development from a business demand perspective. Shi Jie [19] focused on the G75 Lanhai Expressway Taishi service area to explore new pathways for the development of transportation-tourism integration.

In summary, while existing research on tourism space has predominantly focused on the spatial distribution and influencing factors of individual elements such as scenic spots, accommodation, and transportation facilities, there is a noticeable lack of systematic and interrelated studies on the full spectrum of tourism elements. Although the theoretical foundation for transportation-tourism integration is relatively well-established, research has often been confined to highways and highway service areas, leaving a gap in studies and practices within urban contexts. This gap fails to address the need for deep integration, mutual adaptation, and unified development of urban transportation and tourism. This study, centered on Xiamen City, utilizes techniques such as kernel density estimation and synergistic location quotient to conduct a comprehensive analysis of the spatial patterns of tourism elements. Through quantitative analysis of urban road network structural characteristics using space syntax, the study further investigates the relationship between the spatial patterns of tourism elements and the urban road network structure via bivariate analysis. The objective of this research is to provide actionable recommendations for optimizing tourism planning, enhancing the tourism environment, and fully realizing the tourism potential.

2 Study Area and Research Data

2.1 Overview of the Study Area

Xiamen City, situated in the southeastern part of Fujian Province, is a major central city on China's southeastern coast, recognized as a special economic zone and a separately listed city. It governs six districts: Siming, Huli, Jimei, Haicang, Tong'an, and Xiang'an (see Fig. 1). The city spans a total area of 1,699.39 square kilometers, with a GDP of 806.649 billion yuan in 2023 and a resident population of 5.327 million. As one of the first pilot cities for tourism and leisure, Xiamen continues to attract a substantial number of domestic and international tourists. In 2023, Xiamen welcomed nearly 110 million tourists, generating total tourism revenue exceeding 156.7 billion yuan. The city is home to 14 4A-level scenic spots and 2 5A-level scenic spots, offering an abundance of diverse tourism resources. Given the lack of connectivity between the Gulangyu area and the main urban road network, and considering the accessibility characteristics of the urban road network, this study focuses on Xiamen City (excluding the Gulangyu area) as the research area.

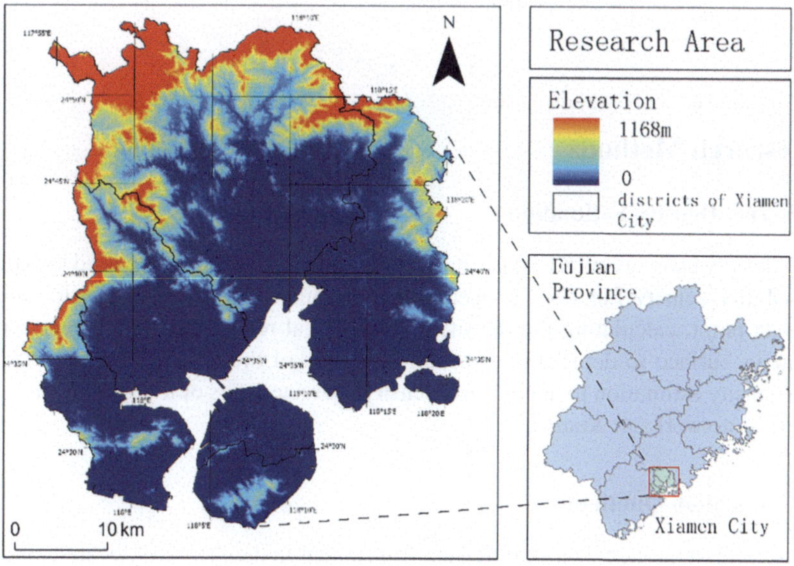

Fig. 1. Research Area

2.2 Data Sources

The tourism POI (Point of Interest) data and urban road network data for Xiamen City used in this study were sourced from the Gaode Open Platform. The 2021 POI dataset encompasses four categories of tourism elements: restaurants, accommodations, transportation, and attractions, resulting in a total of 65,310 entries after data cleaning and correction. The Xiamen urban road network data includes major urban roads, secondary roads, expressways, highways, and pedestrian paths, with a total of 15,460 entries following cleaning and correction, as detailed in Table 1.

Table 1. Types and Quantities of Tourism Elements in Xiamen City

Primary classification	Secondary classification	Poi count	Poi proportion/%
Restaurants	Coffee and tea, chinese restaurant, international restaurant, snacks and fast food, tea house, cake and desserts	46206	70.74
Accommodation	Star-rated hotels, budget hotels, homestays, youth hostels	5475	8.38
Transportation	Subway stations, bus stops, train stations, airports, ferry terminals, parking lots, highway service areas, toll booths	11432	17.50
Attractions	Park, square, religious temples, zoo, botanical garden, memorial hall, natural scenery, cultural relics	2197	3.36

3 Research Methods

3.1 Kernel Density Estimation

Kernel density estimation (KDE) is a non-parametric statistical method used to estimate a probability density function. It operates by defining a smooth local region around each data point, calculating the density of these local regions, and then generating a continuous surface to depict the distribution pattern of the data. This study employs kernel density estimation to analyze the density characteristics of tourism elements and urban road networks in Xiamen.

3.2 Co-location Quotient

The co-location quotient is a significant advancement in the theory of location quotient within economic geography. It is primarily utilized to reveal the synergistic relationships between different types of point features within a specific region. The co-location quotient can be categorized into Global Co-location Quotient (GCLQ) and Local Co-location

Quotient (LCLQ). This study applies the co-location quotient to assess the strength of connections between tourism elements.

The Global Co-location Quotient (GCLQ) is a metric based on nearest neighbors that measures the overall relationship between the observed number of type B points near type A points and the expected number. This quotient reflects the degree to which type A points are attracted to type B points. The calculation formula is as follows:

$$GLCQ_{A\to B} = \frac{N_{A\to B}/N_A}{N_B/(N-1)} \tag{1}$$

$$N_{A\to B} = \sum_{i=1}^{N_A}\sum_{j=1}^{u_i}\frac{f_{ij}(0,1)}{u_i} \tag{2}$$

In the formula: N represents the total number of points; N_A and N_B represent the number of type A and type B points, respectively; $N_{A\to B}$ indicates the number of type A points whose nearest neighbors are type B points; u_i is the number of nearest neighbors for point i; f_{ij} is a binary variable indicating whether the nearest neighbor j of point i is a type B point, with 1 indicating yes and 0 indicating no. If $GCLQ_{A\to B}$ is greater than 1, it suggests that type A points are likely to be attracted to type B points, indicating spatial association between the two types; if $GCLQ_{A\to B}$ is less than 1, it indicates that type A points tend to be distant from type B points; if $GCLQ_{A\to B}$ equals 1, the two types of points tend to be randomly distributed.

The Global Co-location Quotient has certain limitations in its inability to precisely analyze the co-location relationships between point features that vary with space, which poses a challenge in related research fields. In 2014, researchers such as Cromley conducted an in-depth study and introduced the Local Co-location Quotient (LCLQ) to reveal the spatial variation characteristics of the association between two types of point features.

$$LCLQ_{A_i\to B} = \frac{N_{A_i\to B}}{N_B/(N-1)} \tag{3}$$

$$N_{A_i\to B} = \sum_{j=1}^{N}\frac{w_{ij}f_{ij}}{\sum\limits_{j=1(j\neq i)}^{N} w_{ij}} \tag{4}$$

$$w_{ij} = \exp\left(-0.5 \times \frac{d_{ij}^2}{d_{ib}^2}\right) \tag{5}$$

In the formula: $LCLQ_{A_i\to B}$ represents the degree to which type B points are attracted by point A_i of type A;N_B is the number of type B points in the study area; $N_{A_i\to B}$ represents the weighted average number of type B points nearest to point A_i; d_{ij} represents the distance between point A_i and point j; w_{ij} is the Gaussian kernel density function, representing the weight of point j, where the closer the distance, the higher the weight; d_{ib} represents the bandwidth distance near point A_i.

3.3 Space Syntax

Space syntax is a theoretical and methodological framework used for the quantitative analysis of urban spatial structures. Grounded in graph theory, it segments urban spaces and analyzes the morphology and structure of spatial elements through topological relationships. This approach emphasizes understanding urban spaces and buildings from a human visual perspective, exploring the interactions between spatial organization and human activities [21]. The segment model within space syntax incorporates real-world coordinates of road networks and considers the impact of road angles on travel choices, providing a more accurate representation of actual pedestrian and vehicular flows [22]. Therefore, this study utilizes street density, integration, and angular choice variables derived from the segment model to examine the urban road network structure of Xiamen.

(1) Street Density (NC): The number of road segments within the search radius.
(2) Integration (INT): Indicates the accessibility of a spatial node. The higher the integration, the greater the accessibility. The calculation formula is:

$$INT = \frac{NC^2}{ATD} \tag{6}$$

$$ATD(x) = \sum_{i=1}^{n} d\theta(x, i) \tag{7}$$

In the formula: ATD represents the sum of the angular topological distances from a given node to all other nodes within the search radius; d θ (x,i) represents the angular topological distance between node x and node i; n is the number of other nodes within the search radius.

(3) Angular Choice (ACH): The number of times a node is traversed through the shortest angular topological distance. The higher the choice value, the stronger the spatial permeability. The calculation formula is:

$$ACH = \frac{\sum_{i=1}^{n}\sum_{j=1}^{n} \sigma(i, x, j)}{(n-1)(n-2)} \tag{8}$$

In the formula: σ(i,x,j) represents the shortest angular topological distance between points i and j that passes through node x.

(4) Search Radius (n): Refers to the area within which other nodes are considered when calculating the parameters of a given node. The area is defined by taking the node as the center and n as the radius.

3.4 Pearson Correlation Coefficient

The Pearson coefficient is a statistic used to measure the strength and direction of the linear relationship between two variables. It is one of the most commonly used correlation coefficients. In this study, the Pearson coefficient is employed to assess the direction and degree of association between tourism elements and the urban road network.

The research process mainly includes the spatial pattern analysis of tourism elements, the morphological analysis of the urban road network, and the correlation analysis between the two. The technical route is shown in Fig. 2.

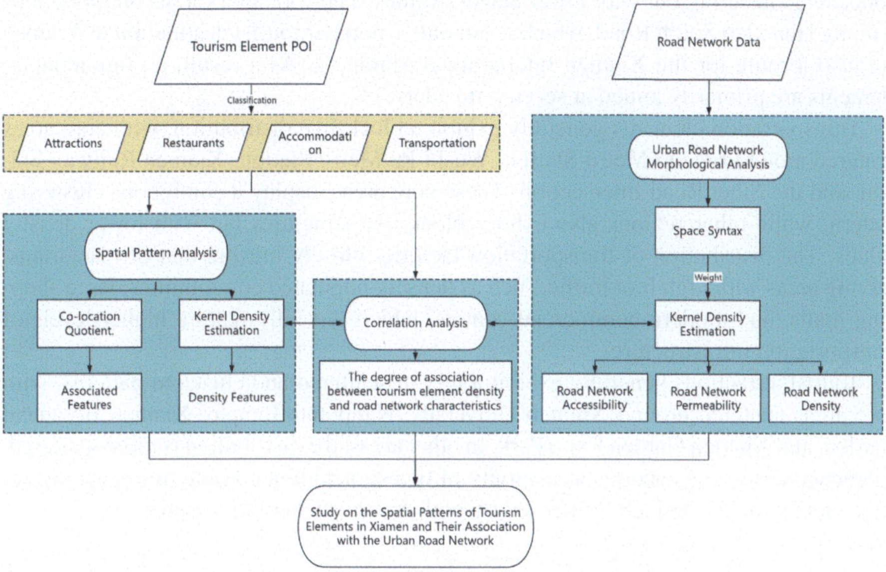

Fig. 2. Technical Route

4 Spatial Patterns of Tourism Elements in Xiamen

4.1 Density Features of Tourism Elements

Using ArcGIS software and the kernel density estimation method, kernel density analyses were conducted on the four selected types of tourism elements—"Restaurants, Accommodation, Transportation, and Attractions"—to explore their spatial distribution in Xiamen. The results revealed that these distributions exhibit spatial non-uniformity (Fig. 3). To determine the optimal search radius for the tourism data, multiple kernel density analyses were performed. After comparison, a search radius of 1,000 m was found to generate a relatively smooth density surface, effectively highlighting the kernel density centers of the tourism elements.

Restaurant elements generally exhibit a block-like distribution, with core areas centered around the Ferry Business District, Mingfa Plaza, Houpu Village, Daliao Village, and Sun Cuo Community. These core areas display a band-like distribution and show a trend of connection along Xiamen-Haixi Road and Jiahe Road. Other regions demonstrate notable clustering but do not exhibit peak values. The distribution of restaurants is significantly influenced by foot traffic, with a tendency to cluster near residential areas and large shopping malls.

Accommodation elements generally exhibit a clustered distribution, with core areas located in Zengcuo'an, Mingyi Golf Club, and Huangcuo Community, forming a "string of beads" pattern connected by Huandao South Road. The surrounding area is an ecological protection zone, where the density value drops sharply. Other regions also show noticeable clustering but with lower density values. The core area for accommodations is along Huandao South Road, which is not only a popular tourist destination in Xiamen but also a route for the Xiamen International Marathon. As a result, accommodation elements are primarily aimed at serving travelers.

Transportation elements generally exhibit a block-like distribution, with core areas centered around LÜcuo Metro Station, Wushi Pu Metro Station, Xiamen Railway Station, and the Sibei Road Intersection. These core areas display a contiguous clustering pattern, while other regions also exhibit block-like structures but with lower density values. The distribution of transportation facilities closely mirrors that of restaurants, serving areas with high foot traffic, such as densely populated communities, large shopping malls, and modern commercial centers, which typically feature higher levels of transportation infrastructure.

Tourist attractions generally exhibit a mix of scattered and clustered patterns, with core areas centered around Xiamen University, Nanputuo Temple, Xiamen Botanical Garden, and Xiamen Garden Expo Park. In other areas, the distribution is more scattered. Xiamen's attractions encompass a variety of types, including coastal views, ecological parks, and temples, and are influenced by both natural and cultural factors.

4.2 Spatial Correlation Characteristics of Tourism Elements.

4.2.1 Overall Spatial Correlation Pattern.

Using the Co-location Analysis tool in ArcGIS Pro, the overall spatial correlation pattern between tourism elements in Xiamen City was analyzed. After conducting 100 Monte Carlo simulations for significance testing and multiple experimental verifications, bandwidths were set to the 1st, 10th, and 25th nearest neighbors. The global co-location quotient reflects the strength of the association between pairs of tourism elements in Xiamen City. As shown in Table 2: ① The global co-location quotient increases gradually with the bandwidth but remains less than 1, indicating that the spatial distribution of tourism elements in Xiamen City generally exhibits a weak association pattern. ② The mean global co-location quotients are 0.7043 for accommodations and restaurants, 0.7018 for transportation and restaurants, with the maximum global co-location quotient occurring at a bandwidth of 25 for transportation and restaurants. This suggests that accommodations, transportation, and restaurants tend to be interdependent to some extent.

4.2.2 Local Spatial Correlation Characteristics.

Using the Local Indicators of Co-location Quotient (LCLQ), the local spatial correlation characteristics between pairs of tourism elements in Xiamen City were examined. Given the large number of elements, data stacking and overlap can occur, making it difficult to discern key patterns. To present the main features more clearly, this study focuses on

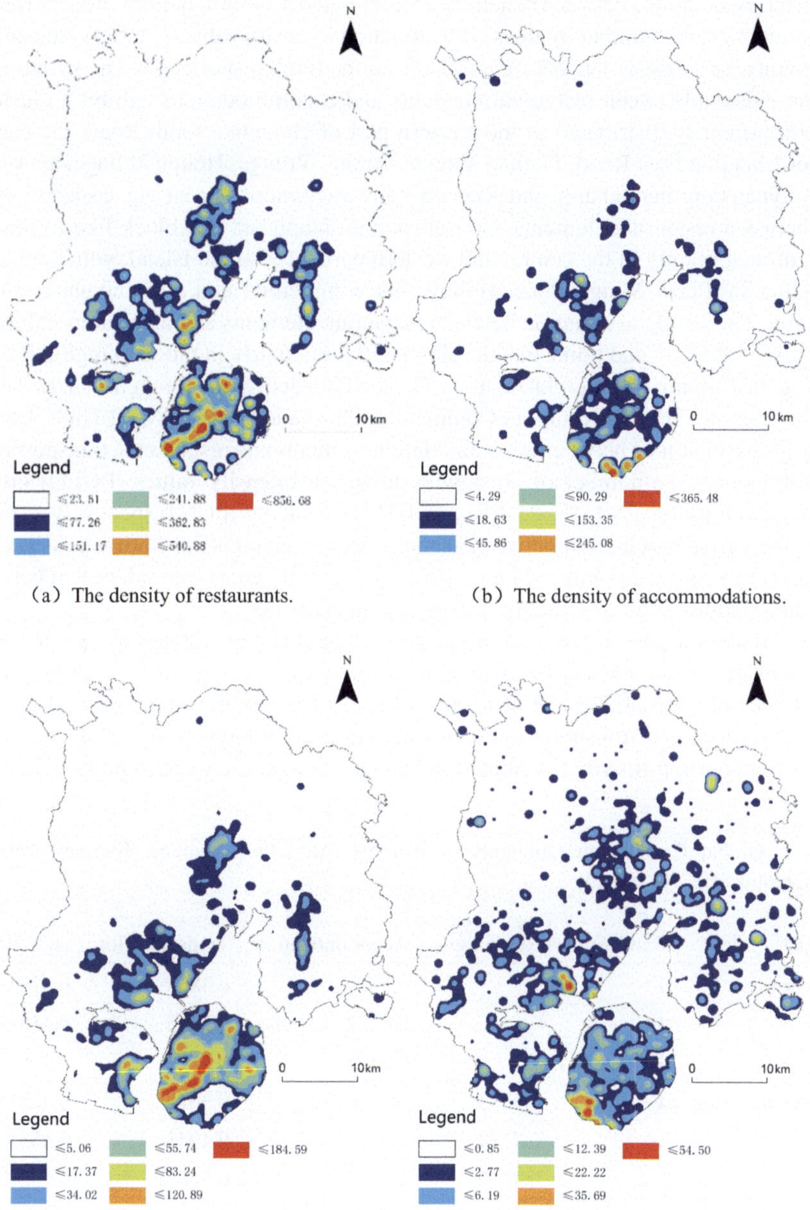

(c) The density of transportation

(d) The density of attractions

Fig. 3. Density Values of Tourism Elements.

displaying areas where the LCLQ is greater than 1. In this context, an LCLQ greater than 2 is considered indicative of a strong association.

In the associations between restaurant elements and accommodations, transportation, and attractions, the number of areas with attraction intensity values (LCLQ) greater than 1 accounts for 21.44%, 19.78%, and 12.40% of the total, respectively. The strong association clusters between restaurant elements and accommodations exhibit a clustered pattern, primarily distributed in the western part of Huandao South Road, the eastern part of Lianqian East Road, Dailiao Village, Andou Village, Houpu Village, as well as the Kaiyuan commercial area and Xiamen Railway Station. The strong association pattern between restaurant elements and transportation appears in a block-like formation, concentrated mainly in the central and western parts of Xiamen Island, with core areas including SM Plaza, Mingfa Plaza, Huli Avenue commercial area, and Kaiyuan commercial area. The strong association between restaurant elements and attractions exhibits a band-like extension and point-like development, particularly in the strip region encompassing the Lujiang commercial district, Xiamen University, and Zengcuo'an, as well as in the areas surrounding Xiang'an Commercial Plaza and Tong'an No. 1 High School.

In the associations between accommodation elements and restaurants, transportation, and attractions, the number of areas with attraction intensity values (LCLQ) greater than 1 accounts for 28.11%, 21.13%, and 16.07% of the total, respectively. Among these, the strong association between accommodation elements and attractions is characterized by scattered points, while no strong association areas were identified between accommodations and restaurant or transportation elements.

In the associations between transportation elements and restaurants, accommodations, and attractions, the number of areas with attraction intensity values (LCLQ) greater than 1 accounts for 24.79%, 21.56%, and 17.75% of the total, respectively. The strong association between transportation elements and accommodation elements exhibits a clustered pattern, primarily distributed in Dailiao Village, the western part of Huandao

Table 2. Global Co-location Quotients of Pairwise Attractions Among Tourism Elements (Bandwidths of 1, 10, and 25)

GCLQ	Bandwidth	Restaurants	Accommodation	Transportation	attractions
Restaurants	1		0.5569	0.3855	0.3731
	10		0.6232	0.5445	0.4584
	25		0.6808	0.6141	0.4982
Accommodation	1	0.6393		0.3610	0.3251
	10	0.7179		0.4945	0.4531
	25	0.7559		0.5482	0.5028
Transportation	1	0.5187	0.3670		0.4290
	10	0.7525	0.5704		0.6048
	25	0.8342	0.6447		0.6758
Attractions	1	0.5557	0.3984	0.5114	
	10	0.7185	0.5882	0.7216	
	25	0.7704	0.6483	0.7722	

South Road, and the central part of the Lujian commercial area. The strong association between transportation elements and attractions follows a road-oriented pattern, particularly evident along Guankou Avenue, Fengnan Road, Min'an Avenue, and Xiang'an South Road. No strong association areas were identified between transportation elements and restaurant elements.

In the associations between attraction elements and restaurants, accommodations, and transportation, the number of areas with attraction intensity values (LCLQ) greater than 1 accounts for 27.54%, 21.96%, and 24.19% of the total, respectively. No strong association areas were identified between attraction elements and restaurant, accommodation, or transportation elements (Fig. 4).

5 Study on the Correlation Between Tourism Elements and Urban Road Networks

5.1 Structural Characteristics of Urban Road Networks

The road network data of Xiamen City was imported into spatial syntax software, and eight key indicators were derived through segment model analysis to evaluate the road network morphology: global integration (INT), local integration (INT5000, INT1600), global choice (CH), local choice (CH5000, CH1600), and local street network density (NC5000, NC1600). Here, 1600 m represents the walking activity range of residents, while 5000 m represents the non-motorized activity range of residents or visitors. These indicators were used as weights in ArcGIS for density estimation, with a search radius of 1000 m for road network morphology analysis (Fig. 5). The density values of choice, integration, and street network density in the density map visually represent the importance of the road network in the area.

The high-value areas of local integration are primarily concentrated around the Jiangtou Interchange, Shigushan Overpass, Haixiang Interchange, Xiamen North Station, and Sun Cuo Community. As the observation scale expands, these high-value areas extend outward, forming a more extensive transportation network. Notably, the Jiangtou Interchange and Shigushan Overpass extend radially, presenting a connected pattern. Meanwhile, the Haixiang Interchange and Sun Cuo Community develop along Xiang'an Avenue and Jimei Avenue as development axes, exhibiting linear radial spread characteristics. Additionally, new core areas have emerged around the Haicang Bridge, Haicang Tunnel, Xiamen Bridge, Xinglin Bridge, Jimei Bridge, and Xiang'an Bridge. These bridges serve as crucial channels for external transportation within Xiamen Island, playing a key role in connecting various areas of Xiamen and acting as central transportation hubs for the city.

The peak of local choice is observed in the Haixiang Interchange area, with high-value zones distributed around the Jiangtou Interchange and the cross-sea bridges connecting various districts of Xiamen Island. As the observation scale expands, these areas become interconnected through major traffic arteries, displaying a block-like pattern. The difference between global and local choice is significant. Global choice reveals the overall structure and network of Xiamen's transportation system, with Xianyu Road traversing the island and connecting with major roads such as Xiang'an Avenue, the

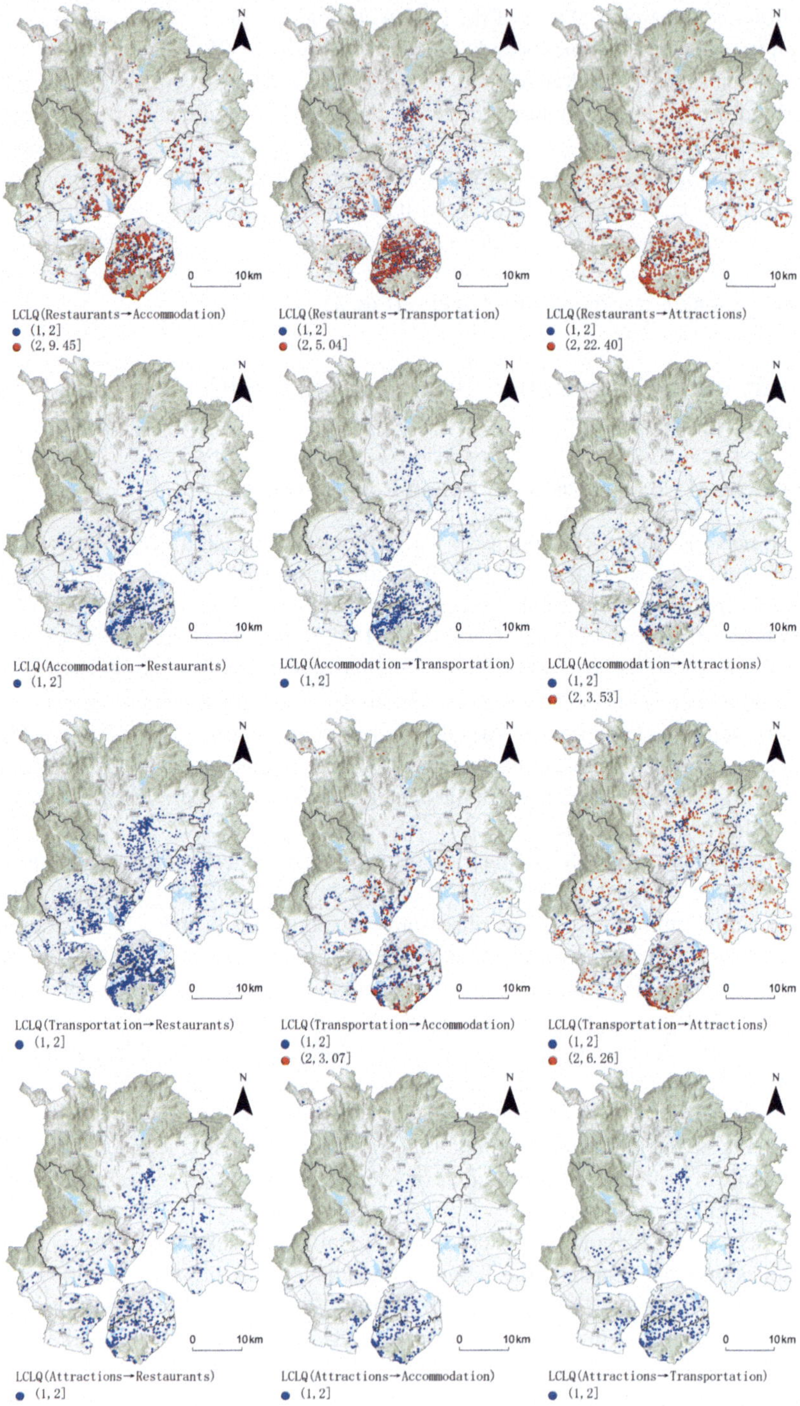

Fig. 4. Local Co-location Quotients of Pairwise Attractions Among Tourism Elements (Bandwidth of 25)

Quanzhou-Xiamen Expressway, and Guankou Road, forming a ring road that links different districts of Xiamen. Additionally, the Shaxia Expressway and County Road 414, serving as external trade corridors, are particularly prominent at the global scale.

The street density at two different scales exhibits a certain similarity and continuity in the overall spatial layout. Core areas are concentrated around key nodes such as the Haixiang Interchange, Jiangtou Interchange, and Shigushan Overpass. As the observation scale expands, these core areas display linear extension characteristics. For example, the Jiangtou Interchange extends linearly along Xiamen Avenue and Jiahe Road, connecting the Sibei Road commercial area, Wenzao commercial area, and Lianban commercial area, and shows a block-like connection trend with the Shigushan Overpass and Gaoqi Interchange. In contrast, the Haixiang Interchange develops along Haixiang Avenue as a development axis, radiating outward.

5.2 Analysis of the Relationship Between Tourism Elements and Urban Road Networks

To investigate the relationship between the spatial distribution of tourism elements and the characteristics of the road network structure, Xiamen City was divided into 100 m × 100 m grid cells. Kernel density estimation was applied to obtain kernel density values for each grid cell, with weights based on the spatial locations of tourism elements and road network indicators, including global integration (CH), local integration (CH5000, CH1600), global choice (I), local choice (I5000, I1600), and local street network density (NC5000, NC1600). The bivariate method was then employed to analyze the direction and degree of association between the spatial distribution of tourism elements and various road network morphology indicators (see Table 3). The results are as follows:

1. There are notable differences in the correlation between integration and the four types of tourism elements across different scales. The correlation between transportation and restaurant elements with integration is consistently moderate to high at all scales, with the strongest correlation observed at the 5000-m scale. This suggests that transportation and restaurant elements are more likely to be located in areas with higher accessibility, particularly at the 5000-m scale, where this association is most pronounced. Transportation elements, such as bus stations, subway stations, and parking lots, are typically situated on streets with good transportation access and high accessibility to meet people's travel needs. Similarly, restaurant elements have high requirements for street location, as accessibility is crucial for attracting customers, especially within the 5000-m range, which is a primary activity area for residents. In contrast, accommodation and attraction elements show weaker correlations with integration across various scales, with the highest correlations appearing at the 1600-m and global scales, both of which are weak. The accommodation industry, when selecting locations, considers not only transportation accessibility but also the scenic environment and clustering effects. As a result, accommodation elements have slightly lower requirements for street accessibility compared to restaurant elements. Attraction elements, such as seaside beaches and ecological parks, due to their unique geographical locations and environmental characteristics, exhibit a relatively low correlation with areas of high integration, as they are less likely to be distributed in areas with higher integration values.

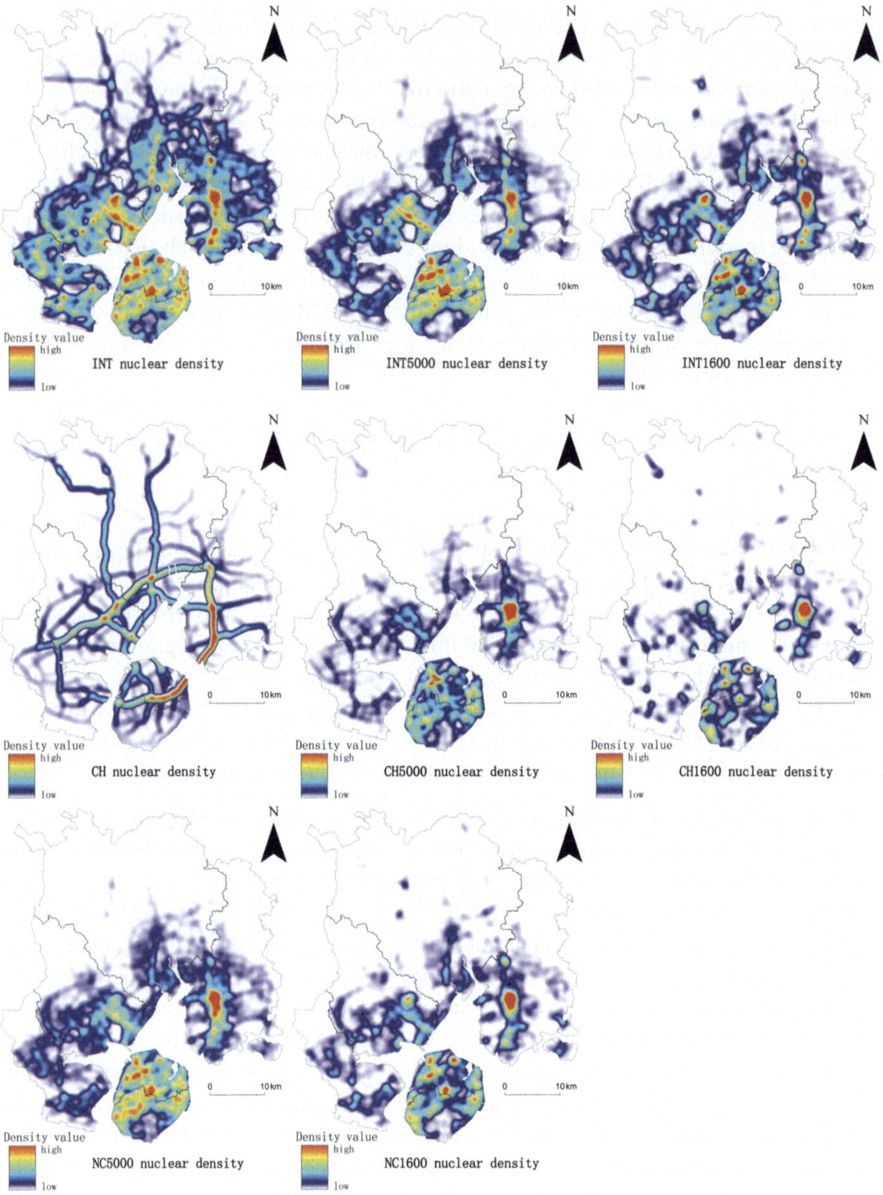

Fig. 5. Kernel Density Distribution of Xiamen's Road Network

2. The correlation between choice and the four types of tourism elements shows significant differences across various scales, particularly at the global and local levels. For restaurant and transportation elements, there is a moderate positive correlation with local choice, but only a low positive correlation with global choice. Accommodation and attraction elements exhibit a low positive correlation with local choice and

almost no correlation with global choice. The four major tourism elements are generally not distributed in areas with high global choice, which are typically important transportation hubs connecting the city internally and externally, primarily serving a "transitional" role within the urban transport network. Due to the mobility characteristics of these areas, tourists and residents are less likely to congregate there, resulting in a relatively lower distribution of tourism elements. However, at the 1600-m and 5000-m scales, restaurant and transportation elements still maintain a moderate correlation with choice. At smaller scales, streets with high choice often serve as preferred routes for residents, making them ideal locations for restaurants and transportation facilities to enhance convenience for residents. In contrast, accommodation and attraction elements are more frequently seen as "endpoints" of travel routes for tourists and residents, rather than transit points. This means these elements are more likely to be located in places that offer specific experiences and functions, rather than on high-traffic streets.

3. The correlation between street density and the four types of tourism elements varies across different scales. Transportation and restaurant elements show a moderate to high positive correlation with street density, particularly pronounced at the 5,000-m scale. This suggests that transportation and restaurant facilities often depend on good street accessibility and a well-developed road network to provide convenient travel conditions for customers. In contrast, the correlation between street density and tourist attractions or accommodation elements is low and exhibits little variation across different scales. Many natural attractions in Xiamen are located in coastal areas and hilly terrains, where the natural landscape is the primary attraction, reducing the need for high street density. Similarly, hotels that cater to tourists tend to be situated in scenic coastal or naturally pleasant areas, where the appeal lies in the unique geographic location and environmental conditions rather than the density of streets.

Table 3. Correlation between Tourism Element Density and Road Network Characteristics.

	Integration			Angular Choice			Street Density	
Scope	R = N	R = 5000	R = 1600	R = N	R = 5000	R = 1600	R = 5000	R = 1600
Restaurants	0.51**	0.59**	0.54**	0.18**	0.53**	0.43**	0.62**	0.55**
Accommodation	0.21**	0.20**	0.22**	0.06**	0.22**	0.20**	0.24**	0.24**
Transportation	0.57**	0.68**	0.63**	0.18**	0.65**	0.53**	0.74**	0.65**
Attractions	0.19**	0.20**	0.23**	0.04**	0.31**	0.36**	0.29**	0.31**

Significantly correlated at the 0.01 level (two-tailed)

6 Conclusions and Discussion

6.1 Conclusion

This study utilizes POI data of tourism elements (restaurants, accommodations, transportation, attractions) and road networks in Xiamen, applying methods such as kernel density estimation, spatial location quotient, and spatial syntax to calculate the density and association characteristics of tourism element spatial patterns, as well as the density, integration, and choice characteristics of the urban road network. The correlation between tourism elements and road network features was analyzed, leading to the following conclusions.

Regarding the spatial pattern distribution characteristics of tourism elements:

1. Density Characteristics: High-density areas for all tourism elements are primarily concentrated on Xiamen Island, with relatively fewer distributions outside the island. Notable exceptions include the Garden Expo Park in Jimei District and the Sun Cuo community, which are relatively prominent among attractions and restaurants. As the city center, Xiamen Island has a higher population density and economic activity. Its well-developed transportation infrastructure and unique island scenery provide favorable conditions for the concentration of tourism resources. This concentration not only promotes the prosperity of tourism but also highlights the imbalance in development between the island and the mainland.
2. Association Characteristics: At the city-wide level, the global spatial location quotient between tourism elements gradually increases with bandwidth but remains less than 1, indicating that the overall spatial distribution of tourism elements exhibits a relatively weak spatial association. However, accommodation, transportation, and restaurants demonstrate relatively high associations, with mean values all greater than 0.7, suggesting that accommodation and transportation have a certain attraction for restaurants, likely due to their intrinsic connections and complementarity. At the local scale, strong spatial associations between restaurants and transportation, accommodation, and attractions manifest as multiple block-like and band-like patterns, with the strongest associations observed between these elements. Strong spatial associations between accommodation and attractions appear only in scattered points, while transportation and accommodation, as well as transportation and attractions, show strong associations in scattered and band-like patterns. No strong association spaces are observed between the remaining elements.

Strip-shaped strong association areas primarily develop along major transportation arteries and coastlines. These regions, due to their linear characteristics, attract and channel substantial flows of people and goods. Block-shaped strong association areas are typically located in commercial centers, where a high concentration of business activities and people creates a pronounced clustering effect. Scattered strong association areas reflect individual locational advantages or inherent strong attractiveness.

Regarding the spatial distribution of road network structure characteristics:

At the smallest local scale, the Jiangtou Interchange, Shigu Mountain Overpass, and Haixiang Interchange emerge as core areas for integration, choice, and street density. These nodes serve as local transportation hubs due to their critical positions within the city's transport network. As the analysis scale broadens, areas with high integration and street density expand outward from these core areas, revealing important global transportation hubs such as Haicang Bridge, Xiamen Bridge, Jimei Bridge, Xiang'an Bridge, and Xiang'an Tunnel. This expansion reflects functional changes in the road network structure at different scales and underscores the significant role of bridges and tunnels in connecting larger areas. High-value areas for choice evolve from small core regions into major traffic arteries that connect various streets and areas, primarily forming a ring road consisting of major routes such as Xianyue Road, Xiang'an Avenue, the Quanzhou-Xiamen Expressway, and Guanxin Road. These roads, as the main arteries of Xiamen's transportation network, constitute the backbone of the city's traffic flow and regional connections.

Regarding the correlation between road network structure characteristics and the distribution of tourism elements:

In the correlation between restaurants, transportation elements, and road network structure characteristics, only global choice shows a low positive correlation, while other correlations are moderate to high, with the strongest association observed at the 5000-m scale. For accommodation and restaurant elements, there is no correlation with global choice, and other associations are weakly positive. This indicates that the distribution of restaurants and transportation elements is significantly influenced by the road network structure, whereas the distribution of accommodation and restaurant elements is less affected. Global choice often serves a "transitional" role within the road network structure and does not typically attract the placement of tourism elements.

6.2 Discussion

Enhancing the connection between major transportation routes and key tourist attractions is crucial for advancing tourism development.However, in practice within Xiamen, there is a noticeable lack of correlation between attractions and the road network structure. Specifically, the association between attractions and the accessibility, permeability, and density characteristics of the road network is weak or nonexistent, which can negatively impact the travel experience of tourists. While road construction and optimization must consider various factors such as economic conditions, terrain, and population distribution, Xiamen, as a major tourist city, must prioritize the accessibility and convenience of attractions in its urban planning. Additionally, in line with the development requirements of building tourism and leisure cities and districts, this study identifies a lack of significant spatial correlation between tourist attractions in Xiamen and key tourism service elements such as restaurants, accommodation, and transportation. This spatial disconnection not only weakens the overall tourist experience but also hampers the development of tourist and leisure cities and districts. To address this issue, Xiamen should focus on enhancing service infrastructure around tourist attractions and connecting key tourist areas to create service-oriented districts. For example, areas along Line 1 of the subway, including Zhenhai Road Station to Lücuo Station, SM Plaza, Mingfa Commercial Plaza, the Cruise Center, Xiamen Railway Station, Xiamen University, and Ring Island South Road, possess abundant tourist resources and have significant potential to improve tourism service efficiency through clustered development. By enhancing the clustering of tourism service elements in these areas and improving the quality and coverage of supporting services, Xiamen can better meet the critical needs of tourism development, thereby promoting the sustainable growth of the tourism industry and increasing the city's attractiveness and competitiveness as a tourist destination.

This study only explores the potential reasons behind the formation of tourism element spatial patterns by examining the correlations between tourism elements and road network structure characteristics, considering too few influencing factors. Future research could build on this by incorporating various aspects such as regional planning, terrain and topography, and tourism behavior to further investigate the factors affecting the spatial patterns of tourism elements. A comprehensive analysis of these factors can provide a more thorough understanding of tourism element distribution patterns, offering deeper insights for the scientific planning and management of tourism, and providing more informed decision support for the optimization of tourism element allocation.

Acknowledgments. The authors would like to thank the anonymous reviewers for their valuable comments on the manuscript.

References

1. The State Council: Notice of the State Council on Printing and Distributing the "14th Five-Year Plan" for Tourism Development [EB/OL]. (2022). http://www.gov.cn/zhengce/content/2022-01/20/content_5669468.htm. Accessed 22 August 2024
2. Gang, C., Yiming, Z.: Mechanisms and Pathways for Tourism Factors to Promote Rural Revitalization. Journal of Lanzhou University of Finance and Economics **04**, 84–91 (2023)
3. Honglian, W., Xiaoli, W., Min, H., Wenzhe, Z., Huiyue, L., Wenjing, N.: Study on the Spatial Pattern and Influencing Factors of County Tourism Elements in the Guanzhong Plain Urban Agglomeration Based on POI Data. Arid Zone Resources and Environment **06**, 200–208 (2024)
4. Lianjiao, S., Qingzhong, M.: Characteristics of Spatial Distribution of Tourism Facilities and Analysis of Their Spatial Misalignment with Tourism Resources. Resources Development, Market **04**, 605–613 (2024)
5. Li Peng, Y., Yingjie, W.: Study on the Spatial Agglomeration Characteristics of 3A and Above Tourist Attractions in China. Geographical Sciences **11**, 1883–1891 (2018)
6. Wei, Z., Xiaomeng, W.: Study on the Spatial Distribution Patterns of 5A and 4A Tourist Resources in China. Statistics and Decision **24**, 103–106 (2012)
7. Agusti, D.P.I.: Territorial distribution of tourist attractions. Comparing projected and perceived image in Uruguay. Economía Sociedad Y Territorio **18**(58), 735–762 (2018)
8. Sena, I.S., Ruchkys, U.D., Travassos, L.E.P., et al.: Geotourism Potential in Karst Geosystems: An example from the Lund Warming Ramsar Site, Minas Gerais, Brazil. CATENA **208**, 105717 (2021)
9. Fei, L., Jiaming, L., He, Z., Tao, L.: Spatial Distribution and Influencing Factors of Homestays in the Yangtze River Delta Region. Geogr. Res. **04**, 950–960 (2019)
10. Wang Chaohui, L., Lin, F.T., Qiaoyun, X.: Evolution of the Spatial Pattern of the Tourism Accommodation Industry in Shanghai During the Expo Construction Period. Acta Geogr. Sin. **10**, 1423–1437 (2012)
11. Zhiqiang, H.: Empirical Analysis of the Effect of Transportation Infrastructure on Regional Tourism Economic Growth—Based on a Spatial Econometric Model of Provincial Panel Data in China. Macroeconomic Research **06**, 118–132 (2018)

12. Shuang, J.: Research on the Spatial Layout Design Methods of Auxiliary Facilities in Integrated Transportation Hubs. Transportation and Transport **S1**, 67–70 (2023)
13. Lili, Z., Jinshan, Z.: Mechanism, Fields, and Future Trends of Integrated Development of Transportation and Tourism. China Highway **05**, 54–56 (2021)
14. Lili, Z., Jinshan, Z.: New Practices in the Integration of Tourism and Tranportation Development. China Highway **12**, 44–47 (2018)
15. Rui, Y., Congdi, S., Ziqun, K.: Exploration of the "Transportation+" Model Under the New Situation of Highway Investment. Transportation Technology and Management **24**, 155–157 (2023)
16. Rongjie, L.: Research on the Big Data Application Framework Based on Transportation-Tourism Integration. China Highway **22**, 94–95 (2020)
17. Fengbo, L., Cheng, Y., Cong, G.: Integration Development of Transportation and Tourism: Issues, Mechanisms, and Pathways. Shandong Social Sciences **04**, 144–149 (2020)
18. Chao, W., Ting, L., Xiaosha, M., Jianqiang, F.: Analysis of Coordinated Development of Tourism Economy and High-Speed Transportation Under "Transportation-Tourism" Integration: A Case Study of the Guanzhong Plain Urban Agglomeration. Arid Zone Geography **06**, 1015–1025 (2024)
19. Jie, S.: Research on Thematic Construction of Transportation-Tourism Integration Projects: A Case Study of the G75 Lanhai Expressway Taishi Service Area. Gansu Science and Technology Review **12**, 70–73 (2023)
20. Zhang, X., Zhang Shizhi, T., Jingyu.: Integrated Development of Transportation and Tourism and Planning Responses. Comprehensive Transportation **06**, 28–32 (2017)
21. Yishu, Q., Meng, Y.: Study on the Impact of Urban Road Network Morphology on Hotel Layout Based on Space Syntax. Tourism Forum **02**, 101–111 (2018)
22. Hillier, B., Yang, T., Turner, A.: Normalising least angle choice in Depthmap and how it opens new perspectives on the global and local analysis of city space. Journal of Space syntax **3**(2), 155–193 (2012)

Methodology for Spatialization of Regional Demographic Data Based on Remote Sensing and Geographic Conditions Census Data

Yahui Chen[1], Xiaoyue Gao[1(✉)], Runliang He[1], Yunzhi Zhang[1], and Chenfei Yu[2]

[1] China Earthquake Networks Center, Beijing 100045, China
{chenyahui,xygao}@seis.ac.cn
[2] Beijing Earthquake Agency, Beijing 100080, China

Abstract. In order to meet the needs of rapid assessment of earthquake disasters and visual expression of spatial distribution of population, this paper takes Luding County of Ganzi Tibetan Autonomous Prefecture in Sichuan Province as the study area. The research employs an object-oriented classification method to extract the shadows of buildings from high-resolution remote sensing images, obtain the length of the shadows, and combine this data with building data of geographic conditions census to determine the number of floors in the building zones. Concurrently, seventh national population census data are integrated with building area to create a population data set based on buildings, thereby completing the high-resolution regional population data spatialization simulation. Furthermore, 50-m grid data is generated through data resampling, which markedly enhances the accuracy of the population grid data, allowing for more precise localization of the population distribution, thus proving advantageous for post-earthquake emergency response operations. The feasibility of this method has been demonstrated, and its potential for broader application is evident.

Keywords: Geographic National Census · Spatialization of Demographic Data · Building Shadows · Classification

1 Introduction

Following the occurrence of destructive earthquakes, it is imperative to undertake a rapid assessment in the shortest possible time. This assessment should produce basic information in the vicinity of the epicentre, estimates of potential economic losses and casualties, and a range of thematic maps. The population data, as a significant information source, serves as the fundamental basis for assessment and mapping. It is instrumental in determining the accuracy and reliability of the assessment results and casualty distribution, and consequently, it affects the decision-making and emergency response efficiency following an earthquake. Currently, population data published by the state is based on administrative divisions as the statistical unit and summarized at various levels through censuses [1, 2]. This statistical method is detached from the geospatial characteristics of population aggregation, and there is a growing interest in research in the integration of demographic data with geographic data to realize the spatialized population model.

G. Luo (Ed.): GFSISD 2024, SPPHY 417, pp. 148–158, 2025.
https://doi.org/10.1007/978-981-96-2490-4_10

At present, population spatialisation models can be classified into five categories: area weighting model, core estimation model, land use type influence model, gravity model and multivariate data fusion model [3–5]. Among these models, the land use type influence model employs a calculation of the relationship between population and land use type to determine the unit population density. Pan Ying et al. [6] selected the zonal modelling method to spatialise the county-level population on the land use data of Fujian Province. The resulting data were found to align with the actual population distribution of counties. Hardin et al. [7] applied the multiple regression method to model the land use data, with the objective of determining the density of building units. Furthermore, numerous scholars at home and abroad have employed 3S technology to support the simulation of population spatial distribution, particularly through the application of remote sensing images for land use classification, the extraction of building outlines and heights [8, 9], the differentiation of building uses and categories [10, 11], the estimation of building numbers and characteristics [12, 13], the establishment of mathematical relationships between population and buildings, in order to facilitate the spatialisation of population.

Since 2013, China has been conducting a geographic national census, which encompasses fundamental geographic environmental elements and geographic composite feature elements [14]. Among these, the building and its subcategories fall within the census's purview. In this paper, we will make use of this refined building data, focusing our analysis on Luding County of Ganzi Tibetan Autonomous Prefecture in Sichuan Province. The number of building floors is calculated through the interpretation of remote sensing images, with statistical population data integrated to achieve a more precise spatialisation of the regional population. This approach provides a foundation for post-earthquake assessment and thematic mapping, offering a more accurate and reliable service for decision-making in earthquake emergency response and rescue.

2 Data Foundation

This paper employs three types of data: remote sensing image, building data and population data of the study area. The remote sensing image is a freely downloadable map tile data set comprising only red, green, and blue bands and a resolution of 0.3 m. The study area is divided into two image strips, both of which are processed separately during the data processing stage. The building data is the result of the initial national geographic census and has been graded and classified according to building density (high/low) and the number of floors (super-high/high/medium-high/low/multi-storey). It should be noted that this classification does not divide each building unit, but instead groups the houses into multiple building zones according to the aforementioned categories. Consequently, in order to ascertain the precise number of floors in each building zone, it is necessary to extract the building shadows by classifying the remote sensing images. The population data are derived from the seventh national census, which allows for the retrieval of resident population data at the county and township levels from official sources. The data are authentic and highly reliable, thereby ensuring the effective application of the research method and the accuracy and credibility of the research results.

3 Shadow Extraction for Buildings

The study area, Luding County, covers an area of 2,165 square kilometers and comprises eight townships. The article extracts more complete shadows of buildings from the images through the processes of image enhancement, image segmentation and threshold classification.

3.1 Image Enhancement

The application of remote sensing images is often limited by the loss of valuable information due to atmospheric or weather-related effects. In order to eliminate some interfering factors and reduce the effects of thin clouds and haze on the clarity of the features in the image, it is necessary to undertake pre-processing of remote sensing images. The objective of image enhancement is to increase the contrast between the target and the surrounding background image, thereby facilitating the interpretation of the image.

The purpose of the remote sensing image enhancement described in this paper is to improve the contrast between the building shadows and the surrounding background image, thereby facilitating the classification and extraction of the shadows. To this end, several common image enhancement transformation methods are compared, including image sharpening, linear transformation, global histogram equalisation, adaptive histogram equalisation, as well as contrast limited adaptive histogram equalization. Revealed that image sharpening is primarily applied to accentuate the edges and contours of features, thereby facilitating edge enhancement. It has been demonstrated that image sharpening is primarily employed to accentuate the edges and contours of the features, thereby facilitating an edge enhancement. The method does not affect the contrast of building shadows. A linear transformation is a method that expands the grey value of an image from a limited range of intervals to a range of 0–255, thereby improving the visual effect of the image and adjusting the contrast and brightness. The histogram equalisation method is used to enhance the contrast of an image by transforming the histogram distribution of the image into a uniform distribution. The adaptive histogram equalisation method is based on this to improve the local contrast of the image and add more detailed information. The contrast limited adaptive histogram equalisation method is further optimised to reduce problems caused by noise amplification by setting a threshold.

In this study, the aforementioned algorithms for image enhancement were implemented in sequence using the Python programming language. As illustrated in Fig. 1, the source image displays a grey tone, presents atmospheric interference, and lacks clarity in the object display. The process of image sharpening results in a progressively deeper greyness, whereas linear transformation leads to an over-brightening of the image. Furthermore, a comparison of the three histogram equalisation methods reveals that they are all effective in enhancing image contrast. However, with regard to image brightness, the global method is on the bright side, the local method is on the dark side, and the restricted contrast method represents a middle ground, capable of accentuating the outlines of features to a greater degree, thereby making the building shadows more apparent and differentiating them from other features. Additionally, the enhanced image exhibits more distinct colors and an appropriate brightness. This also demonstrates that the method is most appropriate for the classification of shadows. Furthermore, it can be observed that the shape of the greyscale histogram is more elongated following the processes of image enhancement (as illustrated in Fig. 2), and that the contrast levels have been significantly increased.

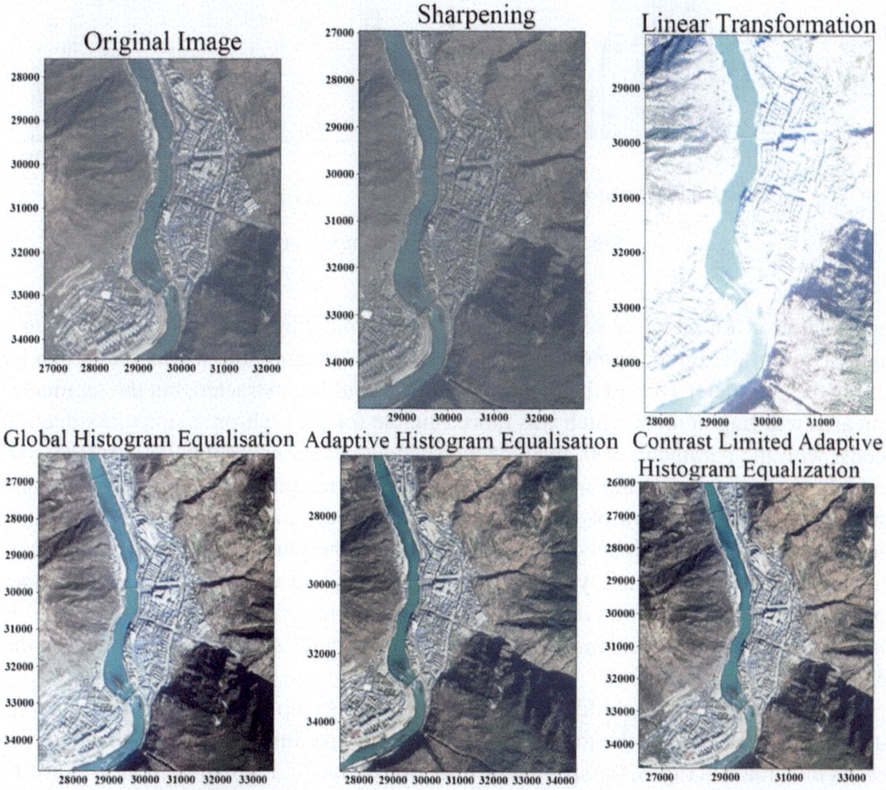

Fig. 1. Comparison of original image and image enhancement results

3.2 Image Classification

The research uses the German eCognition Developer image analysis software to classify the processed remote sensing images, applying an object-oriented classification method. This method is dependent on the spectral characteristics of the features, but is more reliant on their geometric and structural information [15], which makes it more suitable for the processing of map tile data.

Initially, the image is segmented, with image objects exhibiting similar spectral, textural, and other features grouped together within the same region. This paper selects the multi-scale segmentation algorithm, which offers the benefit of synthesizing image information of varying scales, resolving more complex image segmentation issues, thereby

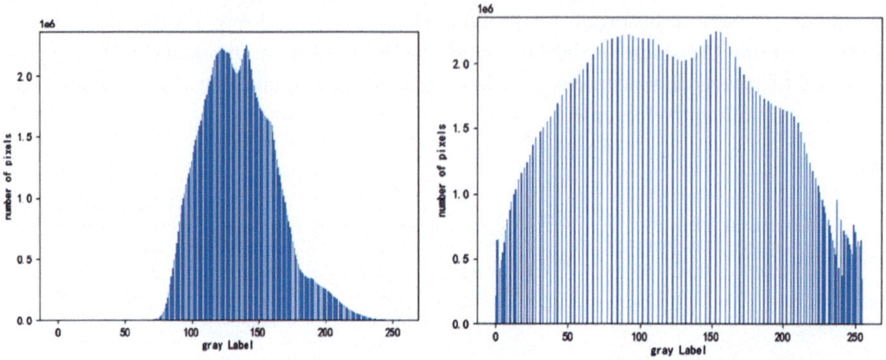

(a) Original Images **(b) Image Enhancement**

Fig. 2. Image histogram comparison

making it more suitable for semi-automatic or automatic image segmentation. After a comparison of the three segmentation scales of 100, 200 and 300, it was found that by setting the parameter 100, all the building shadows could be extracted, but the segmentation area was fragmented, which was more suitable for the high-precision classification needs; with the parameter 300, most of the shadows were incompletely extracted; and with the segmentation scale of 200, the shadow extraction was more complete, and only some of the shadows of the low and dense building zones were integrated into the surrounding feature objects, which remained not finely segmented. Given the inherent limitations in the accuracy of the geographic national conditions data, this paper applies a coarse classification during the process of building shadow classification. This ensures the extraction of at least one shadow for each building zone, thereby setting the segmentation scale to 200.

Subsequently, the threshold classification method is employed to analyse characteristics of the image, select the most appropriate feature parameters and establish the rule set. As illustrated in Fig. 3, the spectral information of the shadow can be more effectively discerned based on the contrast to neighbouring pixels. Furthermore, an comparison of the eigenvalues of the red, green and blue bands on this feature parameter reveals that the shadow exhibits a more pronounced contrast in the green band, facilitating its distinction from other features. To enhance the precision of the extraction, the range of eigenvalues is further delineated to encompass the shadow area. When the characteristic parameter is less than −700, the method is capable of not only extracting the requisite building shadows but also reducing the occurrence of a significant number of misclassifications, avoiding secondary classification, and meeting the demand for coarse classification of building shadows in this study.

Fig. 3. The extraction effect of characteristic parameter of contrast to neighbor pixels.

4 Building Storey Determination

4.1 Calculation of Building Shadow Length

In the calculation of the number of building floors, the traditional method is to use the relationship between the solar azimuth and satellite azimuth, height angle and building shadow at the time of remote sensing image acquisition to deduce the height of the building, thus obtaining the number of building floors. However, the satellite parameters and image time cannot be obtained from the map tile data. Therefore, this paper presents a new calculation method, which measures the sun azimuth angle according to the shadow angle in the map tile data, calculates the shadow length along this angle, and then combines it with panoramic map data or web photos of typical buildings to obtain the actual number of building storeys. Following this, the number of other building floors is calculated according to the ratio of the shadow length.

The image of the study area is divided into two distinct strips. The first strip encompasses Yanzigou Town and Mousi Town in Luding County, while the second strip includes the remaining towns. The solar azimuth angles of the two images have been measured at 117° and 95°. This paper employs the Python programming language for the batch processing of shadow layers, calculating the length of each shadow along the solar azimuth individually and writing the resulting calculations into the layer properties. The primary method involves the creation of numerous parallel lines aligned with the solar azimuth. These lines intersect with the shadows, resulting in a multitude of line segments. However, the presence of buildings and other structures within this area introduces a degree of complexity, as shadows may be obstructed or altered by these structures. To address this issue, the longest line segment is identified as the length of the shadow cast by the building in question.

In the geographic national condition data of building, there are 8,080 building zones in Luding County, and more than 20,000 building shadows have been extracted from the image. In order to facilitate the calculation of shadow length and assignment to the building layer, this paper overlays the shadow data with the building data, associates the ID of the corresponding building zone for each shadow, counts the number of shadows according to the ID code, and takes the average value of the area as the shadow length of the building in that area for the building zone covered by multiple shadows.

4.2 Building Zone Storey Calculation

In order to calculate the number of building floors, it is necessary to collect actual building samples. From these samples, the actual number of floors in each building can be determined, which allows the number of floors in other buildings to be calculated on a proportional basis. The samples are obtained through the network query method. A review of the panorama map revealed that within the entire Luding County, there is only panorama image within a five-kilometer radius of the Luqiao Township government. As a result, the number of layers present in the actual buildings within this area was used to determine the number of layers present in other townships that share an image with Luqiao Township. In the case of Yanzigou Township and Mousi Township, the number of storeys can be determined by combining the data from the images of typical schools, hotels, and other buildings with data from news sources or tourism websites. For example, Fig. 4 (right) illustrates a genuine photograph of the Hailuogou Pearl Garden Hotel in Mousi Town, sourced from a tourism website. By consulting an online map of the hotel's location (depicted in the middle of Fig. 4), the red-roofed structure visible in the downloaded tile map (depicted on the left of Fig. 4) was identified. It is established that the shaded length of the building zone where the hotel is situated is 6.2 m, and the

Fig. 4. Hailuogou Pearl Garden Hotel, Mousi Township

actual number of building floors is 2, which is equivalent to one floor for every 3.1 m. By employing this methodology and acquiring additional data in a similar fashion, it is possible to ascertain the number of floors in all building zones within Mousi Town and Yanzigou Town.

5 Population Spatialisation Simulation

5.1 Spatialisation of Regional Population

The latest census data for Luding County indicates a resident population of 84,204. This paper presents a methodology for the spatial simulation of statistical population data based on buildings. The approach combines the population with the geographic locations of the buildings where they are situated, thereby expressing the spatial distribution of the population and the number of people in each location in an intuitive manner. Furthermore, in order for the data to be applicable within the context of the seismic industry, the data are also gridded on this basis to produce population gridded data.

In the process of creating a spatialized simulation of the population based on buildings, it is necessary to calculate the extended area of the buildings, i.e. the floor area of each building basal area multiplied by the number of floors, divide the total population of the county by the extended area of the building to obtain the population in terms of floor area, and then multiply this value by the extended area of each building zone to obtain the population in each building zone.

$$POP_i = S_i \cdot f_i \cdot \frac{84204}{\sum S_i \cdot f_i}(i = 1, 2, 3, \ldots\ldots, 8080) \tag{1}$$

In Eq. (1), the variable "*Si*" represents the basal area of building zone "*i*", the variable "*fi*" denotes the number of floors in building zone "*i*", and the parameter "*POPi*" signifies the population density in building zone "*i*".

In the case of population data derived from building information, rasterization to the standard 1 km grid will inevitably result in a reduction in data accuracy. Nevertheless, if the data is subjected to further processing to achieve greater refinement in the rasterization process, it can be observed that the smaller the grid division unit, the smaller the damage to the accuracy of the original data. Therefore, it is advisable to avoid providing population data in the grid without the distribution of people. In this paper, we endeavor to generate sample data at a resolution of 50 m and utilize it for the creation of thematic maps.

The grid encompasses the entirety of Luding County, and the grid data is overlaid with the building data to generate a gridded representation featuring the locations of buildings. In each grid, the ratio of the floor area to the area of the building zone in which it is located is calculated, from which the number of people in the gridded buildings can be extrapolated. Subsequently, the value is interpolated in order to obtain the population of the grid in accordance with the ratio of the floor area of the grid to the area of the grid. The resulting population figure corresponds to the 50 m grid, effectively completing the gridding of population data. The processing of regional population data in a gridded format provides a foundation for the subsequent production of high-precision population grid data at the national level. Furthermore, it offers a basis for decision-making in the designation of key rescue zones in the context of an earthquake emergency.

5.2 Error Analysis and Accuracy Validation

To further verify the accuracy of the results of population spatialization, this paper employs the partition statistics function of ArcGIS software to statistically enumerate the population raster data of the 50 m grid in accordance with the boundary range of townships in Luding County. The resulting statistics are compared with the data of the seventh census of the resident population of nine townships in Luding County, and the results of this comparison are presented in Table 1. As can be seen, the total error of the data is 4.26 percent, with an accuracy rate exceeding 95.7 percent. However, the data errors of Yanzigou Township, Mousi Township, and Lan'an Township are higher than 5 percent. This may be attributed to the following reasons: Yanzigou Township has a greater proportion of low-rise buildings, and there are no representative photographs of typical buildings for comparison, which introduces errors in the determination of floor levels. Mousi Township has a greater proportion of high-rise buildings, and there are some errors in the geographic national condition data due to its low precision in delineating building zones. The remote sensing of the northern area of Lan'an Township is affected by cloud cover, and the floor heights in this region are estimated, introducing an error. In conclusion, the method of spatialization of regional population data based on geographic national conditions data used in this paper is a feasible approach with relatively accurate results, which makes it a promising avenue for further research and the potential application of spatialization of population in larger areas.

Table 1. Comparison of regional statistical results with census data

Township Name	Census Population (Person)	Regional Statistical Results (Person)	Difference (Person)	Percentage of Error
Luqiao	37239	37004	235	0.63%
Lengzi	9540	9071	469	4.92%
Xinglong	6611	6587	24	0.36%
Yanzigou	4604	4843	239	5.19%
Mousi	6874	7525	651	9.47%
Dedo	6330	6087	243	3.84%
Pengba	3116	2984	132	4.24%
Dewey	7973	7993	20	0.25%
Lan'an	1917	2097	180	9.39%
Total	84204	84191	Average Error	4.26%

6 Results Application

In examining the data from the Sichuan Luding earthquake (5 September 2022) as a case study, this paper utilises the population grid data derived from building data of geographical conditions census to inform the post-earthquake emergency rapid output maps. Furthermore, it compares this data with the population data from the 1 km grid and the maps are expanded to show extreme seismic intensity. The overall effect of the two thematic maps of population distribution is shown in Fig. 5. The comparison revealed that the unit of population grid data has been upgraded from person/square kilometer to person/square meter, with an accuracy of the meter level. Furthermore, the data accuracy has been significantly enhanced by the refined calculation of the population ratio, which mitigates the resampling error that may arise in the kilometer grid data. Additionally, with regard to population distribution, the data obtained through the spatialization of population using house buildings exhibit greater advantages with respect to location distribution, exhibiting a higher degree of data refinement and more accurate positioning than previously observed. This is of significant assistance to the post-earthquake emergency response operations.

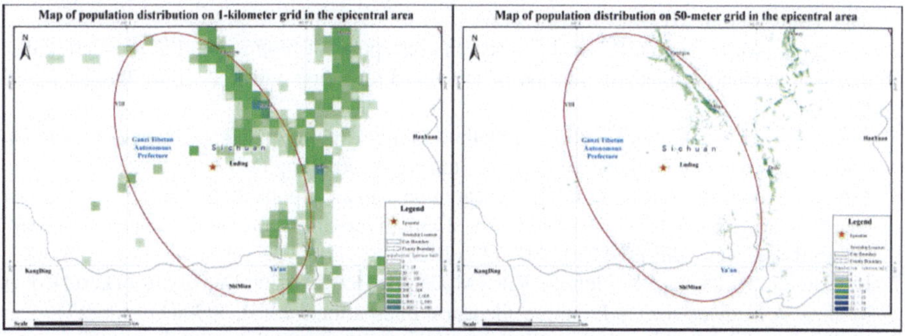

Fig. 5. Comparison of the mapping effect of 1 km versus 50 m population grid data

7 Conclusion and Discussion

This paper focuses on Luding County in China's Sichuan province. It employs high-resolution remote sensing images to extract building shadows and calculate their lengths. Concurrently, the number of all building layers in the study area is accurately determined by combining the fine-grained building data with real photos and panoramic images of building samples. Ultimately, the population census data for Luding County is distributed across the building layers, establishing a quantitative relationship between population distribution and building distribution. This allows for the precise simulation of the spatial distribution of the population in the area of interest. The resulting spatialized population data is highly accurate and the method is both operational and capable of wider data application.

Nevertheless, the building data of the geographical conditions census still require improvement in terms of the accuracy of building zone division. The data divide buildings into zones according to density and height range, which results in the inclusion of roads and a small number of buildings with different heights within the same zone. This increases the error of spatialization of the population. Furthermore, the absence of a distinction between individual buildings in the data precluded the differentiation of building usage in conjunction with high-precision points of interest (POIs), thereby failing to account for the potential influence of industrial buildings on population distribution. Subsequent studies may employ higher precision building data and combine it with POIs to create time-phased population spatialization data.

References

1. Shunbao, L., Zehui, L.: Study on spatialization of population census data based on relation ship between population distribution and land use——Taking Tibet as anexample. J. Nat. Resour. **18**(6), 659–665 (2003)
2. Dong, Y.: Research on spatial fine distribution of urban population based on multi-source data taking Zhengzhou city as an example. Henan University, China (2023)
3. Haiyue, F., Manchun, L., Jun, Z., et al.: Summa of grid transformation models of population data. Hum. Geogr. **21**(3), 115–119 (2006)
4. Xinran, Z., Yujia, L., Yi, S., et al.: Research conception of high resolution population distribution simulation method. Journal of Disaster Prevention and Reduction **39**(02), 76–80 (2023)
5. Zihan, G., Fei, C., Xiaoli, L., et al.: A method to improve the spatial accuracy of population grided data based on land use. Bulletin of Surveying and Mapping **12**, 66–70 (2021)
6. Ying, P., Wanli, H., Shilin, Y., et al.: Research on statistical population spatialization model based on land use type data: An case study of Fujian Province. Journal of Yunnan University (Natural Sciences Edition) **42**(4), 701–711 (2020)
7. Hardin, P.J., Jackson, M.W., Jensen, R.R.: Modelling Housing Unit Density from Land Cover Metrics: A Midwestern US Example. Geocarto International **23**(5), 393–411 (2008)
8. Linlin, L., Huadong, G.: Automatic estimation of urban population with remote sensing imagery based on ArcGIS. Remote Sensing Information **2**, 64–68 (2008)
9. Jing, F.: Urban small area population estimation based on high-resolution remote sensing Data. East China Normal University, Shanghai (2012)
10. Tiantian, F., Jianya, G.: Investigation on small-area population estimation based on building extraction. Remote Sensing Technology and Application **25**(3), 323–327 (2010)
11. Serkan, U., Ejaz, H., Jie, S.: Building Population Mapping with Aerial Imagery and GIS Data. International Journal of Applied Earth Observationand Geoinformation **13**(6), 841–852 (2011)
12. Wu, S.S., Wang, L., Qiu, X.M.: Incorporating GIS Building Data and Census Housing Statistics for Sub-Block-Level Population Estimation. Professional Geographer **60**, 121–135 (2008)
13. Yao, M., Rui, L., Jie, J., et al.: Urban street scale population estimation based on building information. Geomatics and Information Science of Wuhan University **46**(8), 1194–1200 (2021)
14. Haiqin, L.: On classification system for element in geographical condition census. Geomatics & Spatial Information Technology **37**(6), 118–121 (2014)
15. Fenglan, D., Qingjiu, T., Xueqi, X., et al.: Object-oriented image classification analysis andevaluation. Remote Sensing Technology and Application **19**(1), 20–23 (2004)

Multi-target Recognition for Hyperspectral Images Based on Sparse Representation and Multi-scale Analysis

Fangfang Li[2], Kang Sun[1(✉)], Wang Shicheng[2], Chen Jinyong[2], Liang Shuo[2], and Wang Min[2]

[1] School of Geographical Sciences/Hebei Technology Innovation Center for Remote Sensing Identification of Environmental Change, Hebei Normal University, Shijiazhuang 050024, China
skgucas@163.com

[2] The 54th Research Institute of China Electronics Technology Group Corporation, Shijiazhuang 050081, China

Abstract. Hyperspectral images can provide rich spatial and spectral information, providing significant advantages for target recognition. Due to the large number of bands and data in hyperspectral images, the problem of "same object, different spectra" is quite serious in the process of target recognition. In order to solve this problem, this paper proposes two methods: one is a nonlinear mapping method for search recognition, and the other is a recognition method that combines sparse representation and multi-scale analysis for matching recognition. With nonlinear mapping processing, high-dimensional features of small targets can be accurately extracted, achieving precise recognition; The problem of "same object, different spectra" in hyperspectral image target recognition has been effectively solved through sparse representation and multi-scale analysis, improving the accuracy of target recognition. Compared with the classic target recognition methods CEM and MF, the experimental results show that the proposed method can effectively improve the target recognition performance of hyperspectral images. This has promoted the application of hyperspectral images in fields such as land and resources investigation, geological hazard assessment, urban planning, and disaster relief.

Keywords: Sparse representation · Multi-scale analysis · Multi-target recognition · Hyperspectral image

1 Introduction

The target recognition of hyperspectral image is an advanced application of image information, belonging to a more refined application category. The process of recognition requires more stringent information and methods than the process of classification and detection, and the results obtained are more accurate. Target recognition requires a certain amount of prior information [1, 2]. Broadly speaking, object recognition belongs to a binary classification problem, where recognition results can be divided into two

G. Luo (Ed.): GFSISD 2024, SPPHY 417, pp. 159–168, 2025.
https://doi.org/10.1007/978-981-96-2490-4_19

categories: target and background. However, object recognition and binary classification problems are different. The biggest difference between hyperspectral image target recognition and hyperspectral image classification is the source of the sample spectrum. For classification, whether it is supervised classification [3, 4] or unsupervised classification [5–9], the focus of classification is to distinguish the category from other features whose sample spectra are derived from the image itself [10]. And target recognition is a search for specific objects, whose spectral matching samples do not come from the image itself, but from the target database.

For hyperspectral images, target recognition can be divided into two major application scenarios: one is to search for a specified target in the hyperspectral image based on the known features of the target (spectral features, image features, etc.), thus achieving target recognition, which is referred to as search recognition in this article [11–13]; The second is that the target detection results have been obtained, but the attributes of the target are not known. The target recognition method is used to identify the features of the detected target and add identity tags, which is called matching recognition in this article [14, 15] (Fig. 1).

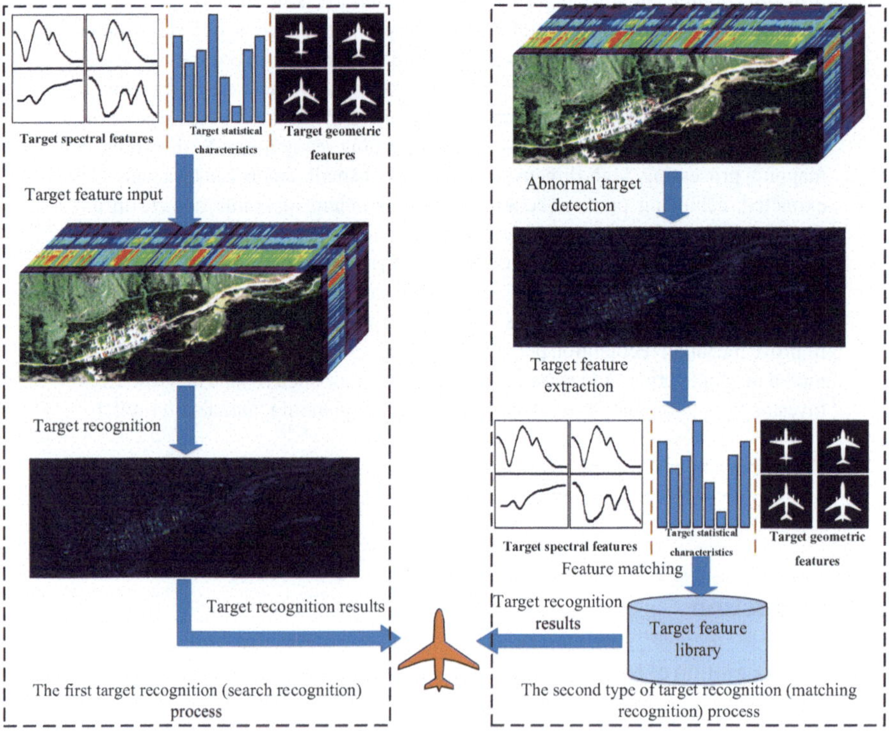

Fig. 1. Two typical application processes for hyperspectral image target recognition

2 Target Search Recognition

The target search and recognition process diagram of this article is shown in Fig. 2, as follows:

(1) Perform spectral feature search and recognition. To improve the non-linear expression ability of features, non-linear mapping techniques were used to perform non-linear feature mapping on the image and target spectrum before the search. This step obtained the target object that satisfies the spectral features;
(2) When the geometric features of the target (size, aspect ratio, shape, etc.) are available, calculate the geometric features of the target and use them for target screening;
(3) If the target statistical features (texture features, moment features, etc.) are available, calculate the statistical features of the target and use them for target screening;
(4) The ultimate goal that meets the conditions is to search for recognition results.

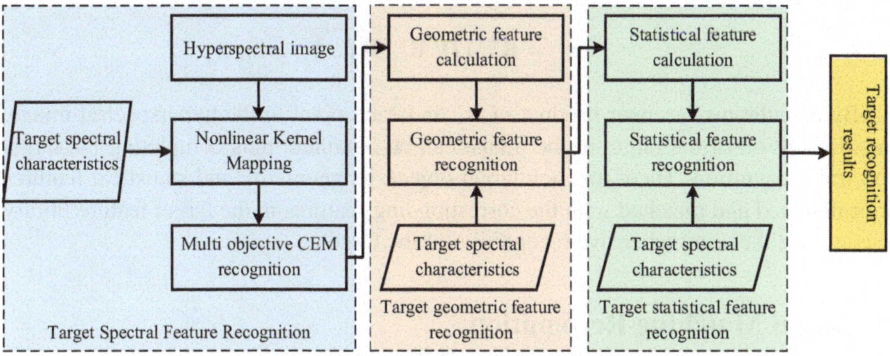

Fig. 2. Target Search and Recognition Process

In the above steps, steps (2) and (3) use ordinary feature matching techniques, among which step (1) has a significant impact on the subsequent results and is a key step in this process. The object of target recognition in this article is multiple targets. Therefore, this article intends to use the following multi-target spectral recognition techniques, as follows.

In general, the spectral characteristics of the ocean background are complex and varied, and can be considered unknown, only knowing the spectrum of the target of interest. Let $\mathbf{D} = \left[\mathbf{d}_1, \mathbf{d}_2, \cdots, \mathbf{d}_q\right]$ be the target spectral matrix, which contains the spectra of interested targets, $\mathbf{d}_1, \mathbf{d}_2, \cdots, \mathbf{d}_q$. The hyperspectral image that needs to be processed is a hyperspectral image of L bands and M pixels obtained, $\mathbf{X} = [\mathbf{x}_1, \mathbf{x}_2, \cdots, \mathbf{x}_L]$, where $\mathbf{x}_i = [x_{i1}, x_{i2}, \cdots, x_{iM}]^T$ is the column vector formed by the i-th band. According to the concept of the single object detection algorithm CEM, object recognition is equivalent to designing an FIR linear filter $\mathbf{w} = \left[w_1, w_2, \cdots, w_L\right]^T$ that minimizes the background output while keeping the target output at 1.That is

$$\min\left\{\frac{1}{M}\sum_{i=1}^{M} y_i^2\right\}, s.t.\mathbf{D}^T\mathbf{w} = \mathbf{1}$$

Among them, $y_i = \sum_{k=1}^{L} w_k x_{ik} = \mathbf{x}_i^T \mathbf{w}$, $\mathbf{l} = \begin{bmatrix} 1 & 1 & \cdots & 1 \end{bmatrix}$. Expanding the objective function of the above equation yields

$$\begin{aligned} \frac{1}{M}\sum_{i=1}^{M} y_i^2 &= \frac{1}{M}\left[\sum_{i=1}^{M} \left(\mathbf{x}_i^T \mathbf{w}\right)^T \left(\mathbf{x}_i^T \mathbf{w}\right)\right] \\ &= \mathbf{w}^T \left(\frac{1}{M}\left[\sum_{i=1}^{M} \mathbf{x}_i \mathbf{x}_i^T\right]\right)\mathbf{w} \\ &= \mathbf{w}^T \mathbf{R} \mathbf{w} \end{aligned}$$

Among them, $\mathbf{R} = \frac{1}{M}\left[\sum_{i=1}^{M} \mathbf{x}_i \mathbf{x}_i^T\right]$ is the autocorrelation matrix of the image. By using Lagrange multiplication, the expression for the filter can be obtained as

$$\mathbf{w}^* = \mathbf{R}^{-1}\left(\mathbf{D}^T \mathbf{R}^{-1} \mathbf{D}\right)^{-1} \mathbf{l}$$

By calculating the inner product of $\mathbf{w}^*$ and each pixel in the hyperspectral image, the distribution of the target in the image can be obtained, thus completing the target spectral recognition. Then, for each target object, its geometric and statistical features are calculated and matched with the corresponding features in the target feature library to complete the comprehensive recognition of the target.

3 Target Matching Recognition

Spectral matching is the process of determining the classification of land features by studying the similarity between two spectral curves. It has become one of the key technologies for identifying ground objects in imaging spectral image processing. The basic principle of hyperspectral target recognition is the quantitative comparative analysis between the spectral vectors of hyperspectral image data and the standard spectral database. By comparing the reflection spectra of known land cover types through waveform or feature matching, the purpose of identifying land cover types can be achieved. The operation mode of spectral matching is as follows: using a spectral database, the standard spectral curves of land features with certain characteristics are compared with remote sensing image pixels as templates, and the most similar pixels are identified and labeled accordingly.

Hyperspectral images have rich spectral information. Therefore, it also has higher resolution capability. Of course, the rich spectral information also brings very complex problems to the processing of hyperspectral data, such as the high processing time complexity of hyperspectral data, the phenomenon of "same object but different spectra" in ground spectra, data uncertainty, curse of dimensionality, and small sample problems, etc. [16].

The sparse representation theory of spectra can solve the above problems. The application of sparse representation theory in hyperspectral remote sensing image target recognition has the following advantages: ① Through the learning process of dictionaries, it is possible to reduce intra class feature differences and expand inter class feature differences; ② The sparse solving process can not only achieve high-precision classification and recognition, but also improve recognition speed; ③ Sparse representation does not require statistical analysis or assumptions about sample distribution. In sparse representation models, different constraints can be added to the objective function as the application scenario changes, thereby increasing the adaptability of the algorithm.

In addition, the target spectrum exhibits significant scale effects, as shown in Fig. 3. There are certain differences in the spectral curves extracted from hyperspectral remote sensing data at different scales, so different spectral features can be extracted using different scales, which is beneficial for accurate identification of target categories.

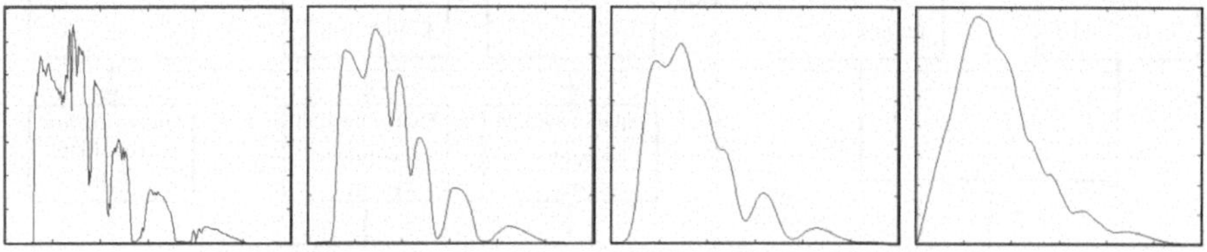

Fig. 3. Spectral curve variation with different scales

Due to interference from sensors and nature, sparse features and similarity between samples can be affected. Therefore, relying solely on the sparse features of spectral signals for recognition can easily lead to false positives and false negatives. Since Gaussian transform can smooth signals and remove some noise, it is equivalent to extracting multi-scale features of spectral signals. Therefore, this paper intends to use a method that integrates multi-scale information and sparse expression for spectral matching recognition. By using Gaussian to obtain information at different scales and using the proposed sparse representation model to identify these different scales, it can be found that the number of correctly identified samples at different scales is roughly the same, but not completely consistent. That is, incorrect samples can be identified at Gaussian scale 1, while positive samples can be identified at Gaussian scale 2. This article intends to adopt a spectral recognition method that combines sparse representation and multi-scale analysis. The specific process is shown in Fig. 4.

The training process of multi-scale dictionary for hyperspectral image target recognition algorithm based on sparse representation is as follows:

(1) Select the scale parameters for the sample and choose the scale where the waveform changes after Gaussian transformation;
(2) At each scale, for Gaussian transformed samples, a dictionary is trained separately for each class of samples;
(3) At each scale, using the trained dictionaries of each category, the maximum sparse recognition algorithm proposed below is used to recognize the training samples. Therefore, for p scales, each training sample has p recognition results;
(4) Train the obtained result matrix using the Adaboost algorithm, select key features that are beneficial for recognition, and use the corresponding Gaussian scale as the key scale. Calculate the weight values of the key scale

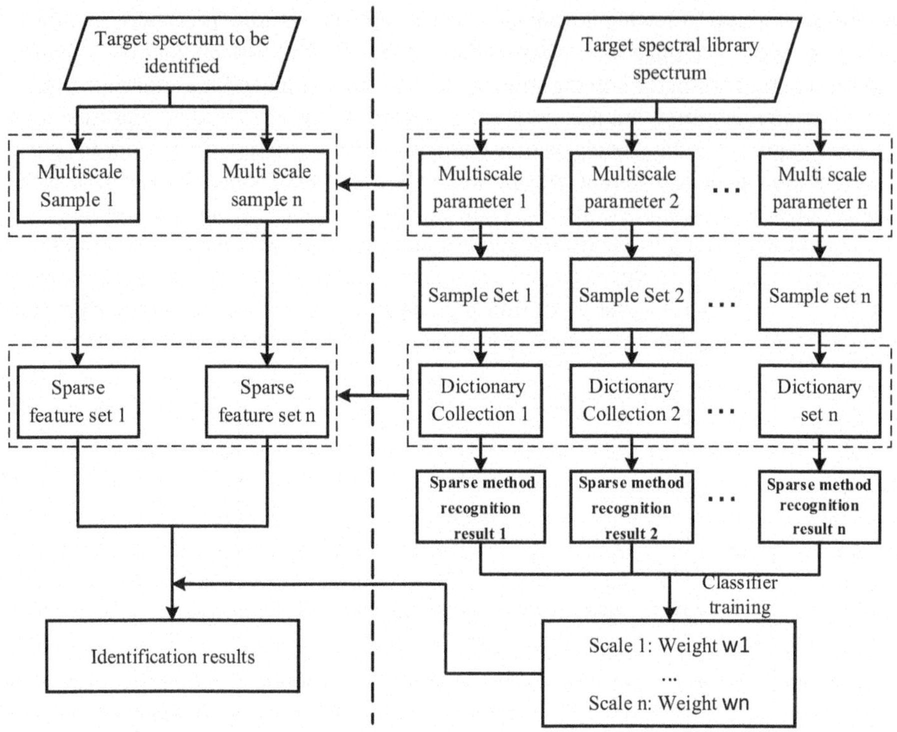

Fig. 4. Spectral recognition based on sparse representation and multi-scale analysis

The recognition process of the target recognition algorithm for hyperspectral images based on sparse representation is as follows:

(1) Using the key scales and feature dictionaries obtained during the multi-scale dictionary training process, at each key scale, the test samples are identified using the training dictionary at that scale, and the maximum sparse recognition algorithm proposed in the previous chapter is used to obtain recognition results;
(2) For each key scale recognition result, combined with the weight values trained using the Adaboost algorithm in the multi-scale dictionary training process, the class of the test sample is calculated and recognized through voting.

The target recognition algorithm for hyperspectral images based on sparse representation and multi-scale information mainly adds multi-scale information on the basis of the maximum sparse recognition algorithm for hyperspectral images based on sparse representation, calculates the weights at these scales, and uses the recognition results and scale weights at each scale for voting to identify the target.

It should be noted that target recognition based on spectral information is based on the differences in target materials, so different targets cannot be distinguished solely by using spectral information if the same material is used. At this point, joint recognition can be performed by combining other features such as geometric and statistical characteristics of the target.

4 Experimental Results

This article uses the Xi'an data obtained from OMIS-II to verify the performance of this method. For comparison, this article compares the performance of classic hyperspectral image target recognition methods CEM [17] and MF [18] under the same data and conditions. The spectrum of the target can be obtained using the endmember extraction method. In this experiment, FGDA [19] was used to extract the target spectrum without dimensionality reduction. In order to quantitatively evaluate the performance of object detection, this experiment uses the Separation Index (SI) to measure, which refers to the degree of differentiation between the target and background. Its definition is as follows [20]:

$$SI = \max_{(i,j)\in R_{tg}} \left\{ \left[\frac{S(i,j) - median_{bkg}(S)}{MAD_{bkg}} \right]^2 \right\}$$

Among them, S is the target detection result, R_{tg} refers to the target area in the image, $median_{bkg}(S)$ and MAD_{bkg} are the median and median absolute deviation (MAD) of the image background, respectively, that is, the sum of the absolute values of the differences between each element and the median. It should be noted that using the separation index as a measurement indicator requires the ground truth value of the target. Due to the lack of pixel level ground truth values in both data sets, this experiment used visual manual production of ground truth values.

This experiment extracts sub regions of pixels from the original Xi'an data, as shown in Fig. 5(A). This region contains two artificial small targets (of the same type), and Fig. 5(B) shows the distribution location of these targets. The target spectrum is obtained from the original image using the endmember extraction algorithm FGDA, corresponding to the pixel spectrum at coordinates (138, 118) in the image. From the figures, it can be seen that the two targets are located in the lower part of the image, with about 30

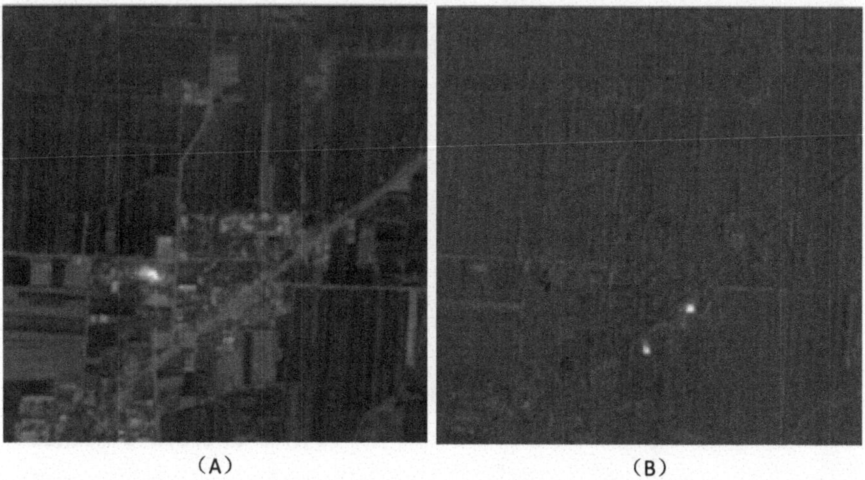

(A) (B)

Fig. 5. Sub regions used in the experiment: (A) 15th band and (B) target distribution location pixels. Due to their small size, it is difficult to visually recognize them in the original image.

This experiment used the ECA band selection method to select 10 bands, and then used CEM, MF, and the method proposed in this paper for target recognition. The results obtained are shown in Fig. 6

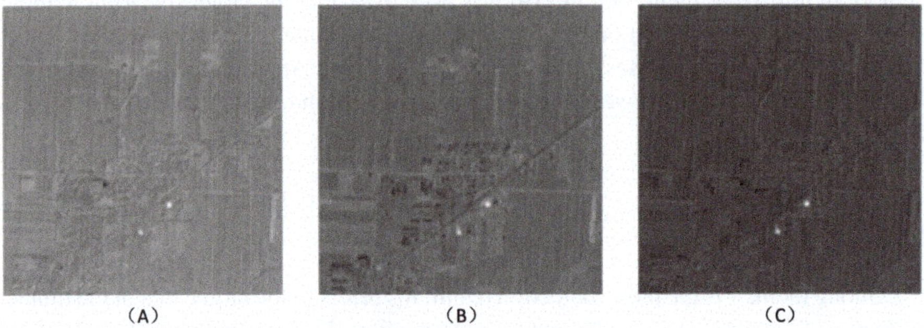

Fig. 6. (A) CEM recognition results, (B) MF recognition results, and (C) target recognition results of our method

From the results, it can be seen that using the target recognition method proposed in this paper for target recognition has better recognition performance compared to CEM and MF, manifested by better background suppression, clearer target recognition results, and effective separation of target and background.

The results of different methods were quantitatively compared using the Separation Index (SI) as a quantitative measure, as shown in Fig. 7. It can be seen that the target

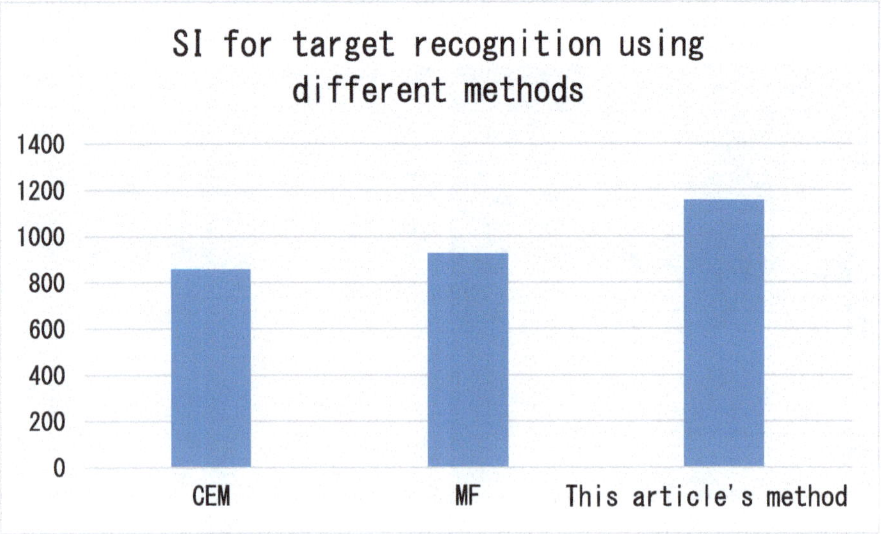

Fig. 7. Quantitative Comparison of Recognition Results from Different Methods

recognition results of this method have the highest separation index, which is the same as the visual situation.

5 Conclusion

This paper proposes a nonlinear mapping method and a recognition method combining sparse representation and multi-scale analysis for the search recognition and matching recognition of hyperspectral image targets. Through nonlinear mapping processing, high-dimensional features of small targets can be accurately extracted to achieve precise recognition; Sparse representation and multi-scale analysis can effectively solve the problem of "same object, different spectra" in hyperspectral image target recognition, and improve the accuracy of target recognition.

References

1. Chang, C.-I.: Target signature-constrained mixed pixel classification for hyperspectral imagery. IEEE Trans. Geosci. Remote Sens. **40**, 1065–1081 (2002)
2. Qingxi, T., Bing, Z., Lanfen, Z.: Hyperspectral Remote Sensing——Principles, Technologies, and Applications. Higher Education Press, Beijing (2006). (in Chinese)
3. Kraut, S., Scharf, L.L.: The CFAR adaptive subspace detector is a scale-invariant GLRT. IEEE Trans. Sig. Process. **47**, 2538–2541 (1999)
4. Geng, X., Ji, L., Sun, K., Zhao, Y.: CEM: more bands, better performance. IEEE Geosci. Remote Sens. Lett. **11**, 1876–1880 (2014)
5. Manolakis, D., Shaw, G.: Detection algorithoms for hyperspectral imaging applications. IEEE Signal Process. Mag.rocess. Mag. **19**, 29–43 (2002)
6. Chang, C.-I.: Hyperspectral Imaging: Techniques for Spectral Detection and Classification. Plenum, New York (2003)
7. Gu, Y., Liu, Y., Zhang, Y.: A selective KPCA algorithm based on high-order statistics for anomaly detection in hyperspectral imagery. IEEE Geosci. Remote Sens. Lett. **5**, 43–47 (2008)
8. Geng, X., Ji, L., Zhao, Y., Wang, F.: A small target detection method for the hyperpectral image based on high-order singular value decomposition (HOSVD). Geosci. Remote Sens. Lett. **10**, 1305–1308 (2013). IEEE
9. Geng, X., Sun, K., Ji, L., Zhao, Y.: A high-order statistical tensor based algorithm for anomaly detection in hyperspectral imagery. Sci. Rep. **4**, 1–7 (2014)
10. Mingyi, H., Wenjuan, C., Shaohui, M.: Advance in feature mining from hyperspectral remote sensing data. Spacecraft Recov. Remote Sens. **34**(1), 1–12 (2013). (in Chinese)
11. Xiurui, G.: Target detection and classification for hyperspectral imagery. Doctoral dissertation, Institute of Remote Sensing Applications, Chinese Academy of Sciences, Beijing (2005). (in Chinese)
12. Bo, D.: Sub-pixel target detection from hyperspectral remote sensing imagery. Doctoral dissertation, Wuhan University, Wuhan (2010). (in Chinese)
13. Jingyang, Z., Haiping, C., Weidong, C., et al.: The discussion of hyperspectral imaging orasis based on micro-satellites. Spacecraft Recov. Remote Sens. **32**(5), 75–79 (2011). (in Chinese)
14. Chang, Y.-L., Liu, J.-N., Han, C.-C., Chen, Y.-N.: Hyperspectral image classification using nearest feature line embedding approach. IEEE Trans. Geosci. Remote Sens. **52**, 278–287 (2014)

15. Geng, X., Ji, L., Sun, K.: Principal skewness analysis: algorithm and its application for multispectral/hyperspectral images indexing. IEEE Geosci. Remote Sens. Lett. **11**, 1821–1825 (2014)
16. Zass, R., Shashua, A.: Nonnegative sparse PCA. In: Advances in Neural Information Processing Systems, vol. 19, p. 1561 (2007)
17. Resende, L.S., Romano, J.M.T., Bellanger, M.G.: A fast least-squares algorithm for linearly constrained adaptive filtering. IEEE Trans. Signal Process. **44**, 1168–1174 (1996)
18. Manolakis, D., Siracusa, C., Shaw, G.: Hyperspectral subpixel target detection using the linear mixing model. IEEE Transactions Geosci. Remote Sens. **39**, 1392–1409 (2001)
19. Sun, K., Geng, X., Wang, P., Zhao, Y.: A fast endmember extraction algorithm based on gram determinant. IEEE Geosci. Remote Sens. Lett. **11**, 1124–1128 (2014)
20. Acito, N., Resta, S., Diani, M., Corsini, G.: Robust technique for anomalous change detection in airborne hyperspectral imagery based on automatic and adaptive band selection. Opt. Eng. **52**, 036202, 1–14 (2013)

Principles and Applications of Nighttime Light Remote Sensing

MHAT: A Mixed High-Order Attention Transformer for DMSP/OLS NTL Image Super-Resolution

Ting Yu[1], Chun Liu[1,2](✉), and Akram Akbar[3]

[1] Urban Mobility Institute, Tongji University, Shanghai, China
{1910961,liuchun}@tongji.edu.cn
[2] College of Surveying and Geo-Informatics, Tongji University, Shanghai, China
[3] College of Electronic and Information Engineering, Tongji University, Shanghai, China
akram@tongji.edu.cn

Abstract. The nighttime light (NTL) data has the unique ability to detect the night light on the earth surface, which provides an indispensable data basis for monitor urban development and characterize the human activities. However, two of the most widely utilized NTL data sources, the Defense Meteorological Satellite Program Operational Line Scanning System (DMSP/OLS) from 1992 to 2013 and the Suomi National Polar-orbiting Partnership-Visible Infrared Imaging Radiometer Suite (NPP/VIIRS) from 2012 to now, were not consistent. To address this problem, we proposed a novel transformer-based DMSP/OLS super-resolution model, namely MHAT, to convert DMSP/OLS data into NPP-like data. It consisted of three modules: pixel embedding, feature extraction, and high-resolution image reconstruction modules. Specifically, we utilized the high-order attention module for deeper information mining on the DMSP/OLS data. The results showed that: (1) Our MHAT model achieved a PSNR of 52.40 dB and an SSIM of 0.97, demonstrating high precision. (2) The DN value distribution histogram of the super-resolution results was very consistent with that of the high-resolution NPP/VIIRS image, which proving the reliability of our MHAT model. (3) We took China as an example and calculated the correlation of SumDN derived from these two images for 385 cities in China. We found that there was a strong correlation between the super-resolution results and the high-resolution NPP/VIIRS image on the city scale ($R^2 = 0.94$).

Keywords: NTL data · Super-resolution · Mixed high-order attention · Transformer · MHAT

1 Introduction

The nighttime light (NTL) data has the unique ability to detect the night light on the earth surface [1, 2]. It has more than 30 years of rich data resources in the world, which provides an indispensable data basis for monitor urban development and characterize the human activities [3, 4]. Two of the most widely utilized NTL data sources, the Defense

G. Luo (Ed.): GFSISD 2024, SPPHY 417, pp. 171–178, 2025.
https://doi.org/10.1007/978-981-96-2490-4_11

Meteorological Satellite Program Operational Line Scanning System (DMSP/OLS) from 1992 to 2013 and the Suomi National Polar-orbiting Partnership-Visible Infrared Imaging Radiometer Suite (NPP/VIIRS) from 2012 to now, has been increasingly employed to monitor urban spatial structure and growth pattern [5, 6], detect social and economic development indicators [7, 8] and evaluate urban public safety [9, 10].

Although the DMSP/OLS and NPP/VIIRS data possess valuable long-term sequence data, their applications have always been limited. The key reasons are three: DMSP/OLS data quality issues, NPP/VIIRS data quality issues, and inconsistency between the DMSP/OLS and NPP/VIIRS data. A large number of studies have corrected and improved the quality of these two data. For example, Li and Zhou [11] proposed a stepwise calibration method to alleviate the inter-year inconsistency of DMSP/OLS data. Cao et al. [12] employed NDVI data to correct the saturation effect of DMSP/OLS data. A logarithmic correction method was proposed by Yu et al. [13] to alleviate the radiation deviation of NPP/VIIRS data. And Zhao et al. [14] employed the empirical mode decomposition method to explore the seasonal change patterns of NPP/VIIRS data. However, the current methods for cross-sensor calibration of the DMSP/OLS and the NPP/VIIRS are limited. In the study of cross-sensor calibration, some studies converted NPP/VIIRS data into DMSP-like data through power function or quadratic polynomial model [4, 10], so as to obtain the NTL data with consistent resolution in long time series at the cost of losing the original fine resolution of NPP/VIIRS data. In addition, since the DMSP/OLS satellite has been retired, the subsequent NTL data are provided by NPP/VIIRS. The conversion of NPP/VIIRS to DMSP-like data is also detrimental to the future use of NTL data. At the same time, some scholars have tried to explore the method of converting DMSP/OLS data into NPP-like data. For example, Sahoo et al. [15] successfully constructed a model of converting DMSP/OLS data into NPP-like data in Uttar Pradesh in India by using machine learning, but the applicability of this algorithm in other regions was unclear. Chen et al. [16] proposed a method combining CNN and auto-encoder model to obtain NTL data of 500 m resolution from 2000 to 2018. However, the attention mechanism commonly utilized in CNN-based methods was rough and could only mine simple and superficial information, and lacked the ability to fully mine detailed information of NTL images, resulting in limited recovery ability. In summary, an accurate and scalable approach for transforming the DMSP/OLS data to NPP-like data is still a challenge.

Recently, transformer-based super-resolution models has shown great potential in image processing [17]. Its aim is to recover high-resolution (HR) images from the corresponding low-resolution (LR) images. Some scholars employed super-resolution methods to restore remote sensing images. For example, Haut et al. [18] proposed a deep generative network for remote sensing images super-resolution. Dong et al. [19] developed a second-order multi-scale super-resolution network for image super-resolution with complex spatial distribution. Using the idea of super-resolution, we regarded the DMSP/OLS image as degraded LR image and the NPP/VIIRS image as HR image, and tried to establish the method of converting DMSP/OLS data into NPP-like data. In addition, the DMSP/OLS and NPP/VIIRS data have two years of repeated observation data from 2012 to 2013, which provides a data basis for the construction of our model.

In this paper, aiming at the inconsistency between the DMSP/OLS and NPP/VIIRS data, we proposed a novel transformer-based DMSP/OLS super-resolution model, namely MHAT, to convert DMSP/OLS data into NPP-like data, so as to provide technical support for using long-term NTL data. Specifically, we utilized the high-order attention (HOA) module for deeper information mining on the DMSP/OLS data.

The main contributions are summarized as follows:

(1) The idea of image super resolution was introduced to solve the inconsistency problem between the DMSP/OLS and NPP/VIIRS data.
(2) The transformer-based framework MHAT was designed to transfer the DMSP/OLS data to NPP-like data.
(3) A novel HOA module was introduced into MHAT, aiming at mining deeper information of DMSP/OLS image.

2 Methodology

The overall architecture of MHAT is shown in Fig. 1, which consisted of three modules: pixel embedding, feature extraction, and HR image reconstruction modules. The pixel embedding layer was utilized to extract features of low-resolution DMSP/OLS data. And the feature extraction was utilized to extract deeper information with the employment of high-order attention network. Finally, the recovery of high-frequency details was in the HR image reconstruction step. More details are described below.

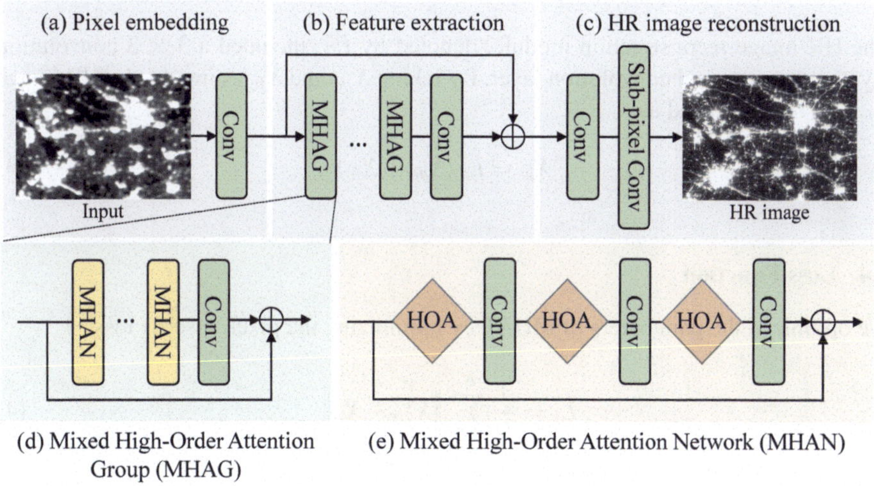

Fig. 1. Overall architecture of MHAT.

2.1 Pixel Embedding

The pixel embedding layer, denoted by L_{pe}, consisted of only a single 3 $\times$ 3 convolution. Its purpose was to transform the low-resolution DMSP/OLS image $I_x \in R^{h1 \times w1 \times 3}$ to the feature embedding layer $X_{pe} \in R^{H\times W\times C}$, where H, W and C represent the channel height, width and number of the intermediate feature respectively. That is:

$$X_{pe} = L_{pe}(I_x) \tag{1}$$

Thus, the channel information of the original input image $I_x \in R^{h1\times w1\times 3}$ was improved and the feature dimension was enriched. Subsequently, the output X_{pe} of this step, which was a hierarchical structure, went to the next layer.

2.2 Feature Extraction

The feature extraction layer, denoted by L_{fe}, was composed of M cascaded mixed high-order attention groups (MHAGs). The final result X_{fe} passes through a single 3 × 3 convolution layer and was concatenated with the original data features. That is:

$$X_{fe} = L_{fe}(X_{pe}) \tag{2}$$

At the heart of MHAT was the MHAG, which was composed of N cascaded mixed high-order attention networks (MHANs) combined with a layer of 3 × 3 convolution. The MHAN was composed of three HOA layers and three 3 × 3 convolution. More details on HOA are referred to Chen et al. [20].

2.3 HR Image Reconstruction

The HR image reconstruction module, denoted by L_{ir}, included a 3 × 3 convolution layer and a sub-pixel convolution layer. By taking X_{pe} and X_{fe} as inputs, the HR image X_{ic} was reconstructed as:

$$X_{ic} = L_{ic}(X_{pe} + X_{fe}) \tag{3}$$

2.4 Loss Function

We optimized the parameters of MHAT by minimizing the pixel-wise L_1 loss:

$$L_1 = \frac{1}{N}\sum_{i=1}^{N}\left\| X_{SR}^{i} - X_{y}^{i} \right\|_1 \tag{4}$$

where, X_{SR}^{i} and X_{y}^{i} represent the super-resolution results and the high-resolution ground truth image respectively, and N denotes the pixel number.

2.5 Evaluation Indexes

We employed the peak signal-to-noise ratio (PSNR) and the structural similarity (SSIM) to evaluate the performance of MHAT. The PSNR evaluated image quality by comparing the signal-to-noise ratio between the original image and the processed image. While the SSIM evaluated structural similarity by combining brightness similarity, contrast similarity and structural similarity. The calculation formula is as follows:

$$PSNR = 10 \times \log_{10}\left(\frac{MAX^2}{MSE}\right) \tag{5}$$

where, MAX represents the maximum possible pixel value of the image, MSE stands for mean square error, which was the average of the square of the pixel difference between the original image and the processed image.

$$SSIM\,(x, y) = \frac{(2\mu_x\mu_y + c_1)(2\sigma_{xy} + c_2)}{(\mu_x^2 + \mu_y^2 + c_1)(\sigma_x^2 + \sigma_y^2 + c_2)} \tag{6}$$

where, x and y represent two images, μ_x and μ_y are the average values of the pixel values of the images x and y, σ_x and σ_y are the standard deviations of the pixel values of the images x and y, σ_{xy} is the covariance between the pixel values of the images x and y, and c_1 and c_2 are constants.

2.6 Settings

The MHAT was constructed based on the products in two temporally overlaid years (i.e., 2012 and 2013) of the DMSP/OLS and NPP/VIIRS data. We utilized the 2013 DMSP/OLS stable light data (version 4) after stepwise calibration [11] and the 2013 annual NPP/VIIRS median data (version 2) data after logarithmic transformation [13] as the training dataset, and the data in 2012 as the validation dataset. We clipped the global NTL images into 480 × 480 for DMSP/OLS and 960 × 960 for NPP/VIIRS data, and inputted them into the network as pairs.

The number of MHAG and MHAN was selected to be 6 in this study. During the experiment, the image size was set to 48 × 48 and the window size was set to 4. Moreover, the Adam optimization method was used to update model parameters. The learning rate was initially set at 10^{-4} and decreased by a factor of 10 per 100 epochs. All experiments were implemented by PyTorch and on NVIDIA GeForce GTX 1660 Ti GPU devices.

3 Results

3.1 Evaluation Indexes Results

The higher the value of PSNR, the smaller the loss of image quality. In general, images with the PSNR value above 30 dB were considered to be high-quality images. For SSIM, its value ranges between −1 and 1. The closer the value is to 1, the more similar the image is. In this study, we defined the super-resolution results of the DMSP/OLS image in 2012 as $I_{SR} \in R^{h1 \times w1}$, and the high-resolution NPP/VIIRS image in 2012 as $I_y \in R^{h2 \times w2}$, where $h_2 = h_1$, $w_2 = w_1$. Our MHAT model achieved a PSNR of 52.40 dB and an SSIM of 0.97 between I_{SR} and I_y, demonstrating high precision.

3.2 Histogram Comparison Results

In order to further compare the similarities and differences between the super-resolution results I_{SR} and the ground truth image I_y, we compared the histogram distribution of pixel DN values of these two images on a global scale. It could be seen that the DN value distribution histogram of I_{SR} was very consistent with that of I_y. The derived histograms of the two data had the rule that the number of pixel distributions decreases with the increase of DN value. However, in the result data I_{SR}, there was still data saturation. This is due to the saturation effect of the original DMSP/OLS data (Fig. 2).

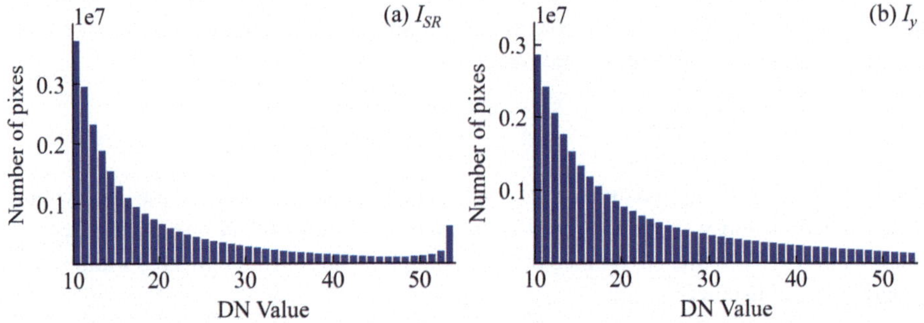

Fig. 2. Histogram distribution of pixel DN values on a global scale between (a) the super-resolution results I_{SR} and (b) the ground truth image I_y.

3.3 Correlation Analysis

To further verify the correlation of the super-resolution results I_{SR} and the ground truth image I_y, we took China as an example and calculated the correlation of SumDN derived from these two images for 385 cities in China. The results showed that there was a strong correlation between I_{SR} and I_y on the city scale ($R^2 = 0.94$) (Fig. 3).

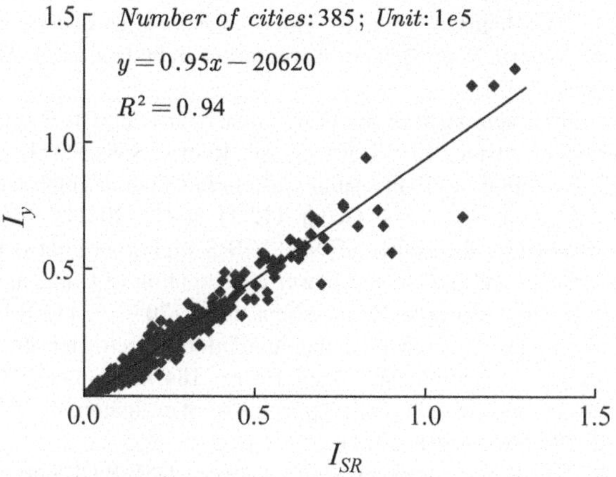

Fig. 3. Correlation of SumDN for 385 cities in China derived from the super-resolution results I_{SR} and the ground truth image I_y.

4 Conclusion

The nighttime light (NTL) data provides an indispensable data basis for monitor urban development and characterize the human activities. Aiming at the inconsistency between the DMSP/OLS and NPP/VIIRS data, we proposed a novel transformer-based DMSP/OLS super-resolution model, namely MHAT, to convert DMSP/OLS data into NPP-like data, so as to provide technical support for using long-term NTL data. Our MHAT model achieved a PSNR of 52.40 dB and an SSIM of 0.97, demonstrating high precision. And the DN value distribution histogram of the super-resolution results was very consistent with that of the high-resolution NPP/VIIRS image. Moreover, we took China as an example and calculated the correlation of SumDN derived from these two images for 385 cities in China. The result showed that there was a strong correlation between the super-resolution results and the high-resolution NPP/VIIRS image on the city scale ($R^2 = 0.94$).

However, there are still some problems that need further study. Our study only converted low-resolution DMSP/OLS data into NPP-like data from the perspective of image recovery, without considering the impact of actual ground objects. Therefore, in future studies, in order to ensure the accuracy of NPP-like data reconstruction, we should also consider the influence of light-sensitive ground objects. For example, areas covered by water are often without lights.

References

1. Elvidge, C.D., Daniel, Z., Baugh, K.E., Tuttle, B.T., Mikhail, Z.: A fifteen year record of global natural gas flaring derived from satellite data. Energies **2**(3), 595–622 (2009)
2. Elvidge, C.D., Baugh, K., Zhizhin, M., Hsu, F.C., Ghosh, T.: VIIRS night-time lights. Int. J. Remote Sens. **38**(21), 5860–5879 (2017)
3. Zhao, M., et al.: Mapping urban dynamics (1992–2018) in Southeast Asia using consistent nighttime light data from DMSP and VIIRS. Remote Sens. Environ. **248**, 111980 (2020)

4. Yu, T., Liu, C., Li, W., Huang, W., Wu, H., Fan, Z.: Characterizing urban actively populated area growth in the Yangtze River Delta using nighttime light data. Int. J. Appl. Earth Observ. Geoinf. **129**, 103857 (2024)
5. Chen, Z., et al.: A new approach for detecting urban centers and their spatial structure with nighttime light remote rensing. IEEE Trans. Geosci. Remote Sens. **55**(11), 6305–6319 (2017)
6. Wang, C., et al.: Evolution of urban spatial clusters in China: a graph-based method using nighttime light data. Ann. Am. Assoc. Geogr. **112**(1), 56–77 (2021)
7. Shi, K., et al.: Evaluating the ability of NPP-VIIRS nighttime light data to estimate the gross domestic product and the electric power consumption of China at multiple scales: a comparison with DMSP-OLS Data. Remote Sens. **6**(2), 1705–1724 (2014)
8. Shi, K., et al.: Detecting spatiotemporal dynamics of global electric power consumption using DMSP-OLS nighttime stable light data. Appl. Energy **184**, 450–463 (2016)
9. Li, X., Li, D.: Can night-time light images play a role in evaluating the Syrian Crisis? Int. J. Remote Sens. **35**(18), 6648–6661 (2014)
10. Li, X., Li, D., Xu, H., Wu, C.: Intercalibration between DMSP/OLS and VIIRS night-time light images to evaluate city light dynamics of Syria's major human settlement during Syrian Civil War. Int. J. Remote Sens. **38**(21), 5934–5951 (2017)
11. Li, X., Zhou, Y.: A stepwise calibration of global DMSP/OLS stable nighttime light data (1992–2013). Remote Sens. **9**(6), 637 (2017)
12. Cao, X., Chen, J., Imura, H., Higashi, O.: A SVM-based method to extract urban areas from DMSP-OLS and SPOT VGT data. Remote Sens. Environ. **113**(10), 2205–2209 (2009)
13. Yu, B., et al.: Urban built-up area extraction from Log-transformed NPP-VIIRS nighttime light composite data. IEEE Geosci. Remote Sens. Lett. **15**(8), 1279–1283 (2018)
14. Zhao, X., Li, D., Li, X., Zhao, L., Wu, C.: Spatial and seasonal patterns of night-time lights in global ocean derived from VIIRS DNB images. Int. J. Remote Sens. **39**(22), 8151–8181 (2018)
15. Sahoo, S., Gupta, P.K., Srivastav, S.: Inter-calibration of DMSP-OLS and SNPP-VIIRS-DNB annual nighttime light composites using machine learning. GIScience Remote Sens. **57**(8), 1144–1165 (2020)
16. Chen, Z., et al.: An extended time series (2000–2018) of global NPP-VIIRS-like nighttime light data from a cross-sensor calibration. Earth Syst. Sci. Data **13**(3), 889–906 (2021)
17. Liang, J., Cao, J., Sun, G., Zhang, K., Van Gool, L., Timofte, R.: SwinIR: image restoration using Swin transformer. In: 2021 IEEE/CVF International Conference on Computer Vision Workshops (ICCVW), pp. 1833–1844 (2021)
18. Haut, J.M., Fernandez-Beltran, R., Paoletti, M.E., Plaza, J., Plaza, A., Pla, F.: A new deep generative network for unsupervised remote sensing single-image super-resolution. IEEE Trans. Geosci. Remote Sens. **56**(11), 6792–6810 (2018)
19. Dong, X., Wang, L., Sun, X., Jia, X., Gao, L., Zhang, B.: Remote sensing image super-resolution using second-order multi-scale networks. IEEE Trans. Geosci. Remote Sens. **59**(4), 3473–3485 (2021)
20. Chen, B., Deng, W., Hu, J.: Mixed high-order attention network for person re-Identification. In: 2019 1EEE/CVF International Conference on Computer Vision (ICCV), pp. 371–381 (2019)

GDP Intelligence Forecasting Models Based on NPP/VIIRS Nighttime Light Imagery

Xiaoli Cai[1,2(✉)], Yunfei Bao[1,2], and Qiaolin Huang[1,2]

[1] Beijing Institute of Space Mechanics and Electricity, Beijing 100094, China
1173138394@qq.com

[2] China Academy of Space Technology, Beijing 100094, China

Abstract. Nighttime remote sensing data are considered an important dataset for measuring human social-economic activities. Due to the strong positive correlation between long time series of nighttime remote sensing data and the degree of economic development, it is possible to construct a Gross Domestic Product (GDP) forecasting model using long time series of nighttime remote sensing data to predict GDP. However, traditional statistical methods struggle to accurately and timely reveal the spatial and temporal information of economic parameters and estimate GDP with insufficient precision. How to eliminate data errors in nighttime remote sensing data and construct reliable long time series is currently a major challenge in research. In addition, different GDP forecasting models have different applicable conditions and forecasting accuracy. How to choose the optimal GDP forecasting model is another major challenge in current research. Thus this paper uses nighttime light imagery data from 2016 to 2022, obtained by the Visible Infrared Imaging Radiometer (VIIRS) carried by the National Polar-orbiting Partnership (NPP) satellite as the data source. Errors generated in the data are processed to obtain long time series data suitable for quantitative analysis. Initially, the Random Forest algorithm is employed for predictive research on the dataset. Subsequently, a regression error estimation based on out-of-bag data is proposed. An improved grid search algorithm is used to tune the parameters of the Random Forest model. The grid search algorithm adjusts parameters step by step according to the step size, traverses the parameter range sequentially to find the parameters with the highest accuracy, and uses 5-fold cross-validation for assessment. Additionally, Bayesian optimization is utilized to tune the hyper-parameters in the Random Forest (RF) model. The Bayesian optimization process mainly includes a prior function and an acquisition function. The prior function is implemented using a Gaussian process, employing nested 5F-CV and estimating the model's generalization ability through an external 5F-CV loop. The internal 5F-CV loop is used to determine the optimal parameters and identify the best parameter model. This optimization method freely combines preset model parameters, conducts a comprehensive search of all parameter combinations, and evaluates the models by analyzing three indicators, comparing the features of different models. The study has established an automatic forecasting system that enables the algorithm model to make accurate predictions automatically based on the input data of the study area. The research results show that after Bayesian optimization and grid

Fund Project: Key Civil Aerospace Projects (D040101)

G. Luo (Ed.): GFSISD 2024, SPPHY 417, pp. 179–188, 2025.
https://doi.org/10.1007/978-981-96-2490-4_12

optimization, the R^2 of the Random Forest algorithm in predicting GDP increased from 0.948 to 0.978 and 0.968, respectively. At the same time, several other evaluation indicators show a significant improvement in forecasting effectiveness. Among them, the Random Forest model improved based on Bayesian optimization has the highest prediction accuracy and the smallest root mean square error value, demonstrating high accuracy and robustness. The research results demonstrate the significant capability of machine learning algorithms and nighttime light index in predicting GDP at the county level. This research framework enriches the application fields and methods of nighttime data and can be used to explore more refined regional development studies in other countries or regions.

Keywords: Nighttime light · Random Forest (RF) model · Bayesian optimization · Time series analysis

1 Introduction

Nighttime Light (NTL) remote sensing images record the night light on the earth's surface through sensors, which can detect the lights from cities, towns and industrial areas, showing good performance in the detection of human social and economic activities. GDP can effectively reflect a country's national strength and financial resources of [1], and accurate pixel-level GDP data is the basis of understanding the regional economic dynamics. Nowadays, it is difficult for the traditional statistical methods to accurately and timely reveal the spatial and temporal information of economic parameters and estimate the GDP. The night light data can make up for the deficiency. Compared with the traditional social and economic census, the remote sensing image has obvious advantages in describing the GDP prediction. As a world, night light data describes the intensity of the earth's economic activity in a unique way. Fan et al. [2] analyzed the relationship between the comparative long time series light sets and GDP by constructing a polynomial prediction model. Zhou et al. [3] quantitatively studied the correlation between NTL and GDP and population, and the results showed a significant positive correlation between night light and GDP and population. The luminous remote sensing data used in the research is grid data, which contains a large amount of data information in each image. The variation and distribution of brightness can be seen from the night light data, thus providing more temporal and spatial information.

At present, there are relatively many studies to estimate economic parameters using data from the Visible Infrared Imaging Radiometer(VIIRS) carried by the Suomi National Polar-orbiting Partnership (NPP) satellite, but few studies have used VIIRS data using machine learning algorithms for GDP prediction. To this end, the long-term night data in the Yangtze River Delta region are used to establish an intelligent GDP prediction model, so as to predict the GDP of the region in a timely and effective manner, and further analyze the economic differences of different regions. The results provide the decision basis for the application field and application mode of luminous data, as well as the sustainable planning and development of the region.

2 Basic Theory

The Random Forest (RF) model is a Bagging ensemble algorithm, characterized by its high flexibility, resistance to over-fitting, For further understanding of the theory behind Random Forest, one can refer to the research by Li Qianqian et al. [4]. The principle of Random Forest involves generating multiple decision trees by randomly sampling with replacement from the dataset and averaging the results of these decision trees as the model output. Consequently, Random Forest has the capability to assess and handle high-dimensional feature data.

The Random Forest has two important parameters: the number of trees and the number of features to split. The core steps of the Random Forest algorithm are as follows:

Step 1: Utilize the Bootstrap method to select replaceable samples, choosing m sets of data as samples. A total of ntree training sets for the trees are generated.
Step 2: Train the ntree trees and produce corresponding models.
Step 3: Select the optimal feature for each split based on the partition index.
Step 4: Each decision tree is independently split, and there is no need for pruning during the splitting process of the decision tree until all training examples belong to the same category. The multiple decision trees form a Random Forest, and the classification or predictive results of each decision tree are integrated.

3 Study Area and Data

3.1 Study Area

The Yangtze River Delta in China, refers to the regions of Shanghai, Zhejiang, Jiangsu, and Anhui. The terrain mainly consists of the Anhui Sub-basin Plain, the Yangtze River Delta Plain, and the Lixiahe Plain. According to the 2019 Yangtze River Delta Regional Integration Development Plan, the Yangtze River Delta urban cluster includes 214 counties. It covers an area of 225,000 square kilometers and has a total population of approximately 243 million people (Fig. 1).

3.2 NPP/VIIRS Data Preprocessing

The original data from the VIIRS uses the WGS1984 coordinate system, to avoid the deformation effects caused by the coordinate system and to facilitate the calculation of image areas, all NPP/VIIRS imagery data are transformed into the Albers Equal Area projection coordinate system. Finally, the images are resampled into a grid size of 0.5 km by 0.5 km. The first edition of the VIIRS monthly composite data situations with background noise, and the light radiation rate values have characteristics of negative values, extremely small values, and extremely high values. The presence of these noise and outlier data can affect the accuracy of data simulation. To address this situation, the following treatments were made according to the method of Shi et al. [5]: values less than zero are set to 0.001 (approximately 0, which has no impact on statistical analysis), background values are set to 0; values greater than 235 are set to 235 to remove some outliers. And by applying the averaging method to the NPP/VIIRS monthly products for yearly synthesis, the annual data for the NPP/VIIRS in the Yangtze River Delta region from 2016 to 2022 was obtained. The study area VIIRS imagery is shown in Fig. 2.

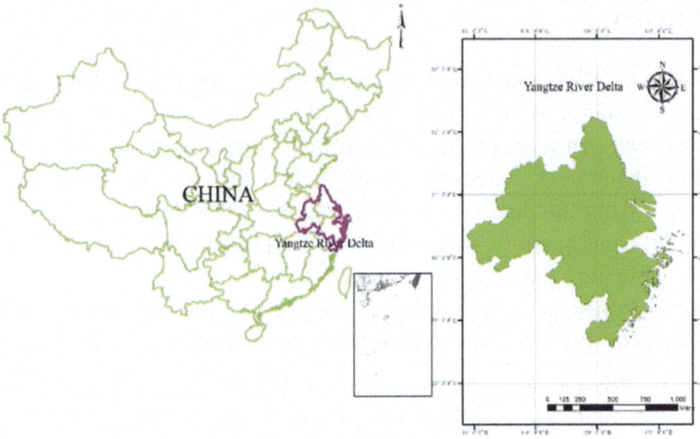

Fig. 1. The location of study area

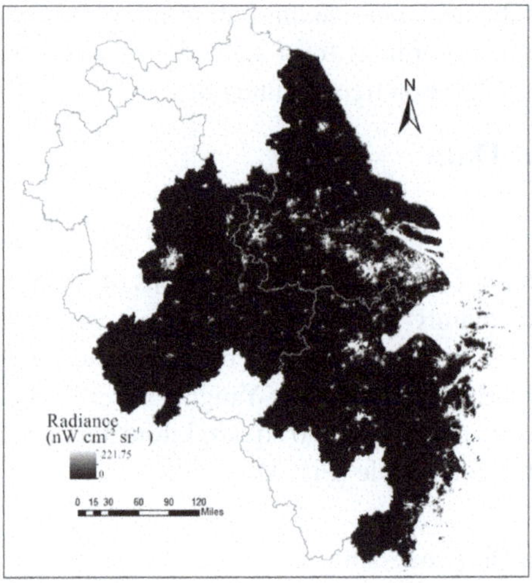

Fig. 2. VIIRS Nighttime Light image for study areas

3.3 Other Auxiliary Data

In addition to the NPP/VIIRS nighttime light data, the study area also employs other auxiliary data such as built-up area, registered population, and total electricity consumption of the whole society. The economic statistical data was retrieved and collected from the "China City Statistical Yearbook" which includes GDP data (Unit: 100 million yuan) for 214 counties in the Yangtze River Delta region from 2016 to 2022. Table 1 lists the data used in the study.

Table 1. Other auxiliary data

Administrative division	2022	National Geomatics Center of China	1:250 000
Economic Statistical Data	2016–2022	Statistical Yearbook and Regional Statistical Reports	District/County
NPP/VIIRS	2016–2022	https://www.ngdc.noaa.gov	0.5 km
Built-Up Area Data	2016–2020	http://data.ess.tsinghua.ed	30 m
End-of-year household registered population	2016–2022	Statistical Yearbook and Regional Statistical Reports	District/County
Electricity consumption	2016–2022	Statistical Yearbook and Regional Statistical Reports	District/County

4 Bayesian Optimization Improved Random Forest Model

The concept of the algorithm is to initially generate a set of candidate solutions, then identify the next point that may be an extremum based on these points, add this point to the set, and repeat this step until the iteration is terminated. Bayesian optimization estimates the mean and variance (i.e., the range of fluctuation) of the true objective function value based on the function values of the points already searched, as shown in Fig. 3.

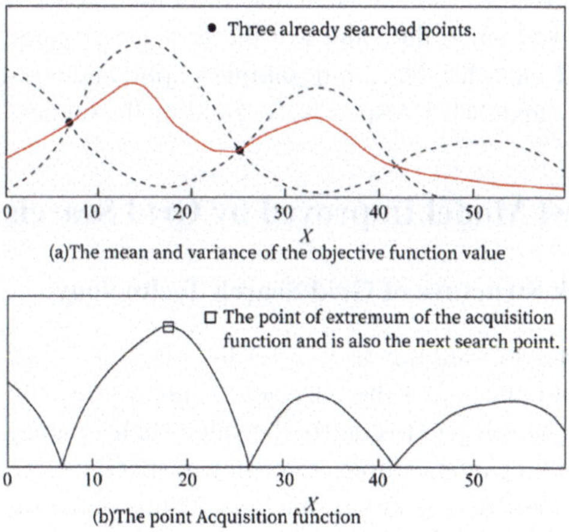

Fig. 3. Bayesian optimization principle

The long time series data from 2016 to 2021 in the experimental dataset is used as the training set, and the data from 2022 is used as the test set. The training set is then trained. Parameters are determined through cross-validation to achieve the best model performance. The algorithm predicts the probability distribution of the function value at any point based on the function values at a set of sampling points, which is implemented through Gaussian process regression. This experiment primarily focuses on Bayesian optimization for the four key parameters: max_depth, min_samples_leaf, n_estimators, and min_samples_split, to enhance the predictive accuracy. The Bayesian optimization process mainly includes the prior function and the acquisition function, The hyper-parameter ranges for constructing the improved Random Forest model with Bayesian optimization are set as follows: max_depth [5, 15], n_estimators [10, 250], min_samples_split [2, 25], and min_samples_leaf [1, 10]. The Bayesian search function in sklearn is used to search the specified ranges for these functions, also using the Root Mean Square Error (RMSE) as the accuracy evaluation metric to identify the optimal parameter combination. The optimal parameter combination is presented in Table 2 as shown below.

Table 2. Optimal parameter combination

Iteration count	Target	max depth	min_samples_leaf	min_samples_split	nestimators
23	0. 963 9	14. 43	1. 289	2. 828	241. 3

The optimization results indicate that the minimum Root Mean Squared Error (RMSE) can be achieved when the values of the four hyper-parameters are set as follows: max_depth, min_samples_leaf, min_samples_split, and n_estimators are taken to be 14.43, 1.289, 2.828, and 241.3, respectively, yielding the highest prediction accuracy.

5 Random Forest Model Improved by Grid Search

5.1 The Framework Structure of Grid Search Technology

The basic principle of grid search is to first set the range of values for the parameters, then divide the feasible interval of the parameters into a grid with a specific step size, and then search through each grid to control the values of the parameters within a certain range. It searches for all the intersections according to specific rules and in this way finds the intersection that optimizes the target function. At this point, the best parameters are the ones at the intersection [6].

5.2 Experimental Validation

To enhance the precision of the forecasting algorithm, parameter optimization is required. A regression error estimation based on out-of-bag data is proposed, and an improved grid search algorithm is used to tune the parameters of the random forest model. The parameter settings are shown in the following Table 3. The grid search algorithm adjusts the parameters incrementally according to a step size, iterating through the parameter ranges to find the set of parameters that yield the highest accuracy, and evaluates them using 5-fold cross-validation. Considering that the number of decision trees and the number of candidate split attributes in the random forest algorithm are discrete values, the improved random forest model based on grid search uses the out-of-bag data estimation error as the objective function value for parameter optimization. Since there is randomness in the construction process of the random forest model, to reduce the impact of uncertainty on parameter selection, the average value of multiple results is taken as the regression error of the random forest model when calculating the regression error.

Constructing a random forest model improved by grid optimization for parameter optimization, the hyper-parameters and their ranges are shown in Table 3. In this experiment, a total of 256 different models were constructed.

Table 3. Set the parameters and ranges that need to be adjusted

Parameters	Range	Optimal parameters
n_estimators	50,100 150,200	150
max_depth	3,4,5,6	6
min_samples_split	2,3,4,5	2
min_samples_leaf	1,2,3,4	1

By searching the interval ranges set forth above to seek the optimal combination of parameters for maximum model accuracy, the relationship between each parameter and the result is shown in Fig. 4. The parameter tuning results indicate that the minimum root mean squared error (RMSE) and the maximum R^2 value are achieved when n_estimators, min_samples_split, max_depth, and min_samples_leaf are set to 150, 2, 6, and 1, respectively. The optimal parameter combination, after adjustment, is input into the model, and the predict function is used for prediction. At this point, the optimal R^2 value obtained is 0.968, the RMSE value is 0.0395, and the mean absolute error (MAE) value is 0.0216.

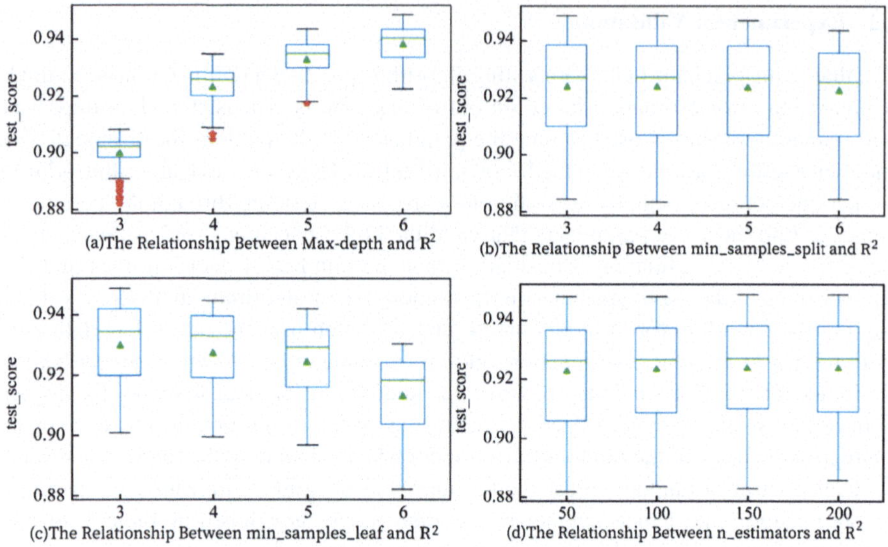

Fig. 4. Relationship between parameters and results

6 Results and Analysis

The model prediction performance comparison is shown in Table 4. After Bayesian optimization and grid optimization, the R^2 values were improved from 0.948 to 0.978 and 0.968, respectively. At the same time, several other evaluation indicators also show a significant improvement in prediction performance. Among them, the random forest model improved by Bayesian optimization has the highest prediction accuracy and a significant increase, with the smallest root mean square error value. The bar chart corresponding to the prediction results is shown in Fig. 5.

Table 4. Comparative analysis of model prediction performance

Model	RMSE	R^2	MAE	T/s
RF	0.047 8	0.948	0.015 2	4.95
Bayesian Optimization Improved RF Model	0.031 6	0.978	0.017 3	3.12
Grid Search Improved RF Model	0.039 5	0.968	0.021 6	3.23

It can be observed that after grid optimization, the running time of the random forest model was improved from the original 4.95 s to 3.23 s. After Bayesian optimization, the running time of the random forest model was further improved from the original 4.95 s to 3.12 s, which is a higher improvement than that achieved by grid optimization, and the parameter search range was significantly expanded.

Comparing the prediction results of the three models before and after improvement with the actual values, as shown in Fig. 6, it can be seen that the prediction results of the

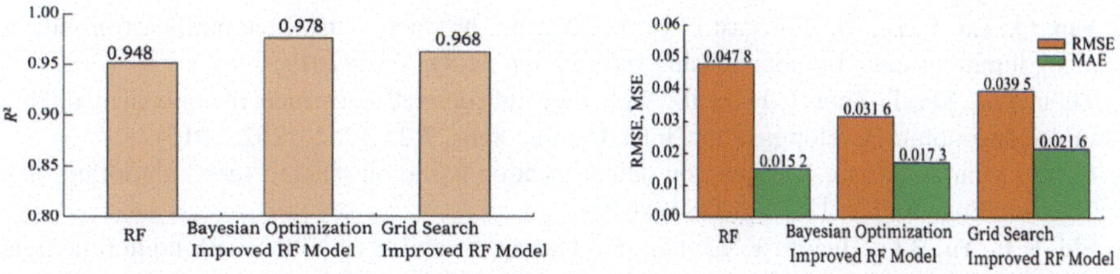

Fig. 5. Histogram of comparison of model predictions before and after improvement

models improved by adjusting hyper-parameters through Bayesian optimization and grid optimization are basically coincident with the zero error line (y = x), indicating that the prediction results of the random forest model improved by Bayesian optimization and grid optimization are basically consistent with the actual values; its prediction accuracy reaches the maximum among all algorithms.

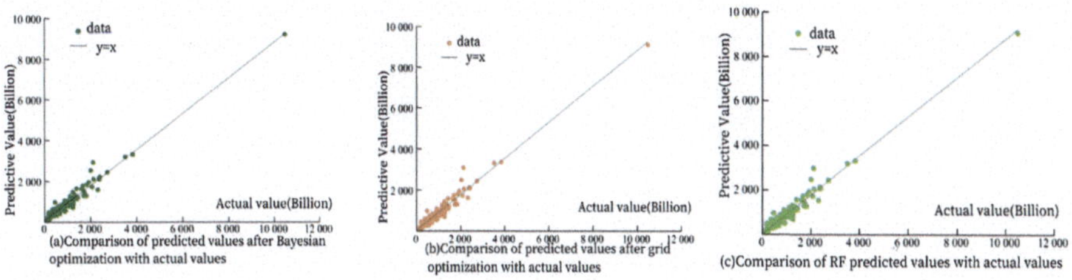

Fig. 6. Correlation between the true value and the predicted value of the test set of different models

7 Conclusion

The research on intelligent forecasting of the GDP in the Yangtze River Delta region employs machine learning algorithms in conjunction with NPP/VIIRS nighttime light data. A series of data processing steps were undertaken, and the random forest algorithm was subjected to Bayesian optimization and grid search for automatic hyper-parameter tuning. The optimization method freely combines preset model parameters and performs an exhaustive search across all parameter combinations. By analyzing three types of indicators, the characteristics of different models are compared, and the models are evaluated. The results show that the random forest model improved by Bayesian optimization has the best predictive performance, with an R^2 score of 0.978, indicating a significant enhancement post-optimization. Future research could focus on expanding the model's training sample set, utilizing longer time series datasets over more extended periods, and refining the current models to make the model more robust and reliable.

References

1. Yuhu, L.: Research on the Development of GDP Inversion at Night Based on Long-Term Sequence of Night Lights. Donghua University, Shanghai (2019)

2. Fan, Q., Lu, J., Li, M.: Forecast of gross domestic product of the three northeast provinces using luminous data. Remote Sensing Information **34**(4), 3–10 (2019)
3. Zhou, Y.K., Ma, T., Zhou, C.H., et al.: Night time light derived assessment of regional inequality of socioeconomic development in China. Remote Sens. **7**(2), 1242–1262 (2015)
4. Li, Q., Mou, Y., Zhao, X.: Function defect location based on random forest algorithm. Sci. Technol. Eng. **20**(32), 13278–13284 (2020)
5. Shi, K.F., Yu, B.L., Huang, Y.X., et al.: Evaluating the ability of NPP-VIIRS night time light data to estimate the gross domestic product and the electric power consumption of China at multiple scales: a comparison with DMSP-OLS data. Remote Sens. **6**(2), 1705–1724 (2014)
6. Yang, J., Xu, T., Bai, Y., et al.: Bearing fault diagnosis based on grid search and support vector machine. Sci. Technol. Eng. **21**(22), 9360–9364 (2021)

Application of Remote Sensing Technology in High Quality and Sustainable Development of Agriculture

Study on the General Sensitivity Index to Subtle Changes of Leaf Chlorophyll Content Under Arsenic Stress

Xuying Chen[1,2], Huitao Gu[1,2], Ruiyin Tang[1,2], Xuqing Li[1,2](✉), Wei Luo[1,2], Qi Wang[3], Bin Wen[1,2], and Yanping Wu[4]

[1] College of Remote Sensing and Information Engineering, North China Institute of Aerospace Engineering, Langfang 065000, China
meililixuqing@163.com

[2] Aerospace Remote Sensing Information Processing and Application Collaborative Innovation Center of Hebei Province, Langfang 065000, China

[3] Hebei Provincial Institute of Geological Surveying and Mapping, Langfang 065000, China

[4] Langfang Agricultural Information Center, Langfang 065000, China

Abstract. China currently faces a severe issue from soil heavy metal contamination in major grain-producing areas, which slightly alters the chlorophyll content of rice leaves. Although soil microbes are unable to break down heavy metals found in the soil, they may readily collect and transform into more hazardous methyl compounds. Some even make their way into human cells through the food chain and build up to potentially harmful levels there, endangering human health. Therefore, the creation of a universal spectrum index that can detect even minute variations in the amount of chlorophyll in leaves is crucial for tracking non-point source pollution in agricultural areas and estimating crop production. In this paper, two cities in the north and south were chosen to provide a matching test field with three distinct types of pollution: non-pollution, moderate pollution, and severe pollution., namely Changchun city and Suzhou city, and 260 groups of ASD spectral data and leaf chlorophyll content data were collected to construct leaf chlorophyll sensitive index. The universal applicability of the model can be proved by the difference of environment and crop varieties in the two regions, and the scientific supply of other conditions to avoid unnecessary stress and other environmental factors. First, three indexes, namely NDSI (normalized difference spectral index), RSI (ratio spectral index) and NVI (new vegetation index), were constructed, and each index was used as an independent variable to deduce the chlorophyll content of leaves. At the same time, 7 vegetation indexes (DattA, MaccioniA, Vogelmann2A, SR3A, SR6A, VogelmannA and RSI(550,800)) which are sensitive to chlorophyll content in leaves are selected as input variables to build a model that can accurately forecast the amount of chlorophyll in rice leaves with a 0.75 accuracy rate. The results showed that NVI(R720, R800, R698) was the best index to evaluate chlorophyll content in leaves. The application effect of this index in two regions with different varieties and environments is good, and the determination coefficient R^2 is 0.83. This index (NVI) is obviously better than other indexes and random forest model. By analysing the spectral properties of rice leaves, it was found that the "red edge" had a slight "blue shift", the height of green peak increased, the absorption depth decreased, and the absorption width

G. Luo (Ed.): GFSISD 2024, SPPHY 417, pp. 191–206, 2025.
https://doi.org/10.1007/978-981-96-2490-4_13

decreased after the moderate and severe pollution of rice. It also explains why the best exponential bands are in the red-edged region. In this paper, a universal index sensitive to the weak change of chlorophyll content in leaves under arsenic stress was found. The index has good application effect in two regions with different varieties and different environments. It may be utilized to field non-point source pollution monitoring by estimating the little variations in chlorophyll concentration in rice leaves under stress from heavy metals in various locales.

Keywords: general sensitivity index · random forest · arsenic stress in rice · chlorophyll content in leaves · Non-point source pollution of farmland

1 Introduction

With the advancement of China's agricultural modernization as well as industrialization, soil heavy metal pollution is a hot issue in today's environmental research. According to the 2018 report, in China's grain-producing regions, the point exceedance rate of soil heavy metals in agriculture was 21.49%, with 13.97%, 2.50%, and 5.02% of the total amount of light, medium, and heavy pollution, respectively [1]. The People's Republic of China's Soil Pollution Prevention and Control Law went into effect on January 1, 2019. The primary challenge in advocating for the bill is tracking the extent of heavy metal contamination in the soils in the primary grain-producing regions [2]. Therefore, China is in urgent need of a method that can conduct large-area monitoring of land contamination in the country to provide the foundation for national food security and the stable development of national livelihoods [3]. One solution is to use remote sensing to monitor soil heavy metal contamination in arable land. Traditional methods generally use field sampling combined with laboratory tests, although the measurement accuracy is high, but time-consuming and labor-intensive, does not have the advantages of fast speed, wide range and low cost, cannot reduce the cost of food production, and is inconsistent with the current development of large-scale food and industrial structure, and cannot better access to spatial and temporal continuous distribution information [4–6]. In contrast, remote sensing as a space technology has the characteristics of broad field of view, rich information, and fast data acquisition, It makes it possible for remote sensing technology to be used more extensively in the study and observation of heavy metal pollution in soil. As a result, it has emerged as one of the best approaches to address the actual issue of heavy metal pollution from surface sources on farms [7–10].

The spectral properties of leaves are mostly determined by the cellular structure, water content, biochemical composition, and pigment content of rice [11]. Heavy metal stress usually leads to a decrease in the amount of chlorophyll [12] and changes in the internal structure of the leaves [13], which react in the visible to near-infrared region of the spectrum [14, 15].

The red edge position is the more obvious spectral characteristic parameter in the reflectance spectrum of green vegetation [16, 17], and it has direct related to the chlorophyll content of crops [18]. Variations in pigment concentration brought on by heavy metal stress are directly linked to variations in visible light reflectance. Water-soluble pigments, such as anthocyanins, and fat-soluble pigments, such as chlorophyll, lutein, carotene, saffronin, and capsaicin, are examples of crop pigments [19]. When a crop faced heavy metal stress, the ions of heavy metals will take the place of Fe2+, Zn2+, and Mg2+ in the chloroplast proteins. This will cause changes in the ionic composition of the chlorophyll protein center and inactivation; additionally, the enzymes required for the synthesis of chlorophyll will be inhibited, which will block the process of chlorophyll formation and reduce the amount of chlorophyll in the crop [20–23]. Finding the ideal index that is more responsive to minute variations in chlorophyll concentration under arsenic stress is the goal of this work.

Leaf chlorophyll content can be measured in the laboratory by chemical solvent extraction, which is a highly accurate but time-consuming and labor-intensive method with limited data acquisition [24, 25]. As technology has advanced, leaf chlorophyll concentration at the regional level can be measured effectively with hyperspectral remote sensing [26]. At present, some research results have been achieved at home and abroad in the estimation of crop chlorophyll content using hyperspectral technology, and the chlorophyll content was estimated by modeling with remotely sensed hyperspectral data. Sun [27] collected imaging hyperspectral data of potato leaves based on the vertical distribution position of leaves, calculated the chlorophyll content of each pixel of potato leaves, and they achieved the visual representation of distribution as well as the non-destructive measurement of the chlorophyll concentration of potato leaves at various leaf locations. Wu [28] utilized the high correlation between chlorophyll content and SPAD value, observed the SPAD and hyperspectral data of rice canopy, and applied the vegetation index to establish a hyperspectral estimation model of SPAD in rice canopy. On the basis of the existing hyperspectral data, the monitoring of chlorophyll content was realized by selecting the chlorophyll spectral characteristic band. Mao [29] Adopted genetic algorithm to optimize the hyperspectral feature parameters of winter wheat canopy, and successfully predicted the chlorophyll content of winter wheat. Jiang [30] applied the RNCA algorithm to realize the chlorophyll spectral characterization and feature band selection of Japonica rice, and constructed the chlorophyll content inversion model of particle swarm optimization and extreme learning machine to realize the detection of chlorophyll content of crops. Yu [31] obtained canopy hyperspectral image data of Japonica rice during the critical fertility period, selected the sensitive bands for the chlorophyll content response in Japonica rice leaves, and constructed the inversion model of Extreme Learning Machine based on Particle Swarm Optimization, with the aim of realizing the fast, accurate, and non-destructive diagnosis of the chlorophyll content of Japonica rice.

There are several issues with using remote sensing to track the presence of heavy metal contamination that have not yet been fully resolved, which are broadly classified into three categories: firstly, direct quantification of the metal content of soil is not possible because most metal ions (with the exception of high concentrations of transition elements) are inactive in the visible-infrared spectral region. Second, research has focused on laboratory or limited field studies, and there is no generalized prediction mechanism for monitoring heavy metal contamination from farmland sources in natural systems of farmland. Finally, in remote sensing, the methods used for estimation in most cases can only estimate relatively severe metal stress. In summary, it is still problematic how mildly contaminated soils can be monitored using remote sensing for natural farmland systems. Therefore, it is necessary to estimate small changes in leaf chlorophyll content under heavy metal stress, the one can be quantitatively monitored with leaf spectral changes, and the leaf chlorophyll index constructed in this paper aims to capture these small changes. The purpose of this study is create a ideal index which is more sensitive to the small changes in chlorophyll content under arsenic stress, and the study attempts to construct a multiregional generalized and universally applicable Neutral Vegetation Index (NVI) to estimate leaf chlorophyll content (LCC) over the entire band of the hyperspectral. A three-band index was constructed by statistically selecting the bands most sensitive to response of chlorophyll content in leaves. The optimal index for assessing leaf chlorophyll content was sought by contrasting it with a random forest model of leaf chlorophyll concentration and other indicators.

2 Materials and Methods

2.1 Overview of the Study Area

Study area I is located in Changchun City (43°51′N, 125°10′E), Jilin Province (Fig. 1),which has an average annual temperature of 5.8 °C and a moderate continental climate with 522–615 mm of precipitation annually. The rice variety is Ji-Japonica 105 series, which has a compact plant with 15 main stem leaves. The rice variety was Jiji-Japonica 105 series, which is a Japonica-type conventional rice, a compact plant with 15 main stem leaves, a transplanting density of 30 cm × 20 cm, a full-life span of 153 days, a plant height of 96.5 cm, and a spike length of 16.8 cm. A total of 150 samples were collected in Changchun City, of which about 20 were used for validation. Study area II is located in Suzhou City (31°24′N, 120°33′E), Jiangsu Province (Fig. 1), which has an abundance of heat and precipitation, a subtropical monsoon climate, 1094 mm of annual rainfall on average, and an average annual temperature of 15.7 °C. The rice cultivar Changyou No. 1 was selected as the first rice cultivar in this area, and was selected as the first rice cultivar in this area. The rice variety was Changyou No. 1, with a compact main stem of 17 slices, transplanting density of about 27 cm × 16 cm, total growing period of about 160 days, plant height of 105 cm, and spike length of 18 cm. A total of 110 samples were collected in Suzhou City, of which about 20 were used for validation.

These two areas have different environments and varieties, and scientific cultivation provides enough irrigation water and fertilizers to prevent needless stress from other environmental concerns, ensuring the applicability of the model's prerequisites. According to the statistical yearbook data of the year, rice in the study area was not affected by other concerns such as water, temperature, and pests and diseases during the growth period.

2.2 Experimental Design for the Study Area

Three sampling fields, designated as A, B, and C, were selected, and three typical rice-growing regions within the research area were designated as experimental areas. Area A was next to a chemical facility, Area B was adjacent to a highway and a little distance from the chemical plant, and Area C was farther away. Some of the samples were selected as experimental samples and some as validation samples. Four to six pieces of rice farmland with an area of about 15 × 15 m^2 in a typical heavy metal contaminated area were selected as the basic experimental area, and in each experimental area a number of sample points were then identified for ground experimental data collection. The various levels of soil heavy metal contamination are chosen as the determinants of biochemical parameter changes in this work, the environmental conditions affecting rice biochemical parameters, such as water and fertility, are essentially the same in the experimental area.

As shown in Table 1, the pollution level of the cultivated land in the three experimental areas was non-polluted (a < soil background value), moderately polluted (soil background value < a < 30 mg/kg), and severely polluted (a > 30 mg/kg) to indicate the corresponding crop growth conditions.

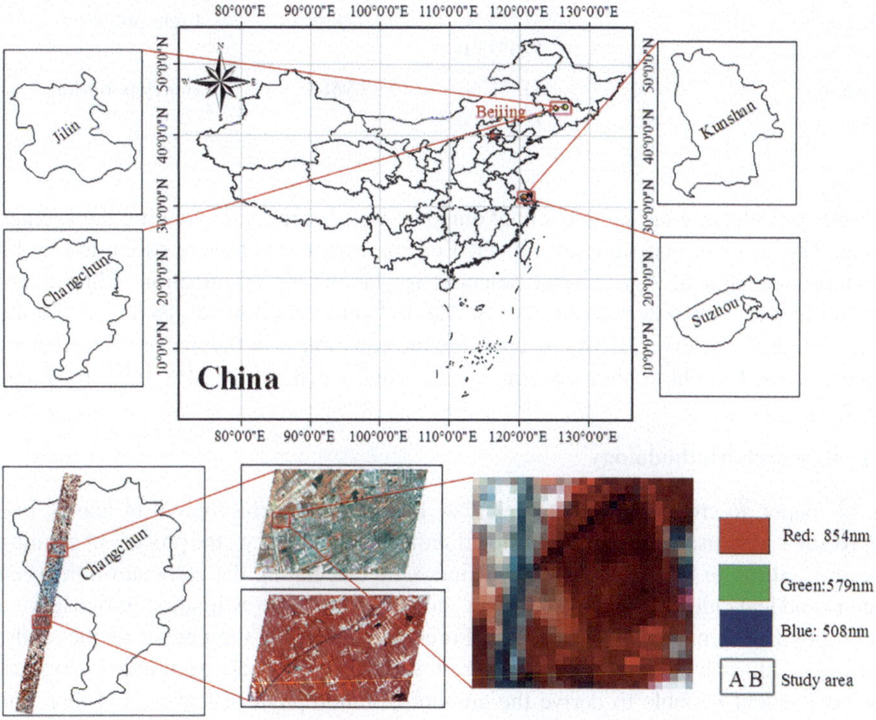

Fig. 1. The study areas in the city of Chang Chun, Jilin and Su Zhou, Jiangsu, China

2.3 Measured Data Collection

In this study, the spectral reflectance of rice leaves was determined using an Analytical Spectral Devices Field Spec 3 spectrometer in the U.S.A. The measured wavelengths ranged from 350 to 2500 nm, with 350–1000 nm light.

The spectral sampling interval was 1.4 nm, and a standard whiteboard calibration was used prior to measurement. Spectral data were automatically collected 10 times

Table 1. Soil heavy metal content in the study area

Study area	Identification of experimental area	Geographical position	Arsenic content	Soil pollution level
Changchun	A	43°42′N, 125°9′E	10.7232	Non-polluting
Changchun	B	43°42′N, 125°9′E	20.4061	Light pollution
Changchun	C	43°43′N, 125°9′E	34.2326	Heavy pollution
Suzhou	A	31°24′N, 120°31′E	8.2971	Non-polluting
Suzhou	B	31°25′N, 120°33′E	18.4858	Light pollution
Suzhou	C	31°21′N, 120°51′E	33.9588	Heavy pollution

for each sample, and the sample's final outcome was determined by taking the average value. The soil and leaf samples collected during this research were submitted to the Chinese Academy of Agricultural Sciences for laboratory examination. Chlorophyll content in this study was derived from SPAD-502 chlorophyll meter readings (Minolta Corporation, New Jersey, USA) by applying the equation: $y = 0.996x - 1.52$, where y represents the leaf chlorophyll content ($\mu g/cm^2$) and x denotes the SPAD-502 readings.

2.4 Research Methodology

In this paper, to create a model to calculate the amount of chlorophyll in leaves, two approaches were used, one is the traditional statistical method, i.e., the process of creating spectral indices to find the best band or index by calculating the association between factors and leaf chlorophyll concentration., and the other is the estimation of rice leaves' chlorophyll content using the Random Forest algorithm. In the course of the study, three leaf chlorophyll content indices were firstly selected, where each index served as an independent variable to derive the amount of chlorophyll in leaves, and then one of the selected vegetation indices was selected to estimate canopy chlorophyll content faced heavy metal stress with higher accuracy, i.e., the one that was more susceptible to variations in the amount of chlorophyll while under severe metal stress.

2.4.1 Statistical Methods

(1) The Normalized Vegetation Index (NVI) formula is defined as

$$\mathrm{NDSI}(x, y) = (y + x)(y - x) \quad (1)$$

where x, y are the values of reflectance of bands i, j i.e. Ri, Rj.

(2) The ratio vegetation index formula is defined as

$$y = 0.1559 + 7.6692x - 0.0650x^2 + 0.0002x^3 \quad (2)$$

where x, y are the values of reflectance of bands i, j i.e. Ri, Rj.

(3) The three-band vegetation index NVI is defined as

$$y = 80.5912 + 2.0844x \quad (3)$$

where x, y and z are the values of reflectance (Ri, Rj and Rk) at i, j and k nm, respectively. They have their own values throughout the band 350–1000 nm.

2.4.2 Random Forest Algorithm

Random Forest Algorithm is an intelligent machine learning algorithm with very fast learning ability and high computational speed, Breiman utilizes OOB to estimate the importance of the variables as well as to estimate the error of the model. Random forest regression algorithm is based on CART (Classification And Regression Tree) regression tree partitioning algorithm, in each tree regression process, the user randomly selects the pattern of each random variable in each tree, the overall prediction data is not included, but rather the difference is indicated by calculating the mean squared deviation of the OOB data and the data in the regression tree,. The random forest algorithm can give the error of the OOB variable for each variable. In this paper, seven vegetation indices sensitive to leaf chlorophyll content were selected as input variables, and the Random Forest algorithm was used to construct a model to predict the value of leaf chlorophyll content of rice and the analysis of model-related accuracy

$$\mathrm{mse} = \frac{\sum_{i=1}^{n} \left(F(x_i) - y_i\right)^2}{n} \quad (4)$$

where xi, yi are the true, predicted values of the sample data.

$$\mathrm{rsp} = \frac{n - \sum_{i=1}^{n} \left(F(x_i) - y_i\right)^2}{\sum_{i=1}^{n} \left(y_i - \bar{y}\right)} \quad (5)$$

where xi, yi are the true, predicted values of the sample data.

Random Forest has three key parameters: (a) the number of forest trees (b) the minimum number of endpoint data points (c) the number of features per node. Typically, the default number of features per node is set to 1/3 of the total independent variables, expressed as mtry = [p/3], where p represents the feature vector of x. In this paper, the node size, n-tree, and feature nodes are set to 5, 2000, and 22, respectively. mse is the mean squared deviation of the vector, and rsq denotes the degree of fit, which is obtained by 1 − mse/Vary(y).

3 Results and Analysis

3.1 Index Modeling

From Fig. 2, it can be seen that NDSI (R780, R730) is the spectral index with the best correlation among the normalized vegetation indices, with r^2 of 0.63 in Changchun and r^2 of 0.59 in Suzhou, and Fig. 3 shows that RSI (R770, R730) has the best correlation with LCC, with r^2 of 0.62 in Changchun and r^2 of 0.58 in Suzhou.

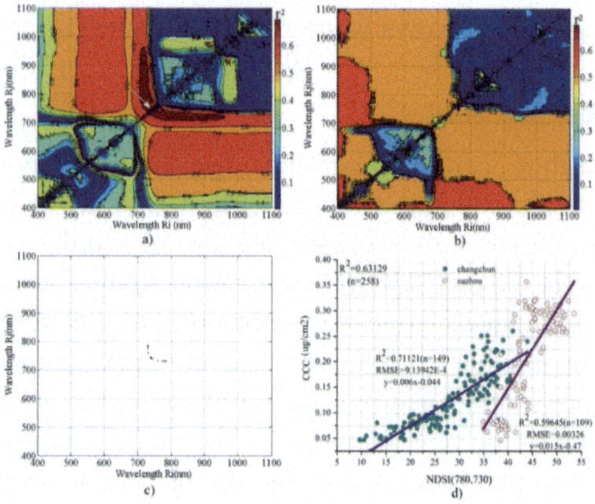

(a) Correlation contour map of NDSI index and chlorophyll content in leaves in changchun city

(b) Suzhou NDSI index and leaf chlorophyll content correlation contour map

(c) intersection of NDSI index and chlorophyll content in leaves of changchun and suzhou

(d) scatter distribution diagram of the correlation between NDSI index and chlorophyll content in leaves in changchun and Suzhou

Fig. 2. r^2 contour map between LCC and NDSI (Ri, Rj) in Changchun and Suzhou

Figure 4 shows the contour plots of r^2 between LCC and NVI (Ri, Rj, Rk) using the complete combination of the three bands at i, j, and k nm from the LCC dataset, respectively. r^2 is the best correlated index with NVI (R_{720}, R_{799}, R_{698}), with 0.83 and 0.77, respectively.

3.2 Construction of Random Forest Model

Seven vegetation indices (DattA, MaccioniA, Vogelmann2A, SR3A, SR6A, VogelmannA and RSI (550, 800)) (Table 2), which are sensitive to leaf chlorophyll concentration, were utilized to predict the value of leaf chlorophyll content in rice. The results showed that the accuracy of the random forest model was high at 0.75.

The distribution of variable importance for the vegetation index is shown in Fig. 5. The greater the variable importance, the more important its characteristics. A reduction in the number of model samples causes a small change in the distribution of variable importance, but the overall change is not significant.

Figure 6 illustrates the optimal ntree and mtry parameters for the random forest regression model developed in this study. a1 and a3 depict the relationship between the

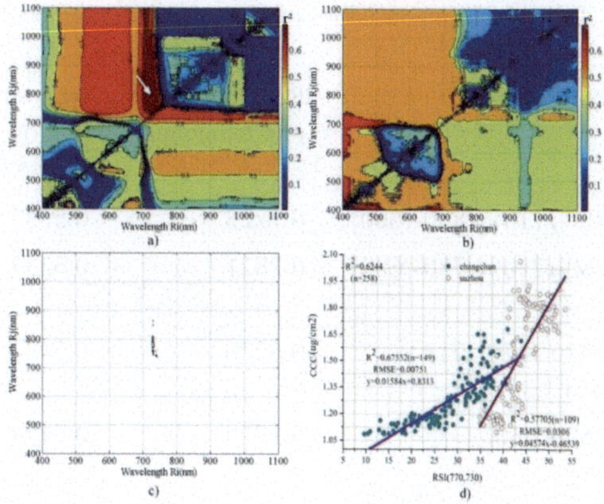

(a) Correlation contour map of RSI index and chlorophyll content in leaves in changchun city

(b) Suzhou RSI index and leaf chlorophyll content correlation contour map

(c) intersection of RSI index and chlorophyll content in leaves of changchun and suzhou

(d) scatter distribution diagram of the correlation between RSI index and chlorophyll content in leaves in changchun and suzhou

Fig. 3. r^2 contour map between LCC and RSI (Ri, Rj) in Changchun and Suzhou

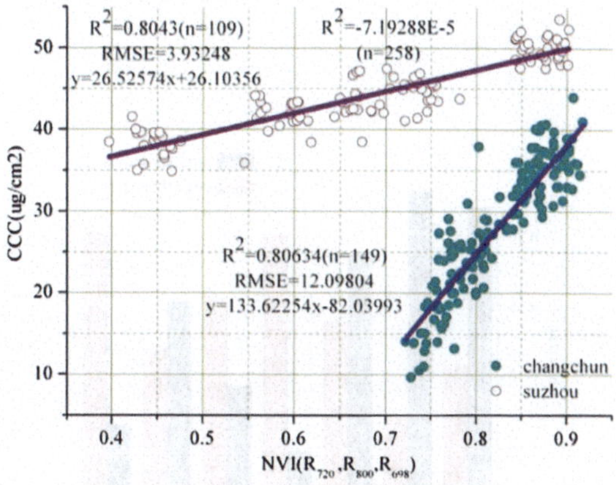

Fig. 4. r^2 contour map between LCC and NVI (Ri, Rj) in Changchun and Suzhou

pseudo-multiple correlation coefficient and the number of random forest trees in the Changchun and Suzhou study areas, respectively. Similarly, a2 and a4 represent this relationship within the test sets for the Changchun and Suzhou regions. b1 and b3 show the relationship between the mean square error (MSE) of the training set and the number of random forest trees, while b2 and b4 highlight the relationship between the MSE of the test set and the number of training trees in the random forest model. c1 and c3 illustrate the relationship between predicted Leaf Chlorophyll Content (LCC) and the measured LCC for the training data, while c2 and c4 compare the predicted and measured LCC for the test sets.

Table 2. Input variables of Random Forests model

Index name	Algorithm	Relevance rate of Changchun	Relevance rate of Suzhou	Quote
Datt[A]	(r850-r710)/(r850-r680)	0.7723	0.5332	Datt (1999)
Maccioni[A]	(r780-r710)/(r780-r680)	0.7821	0.5333	Maccioni et al. (2001)
Vogelmann[2A]	(r734-r747)/(r715 + r726)	0.6276	0.5473	Vogelmann et al. (1993)
SR[3A]	R750/R550	0.5729	0.5128	Gitelson and Merzlyak (1997)
SR[6A]	R750/R710	0.5873	0.4947	Zarco-Tejada and Miller (1999)
Vogelmann[A]	R740/R720	0.6471	0.5534	Vogelmann et al. (1993)
RSI (550, 800)	R550/R800	0.6889	0.5732	Aoki, Yabuki, & Totsuka, 1981

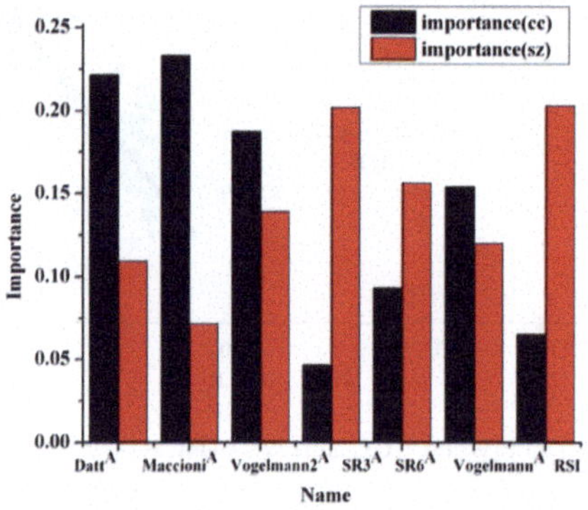

Fig. 5. The variable importance of vegetation index

3.3 Model Comparison

Comprehensively comparing the above indices constructed based on statistical methods and the models constructed based on the random forest algorithm, the results (Table 3) revealed that the highest precision and the best results were obtained by utilizing the triple-band index NVI for estimating the chlorophyll content of leaves under arsenic stress.

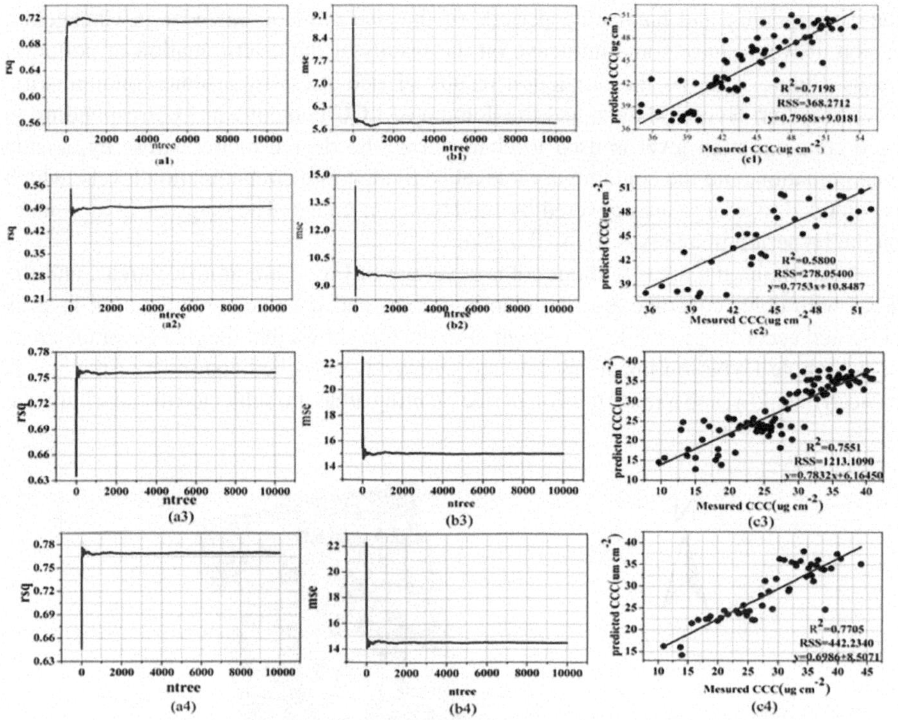

Fig. 6. Comparison of random forests model testing set and training set in Changchun and Suzhou

Table 3. Comparison of model results

Index name	Algorithm	Relevance rate of Changchun	Relevance rate of Suzhou
NDSI	$(R_{780} - R_{730})/(R_{780} - R_{730})$	0.7112	0.5965
RSI	R_{770}/R_{730}	0.6755	0.5770
NVI	$(R_{799} - R_{720})/(R_{799} - R_{699})$	0.8043	0.8063
RF		0.7705	0.5800

3.4 Model Application

The NVI inversion model was applied to the study area through Hyperion imagery by first constructing the NVI three-band index with Hyperion bands, then utilizing the relationship between the LCC and the NVI index in Fig. 4 to obtain the LCC prediction values, and discriminating the situation of the location of the pollution level level based on the domain of the LCC content values corresponding to each pollution level level where these LCC prediction values fall. Since the acquired images were taken during the rice ripening and harvesting periods of the rice climatic calendar in Changchun region, it would have some influence on the results of the discrimination of pollution levels. Figure 7 shows the prediction results of rice stress by arsenic contamination level based on Hyperion data, and the domain of LCC content value corresponding to each contamination level is used to characterize the degree of rice stress by arsenic contamination, and the contamination level is divided into contamination levels, which is extended to a large area covered by the Hyperion image. The range of LCC content on the Hyperion image is -10 to 10 μg/cm^2.

As a whole, the areas without color coverage in Fig. 7 are non-rice areas, and the areas with color coverage are rice planting areas, and the rice contamination level is classified according to the LCC content, and the test shows that the rice coverage areas of the whole Hyperion image are basically correctly classified and most of the rice is basically healthy, and a small portion of rice planting areas are subjected to class 1 arsenic

Fig. 7. LCC concentration of rice and corresponding stress degree of arsenic pollution in Changchun area

contamination (mild contamination) stress and class 2 arsenic contamination. A few areas of rice cultivation were stressed by class 1 arsenic contamination (light contamination) and class 2 arsenic contamination (heavy contamination). It was also found that the closer the cultivated areas were to the residential areas, the more contaminated they were and the higher the level of contamination was.

4 Discussion

Soil heavy metal pollution can lead to small changes in chlorophyll content of rice leaves. In this study, three levels of non-polluted, moderately-polluted and heavily-polluted corresponding experimental plots in the typical study areas of Changchun and Suzhou were selected, and 260 sets of rice leaf chlorophyll concentration data and ASD spectral data were gathered to create leaf chlorophyll sensitivity indices. In the course of the study, three leaf chlorophyll content indices namely Normalized Spectral Index (NDSI), Ratio Spectral Index (RSI), and Three-band Vegetation Index (NVI) were selected, where each of them served as an independent variable to calculate the chlorophyll content of leaves [32]. Among the constructed vegetation indices, an index with higher accuracy in estimating canopy chlorophyll content under heavy metal stress was selected, i.e., an index that was more susceptible to variations in chlorophyll concentration face heavy metal stress. Building a prediction model for chlorophyll content of rice leaveswith the random forest method [33], the accuracy of the model was 0.75. The results showed that NVI was the best index for assessing the chlorophyll content of leaves.

The spectral characterization of rice leaves found that after rice was subjected to moderate and heavy pollution, the "red edge" had a slight "blue shift", the height of the green peak increased, the depth of absorption decreased, and the width of absorption decreased, which also confirmed why the optimal index of each band is in the This also confirms why the best index bands are in the red edge region. Most heavy metals are inactive in the visible-infrared spectral region and it is not feasible to directly quantify the concentration of heavy metals in soil. The methods used for estimation in most cases can only estimate relatively severe metal stress. Therefore, for natural farming systems, how to monitor mildly contaminated soils using remote sensing is still a problem. In this paper, a sensitive universal vegetation index was constructed to capture small changes in leaf chlorophyll concentration under heavy metal stress, and under arsenic stress, this indicator is more responsive to small variations in the amount of chlorophyll, which can be utilized to provide a new method for monitoring mildly contaminated soils using remote sensing to estimate the amount of chlorophyll in rice leaves in different regions.

In order to construct a universal spectral index sensitive to weak changes that chlorophyll concentration of leaves,this work made effort to analyze the correlation of chlorophyll concentration in Changchun and Suzhou by constructing three index models of leaf chlorophyll concentration indices, and concluded that the NVI (R720, R799, R698) was the best correlation index, with the r^2 of 0.83 and 0.77,concurrently, the index was used in conjunction with the remotely sensed data to carry out the corresponding model application in the study area. Meanwhile, the index was combined with remote sensing data to apply the model in the corresponding study area, and the viability of the index was verified based on the LCC predicted values falling in the corresponding LCC content domains of each pollution level, and was successfully extended to a large area. The accuracy of the inversion of chlorophyll content under heavy metal stress was increased [34], and there was a realization of rapid monitoring and assessment of small changes in chlorophyll content under remote sensing [35].

The inversion errors of pollution stress level and LCC in the study mainly come from remote sensing data and inversion methods, and how to improve the inversion accuracy of pollution stress level and LCC will be investigated from the remote sensing inversion methods and data in the subsequent study. The research method in this paper is suitable for the monitoring of arsenic stress in rice where the differences of other environmental factors (PH value, oxides, etc.) are not large, and the relationship between environmental factors and the effects of arsenic stress on the physiological functions of rice under natural conditions needs to be further investigated.

5 Conclusion

In this paper, we constructed four kinds of universal sensitivity indices for small changes in leaf chlorophyll content under arsenic stress, which were NDSI, RSI, NVI and RF models, and compared the four models comprehensively, and found that the validation accuracy of NVI (R720,R800, R698) was 0.83, which was outperformed than the others, and it was the most effective indicator for assessing the amount of chlorophyll in leaves, and it could be applied well in the two regions with different varieties and different environments. The results were good in two regions with different varieties and different environments. These models may be used to calculate the amount of chlorophyll in rice leaves in different regions, and then applied to the monitoring of farmland surface pollution.

Acknowledgments. This research was funded by Supported by Central Guidance Fund for Local Scientific and Technological Development (246Z7401G), Supported by Open Fund of State Key Laboratory of Remote Sensing Science (OFSLRSS202303), Fund of Guangxi Key Laboratory of Arable Land Conservation (23-026-12-24KF7), Graduate Student Innovation Grant Program of North China Institute of Aerospace Engineering (YKY-2024-85).

References

1. Shang, E.P., Xu, E.Q., Zhang, H., et al.: Spatial-temporal trends and pollution source analysis for heavy metal contamination of cultivated soils in five major grain producing regions of China. Environ. Sci. **39**(10), 4670–4683 (2018)
2. Li, X., Zhou, D.: A meta-analysis on phenotypic variation in cadmium accumulation of rice and wheat: implications for food cadmium risk control. Pedosphere **29**(5), 545–553 (2019)
3. Xuqing, L., Xiangnan, L., Meiling, L., et al.: A hyperspectral index sensitive to subtle changes in the canopy chlorophyll content under arsenic stress. Int. J. Appl. Earth Observ. Geoinf. **36**, 41–53 (2015)
4. Zhang, Q., Zhang, H., Zhang, H., et al.: Hybrid inversion model of heavy metals with hyperspectral reflectance in cultivated soils of main grain producing areas. Trans. Chin. Soc. Agric. Mach. **8**(3), 148–155 (2017)
5. Xiong, W., Nie, Y., Wang, Q., et al.: Progress in soil pollution remote sensing and its application prospect. Environ. Sustainable Dev. **42**(6), 51–54 (2017)

6. Zhou, Z.: Remote Sensing Quantitative Inversion of the Chlorophyll of Paddy and Research on the Correlation Between Chlorophyll and Soil Heavy-Metal. Changsha University of Science & Technology, Changsha (2015)
7. Chen, Z., Ren, J., Tang, H., et al.: Progress and perspectives on agriculture remote sensing research and applications in China. J. Remote Sens. **20**(5), 748–767 (2016)
8. Xuqing, L., Xiangnan, L., Meiling, L., et al.: Random forest algorithm and regional applications of spectral inversion model for estimating canopy nitrogen concentration in rice. J. Remote Sens. **23**(4), 923–945 (2014)
9. Liu, M., Liu, X., Li, T., et al.: Analysis of hyperspectral singularity of rice under Zn pollution stress. Trans. Chin. Soc. Agric. Eng. **26**(3), 191–197 (2010)
10. Liu, X., He, C., Pan, Y., et al.: Accuracy assessment of thematic classification based on point and cluster sample. J. Remote Sens. **10**(3), 366–372 (2006)
11. Guyot, G., Baret, F., Jacquemoud, S.: Imaging spectroscopy for vegetation studies. Imaging Spectrosc. 145–165 (1992)
12. Zengin, F.K., Munzuroglu, O.: Effects of some heavy metals on content of chlorophyll, proline and some antioxidant chemicals in bean (Phaseolus vulgaris L.) seedlings. Acta Biol. Crac. Ser. Bot. **47**(2), 157–164 (2005)
13. Slaton, M.R., Hunt, E.R., Smith, W.K.: Estimating near-infrared leaf reflectance from leaf structural characteristics. Am. J. Bot. **88**(2), 278–284 (2001)
14. Schwaller, M.R., Schnetzler, C.C., Marshall, P.E.: The changes in leaf reflectance of sugar maple (Acer saccharum Marsh) seedlings in response to heavy metal stress. Int. J. Remote Sens. **4**(1), 93–100 (1981)
15. Milton, N.M., Ager, C.M., Eiswerth, B.A., et al.: Arsenic- and selenium-induced changes in spectral reflectance and morphology of soybean plants. Remote Sens. Environ. **30**(3), 263–269 (1990)
16. Gao, X., Li, F., Yang, H., et al.: Appropriate calculation method for the use of red edge position to estimate potato nitrogen concentration. J. Plant Nutr. Fertil. **25**(2), 296–310 (2019)
17. Cao, Y., Jang, K., Liu, Y., et al.: Rice chlorophyll inversion based on hyperspectral red edge retrieval. J. Shenyang Agric. Univ. **52**(6), 718–728 (2021)
18. Hanyue, C., Wenjiang, H., Wang, L., et al.: Estimation of LAI in winter wheat from multi-angular hyperspectral VNIR data: effects of view angles and plant architecture. Remote Sens. **10**(10), 1630 (2018)
19. Rosso, P.H., Pushnik, J.C., Lay, M.: Reflectance properties and physiological responses of Salicornia virginica to heavy metal and petroleum contamination. Environ. Pollut. **137**(2), 241–252 (2005)
20. Rahman, M.A., Hasegawa, H., Rahman, M.M., et al.: Effect of arsenic on photosynthesis, growth and yield of five widely cultivated rice (Oryza sativa L.) varieties in Bangladesh. Chemosphere **67**(6), 1072–1079 (2007)
21. Irfan, M., Hayat, S., Ahmad, A., et al.: Soil cadmium enrichment: allocation and plant physiological manifestations. Saudi J. Biol. Sci. **20**(1), 1–10 (2013)
22. Monteiro, M.S., Santos, C., Soares, A., et al.: Assessment of biomarkers of cadmium stress in lettuce. Ecotoxicol. Environ. Saf. **72**(3), 811–818 (2009)
23. Pereira, G.J.G., Molina, S.M.G., Lea, P.J., et al.: Activity of antioxidant enzymes in response to cadmium in Crotalaria. Plant Soil **239**(1), 123–132 (2002)
24. Yang, X., Lu, X., Shi, J., et al.: Inversion of rice leaf chlorophyll content based on Sentinel-2 satellite data. Spectrosc. Spectr. Anal. **42**(3), 866–872 (2022)
25. Yang, X.: Inversion of Rice Leaf Chlorophyll Content Based on Sentinel-2 Satellite Data. Nanjing University, Nanjing (2021)
26. Deng, S., Zhao, Y., Bai, X., et al.: Inversion of chlorophyll and leaf area index for winter wheat based on UAV image segmentation. Trans. Chin. Soc. Agric. Eng. **38**(3), 136–145 (2022)

27. Sun, H., Zheng, T., Liu, N., et al.: Vertical distribution of chlorophyll in potato plants based on hyperspectral imaging. Trans. Chin. Soc. Agric. Eng. **34**(1), 149–156 (2018)
28. Wu, X., Chang, Q., Luo, L., et al.: Hyperspectral estimation model for chlorophyll content of rice canopy. Agric. Res. Arid Areas **37**(3), 238–243 (2019)
29. Mao, B., Li, M., Sun, H., et al.: Optimization of spectroscopy parameters and prediction of chlorophyll content at seeding stage of winter wheat. Trans. Chin. Soc. Agric. Eng. **33**(S1), 164–169 (2017)
30. Jiang, K., An, J., Zhao, Y., et al.: Analysis and inversion of rice chlorophyll spectral characteristics using RNCA-PSO-ELM. Trans. Chin. Soc. Agric. Eng. **38**(8), 178–186 (2022)
31. Yu, F., Feng, S., Zhao, Y., et al.: Inversion model of chlorophyll content in japonica rice canopy based on PSO-ELM and hyper-spectral remote sensing. J. South China Agric. Univ. **41**(6), 59–66 (2020)
32. Zhang, Y., Xiao, J., Yan, K., et al.: Advances and developments in monitoring and inversion of the biochemical information of crop nutrients based on hyperspectral technology. Agronomy (Basel) **13**(8), 2163 (2023)
33. Gao, G., Zhang, L., Wu, L., et al.: Estimation of chlorophyll content in wheat based on optimal spectral index. Appl. Sci. **14**(2), 703 (2024)
34. Li, H., Zhou, B., Feng, X., et al.: Hyperspectral characterization and chlorophyll content inversion of reclaimed vegetation in rare earth mines. Environ. Sci. Pollut. Res. Int. **29**(24), 36839–36853 (2022)
35. Cao, Y., Jiang, K., Jingxian, W., et al.: Inversion modeling of japonica rice canopy chlorophyll content with UAV hyperspectral remote sensing. PLoS ONE **15**(9), e0238530 (2020)

Monitoring and Analysis of Agricultural Non-point Source Risk Sources in Drinking Water Source Based on GF Satellite Data

Xin-zong Xiao[1], Li-li Yin[2,3,4(✉)], Hui Xia[2,3,4], and Cheng Wang[2,3,4]

[1] China South to North Water Diversion Middle Route Corporation Limited, Beijing 100038, China

[2] Changjiang Space Information Technology Engineering Co., Ltd. (Wuhan), Wuhan 430010, China
799571395@qq.com

[3] Changjiang Survey, Planning, Design and Research Co., Ltd., Wuhan 430010, China

[4] Hubei Key Laboratory of Basin Water Security, Wuhan 430010, China

Abstract. This paper takes the drinking water source protection area of Danjiangkou Reservoir (Henan Province) as the research object. Using GF series satellite remote sensing images from spring 2022 (March to May), advanced remote sensing intelligent processing technology and deep learning algorithms are utilized to achieve accurate identification and quantitative analysis of agricultural non-point source risk sources. Through intelligent extraction of agricultural activities such as natural planting, economic forest planting, livestock and poultry breeding, and aquaculture, this study constructed a high-precision spatial distribution map of agricultural non-point source risk sources. The results show that agricultural non-point source risk sources exhibit obvious patchy distribution characteristics around the reservoir, and their distribution is closely related to land use types, terrain conditions, and human activity intensity. The research results have revealed the current situation of agricultural non-point source pollution in the Danjiangkou Reservoir Protection Area, providing a scientific basis for formulating targeted measures for water source protection and agricultural pollution control.

Keywords: Drinking water source protection area · Danjiangkou Reservoir · Remote sensing monitoring · Deep learning · agricultural non-point source risk

1 Introduction

Danjiangkou Reservoir is one of the important drinking water sources in China, and its water quality safety is directly related to the drinking water health of millions of residents and the sustainable development of ecological environment. However, with the increasing frequency of agricultural activities, agricultural non-point source pollution has become one of the main factors threatening the water quality of reservoirs. Agricultural non-point source pollution includes pollutants generated from natural planting, economic forest planting, livestock and poultry breeding, and aquaculture activities.

G. Luo (Ed.): GFSISD 2024, SPPHY 417, pp. 207–216, 2025.
https://doi.org/10.1007/978-981-96-2490-4_14

These pollutants enter water bodies through surface runoff and other means, causing serious impacts on water quality. It is crucial to accurately identify and quantify the spatial distribution of agricultural non-point source pollution in order to effectively manage and control it. Although traditional ground survey methods can provide detailed on-site data, their application in large-scale areas has the problem of high time and labor costs. In recent years, the development of remote sensing technology has provided new means for environmental monitoring, especially the application of high-resolution satellite remote sensing images, which greatly improves the accuracy and efficiency of environmental monitoring.

In recent years, researchers have made some progress in identifying and extracting water sources and their risk sources from remote sensing images. Su Yijun et al. [1] used Landsat8 data to extract forest land, farmland, etc., but due to resolution limitations, there was insufficient recognition of small-scale land such as greenhouses. Yao Yanjuan et al. [2] used high-resolution satellite data and object-oriented methods to locate the risk sources of 25 lakes and reservoirs, and conducted quantitative assessments. Xu Fen et al. identified non-point source risks by calculating risk indices and explored the impact of risk source distribution on pollution. Most of these early studies relied on manual interpretation, which was inefficient. With the development of technology, deep learning, especially convolutional neural networks (CNN), has been introduced into remote sensing image analysis, improving the level of recognition automation. The CNN model proposed by Vakalopoulou et al. [3], combined with a large number of sample learning, achieved high-precision automatic extraction of buildings, and verified the effectiveness of deep learning methods through MRF model optimization results. Yan Zhi et al. [4] applied transfer learning to process Landsat 8 and Sentinel2 data, effectively extracting buildings, but with high requirements for data quality. Cao Qi [5] combined image segmentation and deep learning grading to extract risk sources in the analysis of Gaofen-2 images. Although no risk assessment was conducted, the method covered feature extraction at different scales. Liu Yueming et al. [6] successfully identified aquaculture areas using the RCF model, demonstrating good universality. In summary, from traditional manual interpretation to deep learning driven automation methods, the identification technology of water sources and their risk sources has made significant progress, especially in improving recognition accuracy, expanding identifiable land cover types, and enhancing processing efficiency.

This study takes the drinking water source protection area of Danjiangkou Reservoir (Henan Province) as the research object, and uses GF series satellite remote sensing images from spring 2022 (March to May), combined with advanced remote sensing intelligent processing technology and deep learning algorithms, to achieve accurate identification and quantitative analysis of agricultural non-point source risk sources. Through intelligent extraction of agricultural activities such as natural planting, economic forest planting, livestock and poultry breeding, and aquaculture, this study constructed a high-precision spatial distribution map of agricultural non-point source risk sources, revealing the current situation of agricultural non-point source pollution in Danjiangkou Reservoir Protection Area, and providing scientific basis for formulating targeted measures for water source protection and agricultural pollution control.

2 Research Area and Data Sources

2.1 General Situation

The Danjiangkou Reservoir, as the water source of China's South to North Water Diversion Project [7], spans across Danjiangkou City in Hubei Province and Xichuan County in Nanyang City, Henan Province. Its water quality and safety directly affect the daily water consumption of millions of residents along the route and the health of the regional ecosystem [8]. However, due to the complex industrial and agricultural production activities, potential geological hazards, and urbanization processes within the region, a series of hidden dangers have been brought to the protection of water sources [9]. Henan is the core water source and canal head of the Central Line Project, undertaking the arduous task of ensuring that water sources are not polluted. Efficient and accurate real-time monitoring of risk sources is particularly crucial.

The drinking water source protection area of Danjiangkou Reservoir (Henan Province) is mainly located in Xichuan County, Nanyang City, Henan Province. According to relevant information and policy regulations, its protection area is divided into first level protection zone, second level protection zone, and quasi-protected zone. The specific scope is illustrated in Fig. 1.

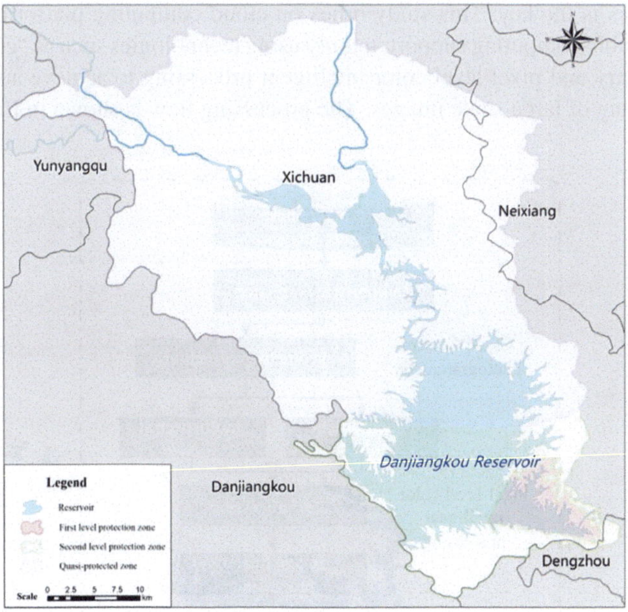

Fig. 1. Schematic diagram of study area

2.2 Image Data and Preprocessing

The data source used in risk source monitoring research in this study is high-resolution series satellite data. The Gaofen series satellites are a series of high-resolution remote sensing satellites independently developed by China, with high spatial resolution, high temporal resolution, multispectral bands, and the ability to cover a large area. They are particularly suitable for remote sensing monitoring of large water bodies such as the Danjiangkou Reservoir and its surrounding areas. The obtained remote sensing image data information is shown in Table 1.

Table 1. Information of satellite imagery in the study area

Serial number	Data source	Resolution ratio/m	Cloud cover	Image time	Data size
1	GF1C	2	<5%	2022.03–2022.05	12.1 GB
2	GF6				19.7 GB

The original remote sensing images have many problems such as geometric distortion and information errors, which require a series of processing [10] to transform the raw and unprocessed remote sensing data into data that can be further analyzed and applied [11]. For large-scale monitoring areas, the efficiency of image processing directly affects the timeliness of image monitoring. How to efficiently and quickly process remote sensing images is the key. This study relies on cloud computing platforms to provide powerful scalable computing support, mainly using technologies such as "cloud control" photogrammetry and pixel level color intelligent processing to achieve automated and rapid processing of large-scale images. The processing flow is shown in Fig. 2.

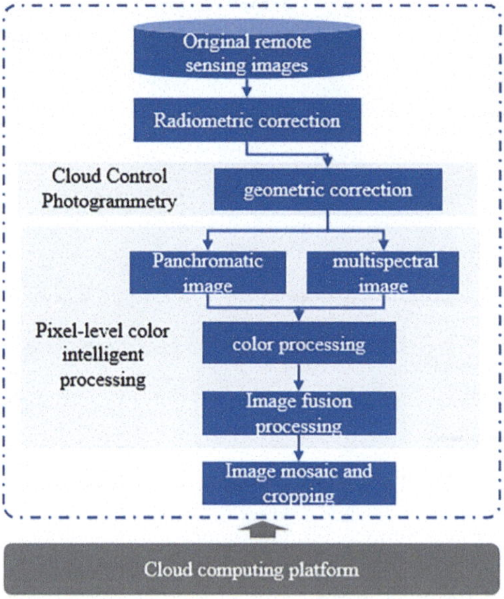

Fig. 2. Intelligent and fast processing flow of large-scale imaging

For the study area, the publicly available SRTM DEM data in the area was used as the basis for control and correction. The "cloud control" intelligent image processing system was used to complete the geometric correction of the image (with an accuracy control of 1–2 pixels), ensuring high spatial accuracy. Afterwards, the Wowei image intelligent processing software was used to complete image preprocessing work such as fusion and color adjustment of the research area images, forming remote sensing orthophoto images suitable for remote sensing interpretation of risk sources in drinking water source areas. Under technological optimization, the entire preprocessing process can be efficiently completed within a few hours, greatly improving the timeliness and practicality of data processing.

3 Research Method

3.1 Monitoring Index

The risk source of the water source may release toxic and harmful substances to the drinking water source, causing pollution sources that deteriorate the water quality of the drinking water source [12]. According to the requirements of agricultural non-point source monitoring in drinking water source protection areas, combined with the information content of high-resolution satellite data, a risk source classification system was constructed, including five types: greenhouse planting, agricultural natural planting, and economic forest planting. Please refer to Table 2 for the risk source monitoring indicator system.

Table 2. Monitoring index system of risk sources

Serial Number	Type	Illustrate
1	Greenhouse planting	The high-density fertilization areas in agriculture are an important source of pollution
2	Agricultural natural planting	Dryland, paddy fields, tea gardens, orchards, etc., residual fertilizers flow into water bodies through surface runoff, seriously affecting the environmental ecology of water source areas
3	Economic forest planting	
4	Livestock and poultry breeding	Mainly includes breeding land for pig, chicken, and cattle farms, including breeding land for pig, chicken, and cattle farms
5	Aquaculture	Mainly includes pond and cage aquaculture, as well as other aquaculture methods that involve feeding on water

3.2 Interpretation Marks

Based on field research and prior knowledge, comparative analysis was conducted with remote sensing images to determine the typical characteristics of various risk sources. As a result, a monitoring indicator system for risk sources was developed, which is suitable for remote sensing interpretation of risk sources in drinking water sources. Table 3 shows the interpretation symbols of some risk sources.

Table 3. Examples of interpretation marks

Type	Image feature	Spatial feature	Latitude	Longitude	Feature description
Aquaculture	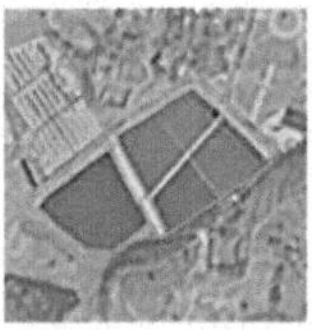		111.3804	33.0032	The geometric features of the image are obvious, the land boundary is regular, the color of the image is light green, the texture of the image is smooth, and the color tone is uniform and consistent

3.3 Risk Source Identification Method

The surface coverage types in water source protection areas are diverse, and the spectral characteristics of land features are complex. This study combines the advantages of remote sensing artificial interpretation technology and artificial intelligence methods to achieve rapid identification and extraction of large-scale regional risk sources.

The identification and extraction of risk sources combines advanced automatic recognition algorithms with professional wisdom of manual interpretation, aiming to more accurately and efficiently discover and distinguish various elements that may pose a threat to drinking water sources from complex and diverse remote sensing images. Firstly, intelligent algorithms such as deep learning and machine learning are utilized to learn the feature patterns of risk sources through a large amount of training data, and to automatically screen out suspected risk areas, such as mining areas, concentrated residential areas, highways, etc. This study used the widely recognized U-Net deep learning model to achieve highly accurate preliminary recognition. For example, the interpretation accuracy for highways and open areas reached about 80%, while in the recognition of concentrated residential areas, the accuracy was improved to about 90% through carefully constructed training and testing sets (Fig. 3.).

After forming preliminary interpretation results, field investigations and verifications will be conducted to improve quality and credibility. This study mainly adopts the method of sampling survey, with a sampling quantity of no less than 10% of the total number of

Fig. 3. Intelligent interpretation effect of aquaculture

spots in the study area, and modifies the problematic spots to control the overall accuracy of remote sensing interpretation of water source risk sources at over 90%.

4 Result and Analysis

4.1 Monitoring Results

Based on the data and research methods described above, potential agricultural non-point source risk sources in the drinking water source protection area of Danjiangkou Reservoir (Henan Province) between March and May 2022 were successfully identified, and indicators such as spatial distribution and area of risk sources were obtained, showing the distribution characteristics and potential threat level of agricultural non-point source risk sources in the study area. The specific information is shown in Table 4.

Table 4. Monitoring results of agricultural non-point source risk sources

Type	First level protection zone/km^2	Second level protection zone/km^2	Quasi-protected zone/km^2
Greenhouse planting	0	0.01	0.57
Agricultural natural planting	5.52	36.9	304.33
Economic forest planting	0.41	5.13	25.45
Livestock and poultry breeding	0	0	0.06
Aquaculture	0.04	0.72	4.64
Total	5.97	42.76	335.05

4.2 Analysis of Spatiotemporal Distribution Characteristics

According to remote sensing monitoring data from March to May 2022, the agricultural non-point source risk sources in the drinking water source protection area of Danjiangkou Reservoir ((Henan Province) show obvious spatial distribution characteristics. The total area of agricultural non-point source in the first level protection zone is 5.97 square kilometers, mainly consisting of natural agricultural planting (5.52 km^2), followed by economic forest planting (0.41 km^2) and aquaculture (0.40 km^2). The total area of agricultural non-point source in the second level protection zone is 42.76 square kilometers, mainly consisting of natural agricultural planting (36.9 km^2), followed by economic forest planting (5.13 km^2) and aquaculture (0.72 km^2), with a smaller greenhouse planting area (0.01 km^2). The total area of agricultural non-point source in the Quasi-protected zone is 335.05 square kilometers, with natural agricultural planting dominating (304.33 km^2), followed by economic forest planting (25.45 km^2) and aquaculture (4.64 km^2). Greenhouse planting (0.57 km^2) and livestock and poultry breeding (0.06 km^2s) have smaller areas (Fig. 4).

Fig. 4. Schematic diagram of the proportion of risk source area

Overall, agricultural natural planting occupies a dominant position in all levels of protected areas, and as the level of protected areas decreases, the area of agricultural non-point source risk sources significantly increases, showing a trend of gradually increasing agricultural activity intensity from first level protection zone to quasi protected areas.

This indicates that agricultural natural planting is the most important form of agricultural activity and also the main potential source of pollution, indicating that the potential threats of agricultural activities (such as pesticide and fertilizer use [13], livestock and poultry breeding, etc.) are the most prominent. In addition, economic forests occupy a larger area in quasi protected areas, while their area is smaller in primary and secondary water source protection areas, but their impact on water quality still needs to be considered. Aquaculture is distributed in all levels of protected areas, and although the area is not large, its impact on water quality cannot be ignored. Greenhouse cultivation and livestock breeding have a small or non-existent area within the first and second level protection zone, but are distributed to a certain extent within the quasi protection zones, and their potential impact on water quality needs to be taken into account.

Through these analyses, it can be seen that agricultural activities (such as pesticide and fertilizer use, livestock and poultry breeding, etc.) pose the most prominent potential threat to water quality. Therefore, strengthening the management and control of agricultural non-point source pollution is particularly important.

5 Conclusion

This study combines advanced remote sensing intelligent processing technology and deep learning algorithms to achieve accurate identification and quantitative analysis of the current situation of agricultural non-point source risk sources in the spring of Danjiangkou Reservoir (Henan Province) drinking water source protection area. A high-precision spatial distribution map of agricultural non-point source risks has been constructed through intelligent extraction of agricultural activities such as natural planting, economic forest planting, livestock and poultry breeding, and aquaculture. The research results show that agricultural natural planting occupies a dominant position in all levels of protected areas, and as the level of protected areas decreases, the area of agricultural non-point source risk sources significantly increases, indicating a trend of gradually increasing agricultural activity intensity. This study reveals the current situation of agricultural non-point source pollution in the Danjiangkou Reservoir Protection Area, providing a scientific basis for formulating targeted measures for water source protection and agricultural pollution control. In addition, the study emphasizes the potential application of remote sensing technology in the field of environmental monitoring, opening up new paths for further improving the water quality management and ecological security level of drinking water sources. Through remote sensing intelligent processing technology and deep learning algorithms, agricultural non-point source pollution can be efficiently and accurately monitored, providing strong technical support for achieving refined management and sustainable development of water source protection areas.

References

1. Su, Y., Li, L., Liu, J., et al.: Potential analysis of GF-5 to monitor the risk source in drinking water source. Yello Wriver **39**(4), 71–77 (2017)
2. Yao, Y., Wang, X., Wu, C., et al.: Non-point source pollution risk mapping and quantitative assessment for surface water source using high spatial resolution satellite data. Res. Environ. Sci. **26**(12), 1349–1355 (2013)

3. Vakalopoulou, M., Karantzalos, K., Komodakis, N., et al.: Building detection in very high resolution multispectral data with deep learning features. In: IEEE International Geoscience and Remote Sensing Symposium, pp. 1873–1876 (2015)
4. Yan, Z.: Application of Transfer Learning in Remote Sensing Monitoring of Risk Sources in Water Source Areas. Henan Polytechnic University (2019)
5. Cao, Q., Zheng, Y., Shen, Q., et al.: Research on risk source extraction of water source areas based on high resolution remote sensing. J. Geomat. **47**(6), 81–85 (2022)
6. Liu, Y., Yang, X., Wang, Z., et al.: Extracting raft aquaculture areas in Sanduao from high-resolution remote sensing images using RCF. Haiyang Xuebao **41**(4), 119–130 (2019)
7. Liu, J., Liu, X., Sun, Z., et al.: Study on the relationship between land use and water quality change in Danjiangkou reservoir. Remote Sens. Technol. Appl. **38**(2), 285–296 (2023)
8. Wang, C., Jia, H., Lei, J., et al.: Evaluation on the effect of ecological control system for non-point source pollution in typical small watershed in Danjiangkou reservoir area. J. Soil Water Conserv. **35**(3), 61–68 (2021)
9. Lin, L., Li, Q., Cao, H., et al.: Challenges and measures for digital twin Danjiangkou reservoir water quality safety construction. China Water Resour. **11**, 32–36 (2023)
10. Sun, Y., Dong, D., Wen, Y., et al.: Extraction of mangrove range in Qi'ao Island based on multi-source remote sensing data. J. Ocean Technol. **41**(5), 5–12 (2022)
11. Wang, H., Wu, X., Yang, J., et al.: Application of remote sensing geological hazard interpretation and identification technology in the Southern network region. China Rural Water Hydropower **12**, 180–184+192 (2020)
12. Environmental Standard Institute. Technical Specification for Remote Sensing Aided Survey of Risk Sources in Centralized Surface Drinking Water Sources: HJ1236–20215. Ministry of Ecology and Environment, Beijing (2022)
13. Gong, S., Xiao, N., Ding, W., et al.: Characteristics of chemical fertilizer application and environmental risk assessment in the core water source area of the Danjiangkou reservoir. Resour. Environ. Yangtze Basin **31**(10), 2259–2271 (2022)

Applications of Oceanic Remote Sensing

Bayesian Meta Pseudo Labeling for Sea Ice Classification: Facilitating Marine Disaster Emergency Management

Xiaoyi Yin[1](✉), Wenyi Zhang[1], Qingwei Chu[1], Jiande Zhang[2], and Guangzuo Li[1]

[1] Aerospace Information Research Institute, Chinese Academy of Sciences, Beijing 100094, China
yinxy@aircas.ac.cn

[2] QiLu Aerospace Information Research Institute, Jinan 250132, China

Abstract. This paper introduces a Bayesian Meta Pseudo Labeling (BMPL) method for optimizing semi-supervised sea ice classification in marine remote sensing imagery. Addressing the challenge of limited labeled data in marine remote sensing, BMPL optimizes the pseudo-labeling process through a Bayesian inference framework and enhances model generalization via meta-learning strategies. Compared to traditional semi-supervised methods, BMPL demonstrates significant advantages in both theoretical foundations and practical performance. Experiments conducted on a comprehensive dataset including open water, new ice, young ice, and first-year ice showcase the method's superior performance, particularly in complex mixed-ice and marginal ice zones. BMPL significantly improves the accuracy and efficiency of sea ice classification, providing reliable technical support for marine environment monitoring, marine resource assessment, and polar navigation safety. This research not only advances marine remote sensing technology but also offers a powerful tool for marine ecosystem protection, climate change studies, and sustainable fisheries management, contributing to multiple ocean-related issues.

Keywords: Sea ice classification · Semi-supervised learning · Meta-learning · Marine remote sensing

1 Introduction

Marine disasters pose significant threats to coastal communities, maritime activities, and ecosystems worldwide [1–3]. The increasing frequency and intensity of these events, exacerbated by climate change, necessitate advanced emergency management strategies. Among the various factors influencing marine disaster preparedness and response, accurate and timely sea ice classification plays a crucial role, particularly in polar and sub-polar regions [4].

Sea ice dynamics significantly impact navigation safety, offshore operations, and coastal erosion patterns [5]. Moreover, changes in sea ice extent and thickness serve as early indicators of potential marine hazards such as storm surges and coastal flooding.

G. Luo (Ed.): GFSISD 2024, SPPHY 417, pp. 219–233, 2025.
https://doi.org/10.1007/978-981-96-2490-4_15

However, the complexity and variability of sea ice formations, coupled with the vast areas to be monitored, present substantial challenges to traditional classification methods [6].

Remote sensing technologies offer a promising solution for large-scale sea ice monitoring. [7–9] Yet, the effectiveness of these technologies is often limited by the scarcity of labeled training data, which is both time-consuming and expensive to acquire in the harsh polar environments. This data limitation hampers the development of robust and accurate sea ice classification models, crucial for timely marine disaster risk assessment and emergency response planning.

To address these challenges, we propose a novel Bayesian Meta Pseudo Labeling (BMPL) method for semi-supervised sea ice classification. Our approach leverages the power of Bayesian inference to optimize the pseudo-labeling process [13], combined with meta-learning strategies to enhance model generalization across diverse sea ice conditions. By effectively utilizing limited labeled data alongside abundant unlabeled data, BMPL aims to significantly improve the accuracy and efficiency of sea ice classification, thereby enhancing our capacity for marine disaster emergency management.

This paper demonstrates the application of BMPL to sea ice classification using a comprehensive dataset encompassing various ice types including open water, new ice, young ice, and first-year ice [10]. We particularly focus on challenging scenarios such as mixed-ice zones and marginal ice areas, which are critical for accurate hazard assessment. Our results show that BMPL outperforms traditional semi-supervised learning methods, offering more reliable sea ice classification in complex environments.

The improved accuracy and efficiency in sea ice classification enabled by BMPL have far-reaching implications for marine disaster emergency management. It enhances our ability to:

1) Provide more accurate early warnings for potential marine hazards.
2) Optimize shipping routes and offshore operation planning in ice-prone areas.
3) Improve resource allocation for disaster preparedness and response.
4) Contribute to long-term climate change monitoring and adaptation strategies.

In the following sections, we detail the BMPL methodology, present our experimental setup and results, and discuss the implications of our findings for advancing marine disaster emergency management through improved sea ice classification.

2 Related Works

Sea ice classification has evolved significantly in recent years, transitioning from traditional manual interpretation of SAR images to more advanced automated methods. Deep learning approaches have shown promising results in automating this process, significantly improving classification accuracy and efficiency [12]. However, these methods often require large amounts of labeled data, which are scarce and expensive to obtain in polar regions.

Semi-supervised learning techniques have emerged as a potential solution to the labeled data scarcity problem. Methods such as pseudo-labeling and consistency regularization have demonstrated effectiveness in various domains with limited labeled data [13]. However, their application to sea ice classification presents unique challenges due to the complex and highly variable nature of sea ice formations.

Recent advancements in meta-learning, particularly approaches like Model-Agnostic Meta-Learning (MAML), have shown promise in few-shot learning scenarios [11]. These methods could potentially be adapted to improve sea ice classification with limited labeled data. Concurrently, Bayesian deep learning methods have gained attention for their ability to quantify uncertainty in predictions, a crucial factor in reliable marine disaster risk assessment [14].

Our proposed BMPL method integrates elements from these various approaches, specifically tailored to address the challenges of sea ice classification in the context of marine disaster management. It aims to leverage the strengths of semi-supervised learning, meta-learning, and Bayesian inference to improve classification accuracy and reliability with limited labeled data.

3 Methodology: Bayesian Meta Pseudo Labeling (BMPL)

In this section, we introduce our Bayesian Meta Pseudo Labeling (BMPL) method for semi-supervised sea ice classification. BMPL integrates Bayesian inference, meta-learning, and pseudo-labeling to address the challenges of limited labeled data in sea ice remote sensing.

3.1 Problem Formulation

Let $(x_{val}, y_{val}) \in \mathcal{D}_{val}$ represent labeled data, $x \in \mathcal{D}_u$ represent unlabeled data, and $\hat{y}$ be the pseudo-label predicted by the teacher model for unlabeled data. We have $x, x_{val} \in \mathcal{X} \subset \mathbb{R}^d$, and $\hat{y}, y_{val} \in \mathcal{Y} \subset \mathbb{R}^C$, where d is the data dimension and C is the number of classes.

The complete dataset is represented as

$$\mathcal{D} = \mathcal{D}_{val} \cup \mathcal{D}_u = \{(x_{val_1}, y_{val_1}), ..., (x_{val_n}, y_{val_n}), x_1, x_2, ..., x_m\},$$

where $m \gg n$, indicating that unlabeled data far outnumber labeled data.

Let W_T and W_S denote the neural network weights of the teacher and student models, respectively. The neural networks are represented as f_{W_T} and f_{W_S}. As the posterior distribution

$$p(y|x, W_T) = \int p(y|f_{W_T}(x))p(W_T|x, y)dW_T \tag{1}$$

is intractable, we use a variational network q_ψ to approximate the original posterior distribution for sampling.

Let q_ψ and p_θ represent the distributions of the teacher and student models, with model parameters ψ and θ, respectively. We have $W_T \sim q_\psi(W_T)$. Let p_ψ be the prior distribution of the teacher model parameters, and $c \in \mathbb{R}$ represent the teacher model's confidence in the pseudo-labels.

We define the cross-entropy as:

$$H(y, \hat{y}) = \sum_i y_i \ln \frac{1}{\widehat{y_i}} = -\sum_i y_i \ln \widehat{y_i} \tag{2}$$

The teacher model is a Bayesian network. Following [15], we model the network parameters as Gaussian, i.e., $\psi \sim \mathcal{N}(\mu, \sigma^2)$. The loss function generated by the Bayesian neural network can be formalized as:

$$\mathcal{F}(\mathcal{D}, \psi) = \mathrm{KL}[q_\psi(W_T)||p_\psi(W_T)] - \mathbb{E}_{q_\psi}[\ln p_\psi(\mathcal{D}|W_T)] \tag{3}$$

For the Bayesian neural network model under semi-supervised settings, substituting the data $\mathcal{D}$, we get:

$$\begin{aligned}\mathcal{F}(\mathcal{D}, \psi) &\approx \sum_{i=1}^{n} \mathrm{KL}[q_{\psi_i}(W_T)||p_\psi(W_T)] - \ln q_{\psi_i}(y_{val}|x_{val}, W_T) \\ &+ \sum_{j=1}^{m} \mathrm{KL}[q_{\psi_j}(W_T)||p_\psi(W_T)] - \ln q_{\psi_j}(\hat{y}|x, W_T)\end{aligned} \tag{4}$$

The first line of this equation deals with labeled data, while the second line deals with unlabeled data, reflected in the inner loop and additional steps, respectively.

3.2 Meta-Learning Framework

We employ a meta-learning approach to enhance the model's ability to generalize across different sea ice conditions. The meta-learning process involves two nested loops:

Inner loop: Update student model parameters θ using the current teacher model;

Outer loop: Update teacher model parameters ψ based on the performance of the updated student model.

In the inner loop, the main objective is for the student model to learn from pseudo-labels. Assuming we are at iteration t, the constraint on the teacher model comes from the distance between the sampled parameters $q_\psi(W_T)$ and the prior distribution $p_\psi(W_T)$:

$$\mathcal{L}_T^{\text{inner}} = \mathrm{KL}\left[q_{\psi^{(t)}}(W_T^{(t)})||p_\psi(W_T^{(t)})\right] \tag{5}$$

The confidence function is defined as

$$c^{(t)} := B/\mathrm{KL}\left[q_{\psi^{(t)}}(W_T^{(t)})||p_\psi(W_T^{(t)})\right]$$

where B represents the batch size.

The constraint for the student model is:

$$\mathcal{L}_S^{\text{inner}} = \mathbb{E}_x\left[c^{(t)} \cdot H\left(\delta(q_{\psi^{(t)}}(x)), p_{\theta^{(t)}}(x)\right)\right] \tag{6}$$

Here, δ represents the vector sharpening function.

In the outer loop, the main objective is to adjust the pseudo-label related parameters ψ of the teacher model:

$$\mathcal{L}_S^{\text{outer}} = \mathbb{E}_{x_{val}}\left[H\left(y_{val}, p_{\theta^{(t+1)}}(x_{val})\right)\right] \tag{7}$$

where $\theta^{(t+1)}$ is a function of ψ.

BMPL adds an additional step to enhance algorithm performance:

$$\mathcal{L}_T^{\text{extra}} = \mathbb{E}_{x_{val}}\left[H\left(y_{val}, q_{\psi^{(t+1)}}(x_{val})\right)\right] \tag{8}$$

The total loss functions for the inner loop, outer loop, and additional step are:

$$\mathcal{L}^{\text{inner}} = \mathcal{L}_T^{\text{inner}} + \mathcal{L}_S^{\text{inner}} \tag{9}$$

$$\mathcal{L}^{\text{outer}} = \mathcal{L}_S^{\text{outer}} \tag{10}$$

$$\mathcal{L}^{\text{extra}} = \mathcal{L}_T^{\text{extra}} + \mathcal{L}_S^{\text{extra}} \tag{11}$$

3.3 BMPL Algorithm

The BMPL algorithm proceeds as follows:

Algorithm 1 BMPL Training procedure

```
Input dataset D, teacher model sampling number K, iteration number τ.
Initialize student model parameters θ^(0) and teacher model parameters ψ^(0).
for t = 1,3,5,...,2τ−1 do
    Sample mini-batches of unlabeled and validation data.
    Generate pseudo-labels using the current teacher model.
    Update student model parameters using L_inner.
    Update teacher model parameters using L_outer.
    Perform extra update steps using L_extra.
end for
Output final model parameters ψ and θ.
```

In the inner loop, the algorithm first samples from the teacher model's distribution $q_{\psi^{(t)}}$ to generate the current parameters of the teacher model. Following [15], we set $\psi = \{\mu, \rho\}$, where μ represents the mean of the network weights, and ρ is the parameter for the standard deviation $\sigma = \ln(1 + \exp(\rho))$, ensuring $\sigma \geq 0$.

Using the reparameterization trick, we get:

$$W_T = \mu + \ln(1 + \exp(\rho)) \cdot \in \tag{12}$$

where $\in \sim \mathcal{N}(0, I)$.Let K be the number of samples. We take the average of K samples as the parameters of the teacher model:

$$W_T = \frac{1}{K}\sum_{k=1}^{K} W_T^k \tag{13}$$

Assuming we are at iteration t, the gradient of the loss function with respect to the teacher model parameters is $\nabla_\psi \mathcal{L}^{\text{inner}}$. Since $\nabla_\psi \mathcal{L}^{\text{inner}} = \{\nabla_\mu \mathcal{L}^{\text{inner}}, \nabla_\rho \mathcal{L}^{\text{inner}}\}$, we get:

$$\nabla_\mu \mathcal{L}^{\text{inner}} = \frac{\partial \mathcal{L}^{\text{inner}}}{\partial W_T} + \frac{\partial \mathcal{L}^{\text{inner}}}{\partial \mu} \tag{14}$$

$$\nabla_\rho \mathcal{L}^{\text{inner}} = \frac{\partial \mathcal{L}^{\text{inner}}}{\partial W_T} \frac{\epsilon}{1+\exp(-\rho)} + \frac{\partial \mathcal{L}^{\text{inner}}}{\partial \rho} \tag{15}$$

The training process for the student model's parameters is:

$$\begin{aligned} \theta^{(t+1)} &= \theta^{(t)} - \eta_S \nabla_\theta \mathcal{L}^{\text{inner}} \\ &= \theta^{(t)} \eta_S \nabla_\theta \mathbb{E}_x \left[c^{(t)} \cdot H\left(\delta(q_{\psi^{(t)}}(x)), p_{\theta^{(t)}}(x)\right)\right] \end{aligned} \tag{16}$$

In the outer loop, the learning process converts the student model's performance on the validation set into a second-order gradient for the teacher model parameters, adjusting the parameters related to pseudo-labels and confidence:

$$\begin{aligned} &\nabla_\psi \mathcal{L}^{\text{outer}}\left(\theta^{(t+1)}\right) \\ &= \nabla_\psi \mathcal{L}^{\text{outer}}\left(\theta^{(t)} - \eta_S \nabla_\theta \mathcal{L}^{\text{inner}}(\theta^{(t)}, \psi^{(t)})\right) \\ &= \nabla_\psi \mathcal{L}^{\text{outer}}(\theta^{(t)}) - \eta_S \nabla_\psi \nabla_\theta \mathcal{L}^{\text{inner}}\left(\theta^{(t)}, \psi^{(t)}\right) \end{aligned} \tag{17}$$

where η_S represents the learning rate or step size of the student model.

The teacher model stores gradient values in the inner loop without updating, and updates all at once in the outer loop:

$$\psi^{(t+1)} = \psi^{(t)} - \eta_T (\nabla_\psi \mathcal{L}^{\text{inner}} + \nabla_\psi \mathcal{L}^{\text{outer}}) \tag{18}$$

where η_T represents the learning rate of the teacher model.

The additional learning process has two main steps:

(1) Use the updated teacher model to recalculate pseudo-labels, and update the student model parameters under the guidance of new pseudo-labels;
(2) Update the teacher model parameters using labeled data.

For the student model:

$$\begin{aligned} \theta^{(t+2)} &= \theta^{(t)} - \eta_S \nabla_\theta \mathcal{L}^{\text{extra}} \\ &= \theta^{(t)} \eta_S \nabla_\theta \mathbb{E}_x \left[c^{(t+1)} \cdot H\left(\delta\left(q_{\psi^{(t+1)}}(x)\right), p_{\theta^{(t)}}(x)\right)\right] \end{aligned} \tag{19}$$

Here, $\psi^{(t+1)}$ and $c^{(t+1)}$ are treated as constants and do not compute gradients.

For the teacher model:

$$\begin{aligned} \psi^{(t+2)} &= \psi^{(t+1)} - \eta_T \nabla_\psi \mathcal{L}^{\text{extra}} \\ &= \psi^{(t+1)} - \eta_T \nabla_\psi \mathbb{E}_{x_{val}} \left[H\left(y_{val}, q_{\psi^{(t+1)}}(x_{val})\right)\right] \end{aligned} \tag{20}$$

where $\psi^{(t+1)}$ is no longer a constant and requires gradient computation.

3.4 Theoretical Analysis

We provide theoretical guarantees for the convergence and generalization of BMPL. Our analysis builds on concepts from Lipschitz continuity and smoothness, and extends to the PAC (Probably Approximately Correct) generalization bounds.

We first define Lipschitz continuity, Lipschitz smoothness, and gradient boundedness:

Definition 1 (Lipschitz Continuity): A function $f(x) : \mathbb{R}^d \rightarrow \mathbb{R}$ is Lipschitz continuous if there exists a constant L such that:

$$|f(x) - f(y)| \leq L|x - y|, \forall x, y \in \mathbb{R}^d \tag{21}$$

Definition 2 (Lipschitz Smoothness): A function $f(x) : \mathbb{R}^d \rightarrow \mathbb{R}$ is Lipschitz smooth if there exists a constant L such that:

$$|\nabla f(x) - \nabla f(y)| \leq L|x - y|, \forall x, y \in \mathbb{R}^d \tag{22}$$

Definition 3 (ρ-Gradient Boundedness): A function $f(x)$ is ρ-gradient bounded if for $\forall x \in \mathbb{R}^d$, we have:

$$|\nabla f(x)| \leq \rho \tag{23}$$

Based on these definitions, we can establish the convergence of BMPL:

Theorem 1 (BMPL Convergence): Let B be the batch size and T be the number of iterations. Assume the student model satisfies Lipschitz smoothness with constant L on the labeled data $\mathcal{D}_{val}$, and both the teacher and student models satisfy ρ-gradient boundedness. If the learning rates satisfy $\eta \leq \frac{2B}{L\rho^2}$, then:

$$\mathcal{L}^{\text{outer}}\left(\psi^{(t+2)}, \theta^{(t+2)}\right) \leq \mathcal{L}^{\text{outer}}\left(\psi^{(t)}, \theta^{(t)}\right) \tag{24}$$

Moreover, let $u > 0$, $v > 0$ be constants satisfying $\frac{u}{T} \leq 1$ and $\frac{v}{\sqrt{T}} \leq L$. If the learning rates satisfy $\eta_S = \min\{1, \frac{u}{T}\}$, $\eta_T = \min\left\{\frac{1}{L}, \frac{v}{\sqrt{T}}\right\}$, then:

$$\min_{0 \leq t \leq T} \mathbb{E}\left[\left|\nabla \mathcal{L}^{\text{outer}}\left(\psi^{(t)}, \theta^{(t)}\right)\right|_2^2\right] \leq O\left(\frac{v}{\sqrt{T}}\right) \tag{25}$$

Let $\epsilon = \sqrt{\frac{\sqrt{T}}{v}}$ be the error control parameter. The convergence rate of BMPL is $O\left(\frac{1}{\epsilon^2}\right)$.

We now present two generalization bounds for BMPL, which provide theoretical guarantees on its performance on unseen data.

Theorem 2 (Generalization Bound for the Teacher Model): Let q_ψ be the posterior distribution of the teacher model parameters ψ, and p_ψ be its prior distribution. Let $R(q_\psi)$ and $\hat{R}(q_\psi)$ be the true and empirical risk of the posterior distribution, respectively. For

$\delta \in (0, 1]$, with probability at least $1 - \delta$, the following generalization bound holds:

$$R(q_\psi) \le \frac{1}{T}\sum_{i=1}^{T}\hat{R}(q_\psi, \mathcal{D}_i) + \left(\sqrt{\frac{1}{2(B-1)}} + \sqrt{\frac{1}{2(T-1)}}\right)\sqrt{\mathrm{KL}[q_\psi||p_\psi] + \ln\frac{T(B+1)}{\delta}} \tag{26}$$

This bound shows that the generalization risk is negatively correlated with the number of iterations and batch size, and positively correlated with the KL divergence between the parameter posterior and prior distributions.

Theorem 3 (Generalization Bound for the Student Model): Let θ be the student model parameters, and c be the confidence. Let $R(p_\theta(c))$ and $\hat{R}(p_\theta(c))$ be the true and empirical risk of the student model, respectively. If $p_\theta(c)$ is λ-Lipschitz continuous with respect to c, then for $\delta \in (0, 1]$, with probability at least $1 - \delta$, the following generalization bound holds:

$$R(p_\theta(c)) \le \hat{R}(p_\theta(c)) + \frac{3\lambda + \sqrt{4d\ln(n) + 8\ln(2/\delta)}}{\sqrt{n}} \tag{27}$$

This bound shows that the generalization risk is negatively correlated with the number of labeled samples n, and positively correlated with the data dimension d, with a scale of $O(\sqrt{d\ln(n)/n})$.

These theoretical results demonstrate that BMPL has strong convergence properties and favorable generalization bounds, supporting its effectiveness in semi-supervised sea ice classification tasks with limited labeled data.

4 Experiments and Analysis

4.1 Dataset and Experimental Setup

We evaluate our BMPL method on a comprehensive sea ice dataset compiled from satellite sources including Synthetic Aperture Radar (SAR) imagery from Gaofen-3. The dataset covers various sea ice types including open water, new ice, young ice, and first-year ice, with a particular focus on complex mixed-ice zones and marginal ice areas. The raw images are collected in full polarization strip mode with the RGB channels representing the polarization states HH (Horizontal-Horizontal), HV (Horizontal-Vertical), and VV (Vertical-Vertical), respectively. We visually annotated the ground truth for these images. During the process of sliding a window across the annotated map, if the proportion of pixels belonging to the same category exceeds a certain threshold (in our experiments, we used 0.9), then the corresponding location in the original image is stored as a sample, with the label being the category from the annotated map.

For all experiments, we employ a modified ResNet18 architecture as our base model, adapted to handle the multi-modal nature of sea ice data. The network could accept input from both SAR or optical satellite imagery. The final fully connected layer is adjusted to output 4 classes corresponding to our sea ice types. We initialize the network

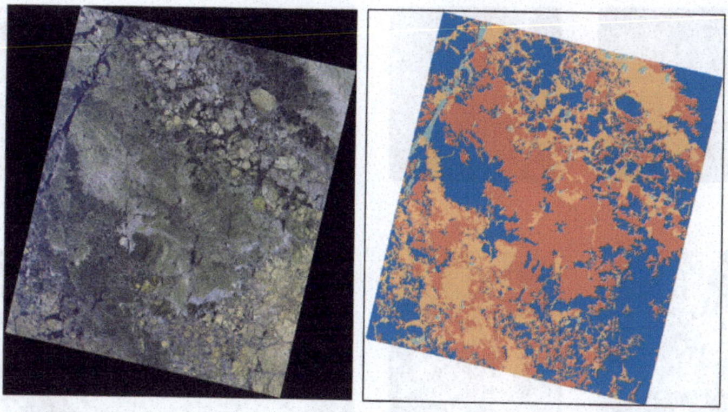

Fig. 1. Examples of original SAR imagery and segmentation labeling

weights using He initialization [16], and for the Bayesian components in BMPL, we use a Gaussian prior with zero mean and unit variance. The training configuration uses Stochastic Gradient Descent (SGD) with a momentum of 0.9, weight decay of 5e−4, and an initial learning rate of 0.01. We employ a cosine annealing learning rate schedule with warm restarts.

The dataset consists of 50,000 image patches, each 64 × 64 pixels, with 10,000 patches reserved for testing. Figure 1 visualizes some example classifications, with subplots (a), (b), (c), and (d) representing open water, new ice, young ice, and first-year ice, respectively as [20]. We create semi-supervised learning scenarios by randomly selecting a small portion of the training data as labeled and treat the rest as unlabeled. The batch size is set to 32, with 4 labeled and 28 unlabeled samples for semi-supervised methods. We train for 200 epochs, applying data augmentation techniques including random horizontal and vertical flips, random rotation (−10 to 10 degrees), and random cropping. For BMPL-specific hyperparameters, we use 10 Monte Carlo samples, a confidence threshold of 0.95 for pseudo-labeling, and 5 inner loop steps for meta-learning. All models are implemented in PyTorch and trained on NVIDIA Tesla V100 GPUs, with a fixed random seed (42) for reproducibility. We report the average performance over 5 runs with different data splits. To ensure fair comparison, all baseline methods use the same network architecture and training setup, with method-specific modifications applied as described in their respective papers. We evaluate the performance using Average Accuracy.

4.2 Baselines Methods

We compare BMPL with the following baselines:

Supervised Learning: This method uses only the labeled data for training. It serves as a lower bound for our experiments, demonstrating the performance achievable without leveraging unlabeled data. In the context of sea ice classification, this approach is limited by the scarcity of labeled data, especially for rare ice types or complex ice conditions.

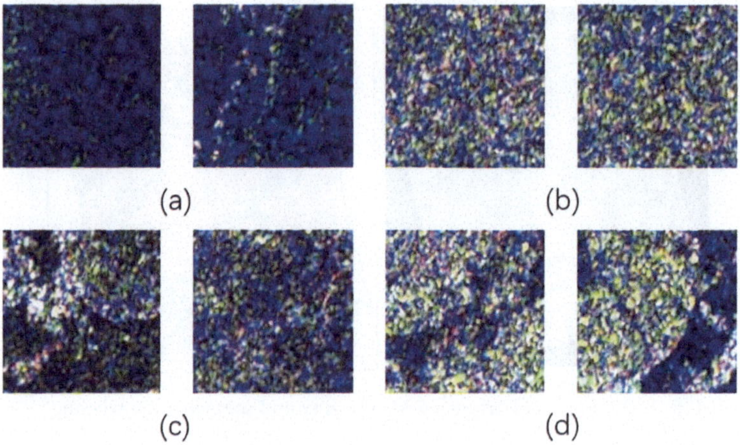

Fig. 2. Some examples for sea ice classification

Pseudo-Labeling [13]: This is a simple yet effective semi-supervised learning technique. It first trains a model on labeled data, then uses this model to generate pseudo-labels for unlabeled data. The model is then retrained on both labeled and pseudo-labeled data. While effective, it can suffer from confirmation bias, where the model reinforces its own mistakes. In sea ice classification, this could lead to propagation of errors, especially in ambiguous areas like marginal ice zones.

Mean Teacher [17]: This method uses temporal ensemble, where the model weights are averaged over training steps. It maintains a teacher model (the temporal ensemble) and a student model. The consistency between teacher and student predictions on unlabeled data provides an additional training signal. This approach can be particularly useful for sea ice classification as it can help smooth out inconsistencies in predictions over similar ice formations.

MixMatch [18]: MixMatch unifies several semi-supervised learning techniques, including data augmentation, mixup, and entropy minimization. It generates pseudo-labels for unlabeled examples using data augmentation and sharpening, then uses mixup to combine labeled and unlabeled data. This holistic approach could be beneficial for sea ice classification by creating a smoother decision boundary between different ice types.

FixMatch [19]: FixMatch combines consistency regularization and pseudo-labeling. It uses weak augmentations to generate pseudo-labels and strong augmentations to enforce consistency. This method has shown state-of-the-art performance on various image classification tasks. For sea ice classification, the strong augmentations could help the model learn invariances to different imaging conditions (e.g., different sun angles, partial cloud cover).

4.3 Performance Comparison

To evaluate BMPL's effectiveness in challenging scenarios, we specifically test its performance in mixed-ice zones and marginal ice areas. BMPL shows superior performance in these complex areas, accurately distinguishing between different ice types even in ambiguous situations. This is particularly evident in the marginal ice zones, where the transition between ice types is gradual.

Table 1 shows the performance comparison of BMPL with baseline methods under different labeled data ratios.

Table 1. Performance comparison of different methods on sea ice classification task (5% labeled data)

Method	Open Water	New Ice	Young Ice	First-Year Ice	Average Accuracy
Supervised	82.3%	70.1%	64.2%	72.5%	72.3%
Pseudo-Labeling	87.6%	75.6%	69.8%	77.9%	77.7%
Mean Teacher	89.5%	78.2%	72.9%	80.3%	80.2%
MixMatch	91.0%	80.9%	76.1%	82.5%	82.6%
FixMatch	92.3%	83.4%	78.8%	84.7%	84.8%
BMPL (Ours)	**93.1%**	**85.2%**	**80.9%**	**86.3%**	**86.3%**

In Table 1, we observe a clear progression in performance across different methods. The supervised learning baseline, using only 5% labeled data, achieves an average accuracy of 72.3%. This relatively low performance underscores the challenge of sea ice classification with limited labeled data. The simple Pseudo-Labeling method shows a significant improvement, reaching 77.7% average accuracy, which highlights the value of leveraging unlabeled data in this domain.

More advanced semi-supervised methods like Mean Teacher, MixMatch, and FixMatch further improve the results, with FixMatch achieving 84.8% average accuracy. However, our proposed BMPL method outperforms all baselines, reaching 86.3% average accuracy. This represents a substantial improvement of 14 percentage points over the supervised baseline and one percentage point over the next best method, FixMatch. From the performance across different ice types, we see that BMPL consistently outperforms other methods. The improvement is particularly noticeable for more challenging ice types such as Young Ice, where BMPL achieves 80.9% accuracy compared to 78.8% for FixMatch and only 64.2% for the supervised baseline.

To assess BMPL's robustness to potential distribution shifts and its performance with varying amounts of labeled data, we conducted experiments with labeled data percentages ranging from 1% to 30%. Figure 2 illustrates the error rates of various semi-supervised learning methods as the percentage of labeled data increases. BMPL consistently outperforms other methods across all labeling ratios, particularly in low-label regimes. At 1% labeled data, BMPL achieves an error rate of 22%, significantly lower than the supervised baseline (35%) and other semi-supervised methods. This advantage is maintained as the labeled data increases, with BMPL showing the lowest error rate of 7% at 30% labeled data. Notably, BMPL's performance improvement is more pronounced in scenarios with limited labeled data. This is evident from the steeper initial descent of its learning curve compared to other methods. As more labeled data becomes available, the performance gap narrows, but BMPL maintains its lead (Fig. 3).

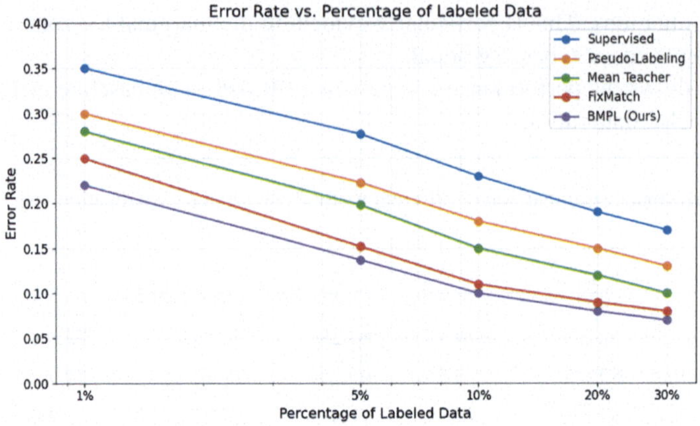

Fig. 3. Error Rates of Semi-Supervised Learning Methods for Sea Ice Classification at Different Labeling Ratios

4.4 Ablation Study

We conduct an ablation study to analyze the contribution of different components in BMPL. Table 2 presents the results.

Table 2. Ablation study of BMPL components (5% labeled data)

Method	BMPL w/o Bayesian	BMPL w/o Meta-learning	BMPL w/o Confidence	Pseudo-Labeling	BMPL (Full)
Average Accuracy	85.2%	84.7%	86.1%	78.6%	86.3%

Table 2 provides insights into the contribution of each component of BMPL through an ablation study. The full BMPL model achieves the highest accuracy at 86.3%. Removing the Bayesian component results in a drop to 85.2%, indicating its importance in handling uncertainty in sea ice classification. The meta-learning component also proves crucial, as evidenced by the performance drop to 84.7% when it is removed. Interestingly, removing the confidence weighting mechanism has a less dramatic but still significant impact, with accuracy decreasing to 86.1%. The results demonstrate that each component of BMPL contributes meaningfully to its performance, with the Bayesian and meta-learning aspects providing the most substantial improvements. The synergy between these components allows BMPL to effectively leverage unlabeled data and adapt to the specific challenges of sea ice classification, resulting in a robust and high-performing semi-supervised learning method for this critical application.

4.5 Shipping Route Safety Assessment

To evaluate the practical benefits of BMPL in marine disaster relief, we designed an experiment focused on Arctic shipping route safety assessment. The study utilized satellite imagery of a frequently used Arctic shipping route over a three-month summer period, collecting a total of 90 daily images. To simulate real-world data scarcity, only 10 images (approximately 11%) were randomly selected for expert annotation, categorizing the route conditions as "safe," "caution required," or "dangerous." The remaining 80 images were left unlabeled.

We trained the BMPL model alongside two benchmark methods, supervised learning and FixMatch, aiming to classify the shipping route status into these three categories. The models were then tested on a separate set of 20 images. We evaluated their performance based on overall accuracy, recall rate (with particular attention to the "dangerous" category), and processing time per image.

To simulate a real-world emergency scenario, we selected a single day's actual route image and had each model perform a rapid classification. Experienced ship captains were then asked to assess the practical utility of each model's output. This approach allowed us to not only compare the quantitative performance of BMPL against other methods but also to gauge its real-world applicability in time-sensitive situations. This streamlined experimental design directly addresses the critical aspect of safety assessment in Arctic shipping routes, a key factor in marine rescue operations (Table 3).

Table 3. Model Performance Comparison

Model	Accuracy	Recall (Dangerous)	Avg. Processing Time (s)
Supervised	72.5%	68.0%	**2.5**
FixMatch	81.0%	79.5%	3.8
BMPL	**87.5%**	**86.0%**	3.2

Ship captains rated BMPL's output as the most useful, citing its high confidence in the "dangerous" classification and additional uncertainty metrics as particularly valuable for decision-making. They noted that while BMPL took slightly longer to process, the additional information it provided justified the small time increase. The captains emphasized that in real-world scenarios, the higher accuracy and recall rates of BMPL, especially for the "dangerous" category, could be crucial for preventing disasters.

5 Conclusion

This paper introduced Bayesian Meta Pseudo Labeling (BMPL), a novel semi-supervised learning approach for sea ice classification in marine disaster emergency management. By integrating Bayesian neural networks, meta-learning, and pseudo-labeling techniques, BMPL effectively leverages limited labeled data alongside abundant unlabeled satellite imagery.

Our experimental results demonstrate BMPL's consistent superior performance over existing semi-supervised learning methods across various labeling ratios, with particularly significant improvements in low-label scenarios. BMPL achieves higher classification accuracy, especially for challenging ice types such as new ice and young ice, and shows better generalization to unseen data in cross-season experiments. The ablation study confirms the importance of each BMPL component, with the Bayesian framework and meta-learning strategy contributing significantly to its performance. The implications of BMPL for marine disaster relief are significant. Accurate sea ice classification is crucial for safe navigation in polar regions, enabling emergency response vessels to plot optimal routes through ice-filled waters. This can dramatically reduce response times during critical situations such as oil spills, ship accidents, or search and rescue operations. Furthermore, BMPL's ability to accurately identify different ice types, including thin and unstable ice, can help prevent disasters by alerting ships to dangerous areas. To demonstrate BMPL's practical value in marine disaster relief, we conducted an experiment on Arctic shipping route safety assessment. The results showed that BMPL significantly outperformed traditional methods in accuracy and recall, particularly in identifying dangerous conditions, with experienced ship captains confirming its enhanced utility for real-world emergency decision-making.

BMPL's robustness to seasonal variations is particularly valuable for year-round disaster preparedness. It allows for consistent monitoring of ice conditions, facilitating early warning systems for potential hazards such as ice shelf collapses or rapid sea ice breakup that could endanger coastal communities or offshore operations. Moreover, the method's efficiency in handling limited labeled data makes it adaptable to rapidly changing Arctic environments, where new ice formation patterns may emerge due to climate change. This adaptability ensures that disaster response strategies can be updated quickly based on the most current ice conditions.

References

1. Guo, L.-Z. et al.: Safe deep semi-supervised learning for unseen-class unlabeled data. In: International Conference on Machine Learning. PMLR (2020)
2. Shokr, M., Mohammed D.: Polarimetric SAR applications of sea ice: a review. IEEE J. Sel. Top. Appl. Earth Observ. Remote Sens. (2023)
3. Zhao, L., et al.: Sea ice detection from RADARSAT-2 quad-polarization SAR imagery based on co- and cross-polarization ratio. Remote Sens. **16**(3), 515 (2024)
4. Han, Y., et al.: Sea ice image classification based on heterogeneous data fusion and deep learning. Remote Sens. **13**(4), 592 (2021)
5. Huang, R., et al.: DF-UHRNet: a modified CNN-based deep learning method for automatic sea ice classification from Sentinel-1A/B SAR images. Remote Sens. **15**(9), 2448 (2023)
6. Park, J.-W., et al.: Classification of sea ice types in Sentinel-1 synthetic aperture radar images. Cryosphere 14(8), 2629–2645 (2020)
7. Boulze, H., Anton, K., Julien, B.: Classification of sea ice types in Sentinel-1 SAR data using convolutional neural networks. Remote Sens. **12**(13), 2165 (2020)
8. Montpetit, B., et al.: Assessing the parameterization of RADARSAT-2 dual-polarized ScanSAR scenes on the accuracy of a convolutional neural network for sea ice classification: case study over Coronation Gulf, Canada. Can. J. Remote. Sens. **49**(1), 2247091 (2023)

9. Malmgren-Hansen, D., et al.: A convolutional neural network architecture for Sentinel-1 and AMSR2 data fusion. IEEE Trans. Geosci. Remote Sens. **59**(3), 1890–1902 (2020)
10. Song, W., et al.: Automatic sea-ice classification of SAR images based on spatial and temporal features learning. IEEE Trans. Geosci. Remote Sens. **59**(12), 9887–9901 (2021)
11. Finn, C., Pieter A., Sergey L.: Model-agnostic meta-learning for fast adaptation of deep networks. In: International conference on machine learning. PMLR (2017)
12. Huang, W., et al.: Sea ice extraction via remote sensing imagery: algorithms, datasets, applications and challenges. Remote Sens. **16**(5) (2024)
13. Arazo, E., et al.: Pseudo-labeling and confirmation bias in deep semi-supervised learning. In: 2020 International Joint Conference on Neural Networks (IJCNN). IEEE (2020)
14. Fan, Y., et al.: Revisiting consistency regularization for semi-supervised learning. Int. J. Comput. Vis. **131**(3), 626–643 (2023)
15. Wang, H., Dit-Yan, Y.: A survey on Bayesian deep learning. ACM Comput. Surveys **53**(5), 1–37 (2020)
16. Marcot, B.G., Trent, D.P.: Advances in Bayesian network modelling: integration of modelling technologies. Environ. Model. Softw. **111**, 386–393 (2019)
17. Tarvainen, A., Harri V.: Mean teachers are better role models: weight-averaged consistency targets improve semi-supervised deep learning results. Adv. Neural Inf. Process. Syst. **30** (2017)
18. Berthelot, D., et al. Mixmatch: a holistic approach to semi-supervised learning. Adv. Neural Inf. Process. Syst. **32** (2019)
19. Sohn, K., et al.: Fixmatch: simplifying semi-supervised learning with consistency and confidence. Adv. Neural Inf. Process. Syst. **33**, 596–608 (2020)
20. Johansson, A.M., et al.: X-, C-, and L-band SAR signatures of newly formed sea ice in Arctic leads during winter and spring. Remote Sens. Environ. **204**, 162–180 (2018)

Stream Processing Techniques for Fast Processing of Sea Surface Wind Field

Qingwei Chu[1], Wenyi Zhang[1(✉)], Guangzuo Li[1], Ninghui Diao[2], and Hu Shuo[3]

[1] Aerospace Information Research Institute,Chinese Academy of Sciences, Beijing, China
wyzhang@alrcas.ac.cn
[2] National Satellite Ocean Application Service, Beijing 100081, China
[3] QiLu Aerospace Information Research Institute, Jinan 250132, China

Abstract. The shipboard processing system is unable to provide high timeliness Sea surface wind field information for polar voyage, if traditional file-driven and data transmission method is used. This paper proposes a streaming processing platform and architecture that can be used for shipboard processing system, which constructs sea surface wind field product generation as independent plugins. The satellite data are interacted and transferred between plug-ins in the form of data flow. This method can effectively improve the data transmission efficiency and reduce the data processing delay. With this method, rapid processing of Sea surface wind field products can be carried out while satellite data being received. After experimentation, the method is able to shorten the processing time for L2 products of sea surface wind fields for Arctic regional data, which is from the traditional 180 min to within 30 min.

Keywords: Shipboard processing system · Stream processing · Fast processing · Sea surface wind field

1 Introduction

During polar voyages, polar research vessels need to cross ice floes several times and alternate between large areas of ice-free water and ice floes. During the voyage, the ship will be affected by environmental factors, human factors and means of transportation. Among the environmental factors, sea ice floes will greatly hinder the ship's speed control and heading adjustment, which is the biggest threat to the polar research ship's navigation in the ice area.

For example, China's polar research vessel Xuelong has a shipborne X-band remote sensing satellite reception and marine information service system, which has a certain degree of marine remote sensing satellite data reception and processing capability. However, the shipbord data processing system still faces with the problem of information and data interaction by means of traditional documents after data reception, and there is the outstanding problem of insufficient timeliness, precision and temporal and spatial resolution of polar protection products, which makes it impossible to provide high timeliness data and information protection capability for polar navigation.

G. Luo (Ed.): GFSISD 2024, SPPHY 417, pp. 234–240, 2025.
https://doi.org/10.1007/978-981-96-2490-4_16

Based on the streaming processing framework, this paper builds a streaming data processing platform, constructs independent plug-ins for data pre-processing, geo-localization, radiation calibration, and product generation of surface wind field, etc. In order to form the capability of polar downlinking of ocean remote sensing satellite data, real-time transmission of raw data, and fully automatic streaming fast processing, the data processing workflow is designed, and distributed real-time processing is carried out under the unified scheduling of the platform.

2 Streaming Computing Technology

Streaming computing refers to real-time computation of arriving data streams in memory, so streaming computing is characterized by low latency, high throughput and continuous operation [1]. Common streaming computing frameworks are Twitter Storm [2], Spark Streaming [3], Yahoo S4 [4], etc. It has become a research hotspot in recent years and is widely used in the fields of Internet, transportation, finance, medical care, etc. It is also applied to the ground processing of satellite data, for example, Sun et al. [5] proposed a real-time processing method for space science satellite data based on a streaming computing framework, which improved the real-time performance of the space science satellite data processing system.

From an implementation point of view, the following figure shows the stream computing mode of operation of an asynchronous framework, which is a set of threads that take messages from their input queues for processing, and then output them to the downstream input queues for the downstream threads to continue to read and process them, in a way that brings about a performance improvement in the processing [6] (Fig. 1).

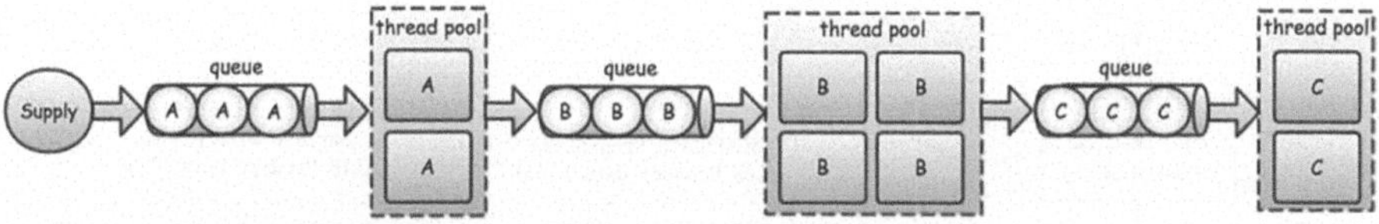

Fig. 1. Stream Computing Operational Model for Asynchronous Frameworks.

3 Ocean Remote Sensing Satellite Data Processing Flow

In this paper, we develop data preprocessing techniques for Microwave Scatterometer (SCAT) and Microwave Scanning Radiometer (SMR) for Oceanic 2B Satellite (HaiYang-2B, or HY-2B) data to generate level 0–2 products. The data preprocessing techniques of SCAT and SMR are used to generate 0–2 level products. It mainly involves the processes related to raw data unpacking, satellite microwave scatterometer data preprocessing, and fast processing of Arctic sea surface wind field. In order to improve the processing efficiency, the process was optimized using the streaming processing method.

3.1 Raw Data Parse

For the raw data of HY-2B satellite received by the ground station, AOS frames are extracted in real time, descrambled and RS decoded, and the data source packets of different payloads are extracted by virtual channel identification to generate L0A products for SCAT and SMR payloads. Perform data calibration and sub-orbital and sub-PASS processing for SCAT and SMR to generate L0B products.

3.2 SCAT Data Preprocess

The L0B product of the HY-2B satellite microwave scatterometer is processed to generate the L1B product of the microwave scatterometer through the processes of format parsing, physical quantity conversion, ephemeris interpolation, geo-localization and backward scattering coefficient calculation.

3.3 Fast Processing of Arctic Surface Wind Fields

For the L1B product of the HY-2B satellite microwave scatterometer, processes such as surface element matching, sea-land identification, sea-ice identification, and rainfall identification were performed to generate the L2A product of the scatterometer in HDF format and metadata in XML format; and then processes such as wind vector inversion and fuzzy deconvolution were performed to generate the L2B product of the scatterometer in HDF format and metadata, which were finally outputted (Figs. 2 and 3).

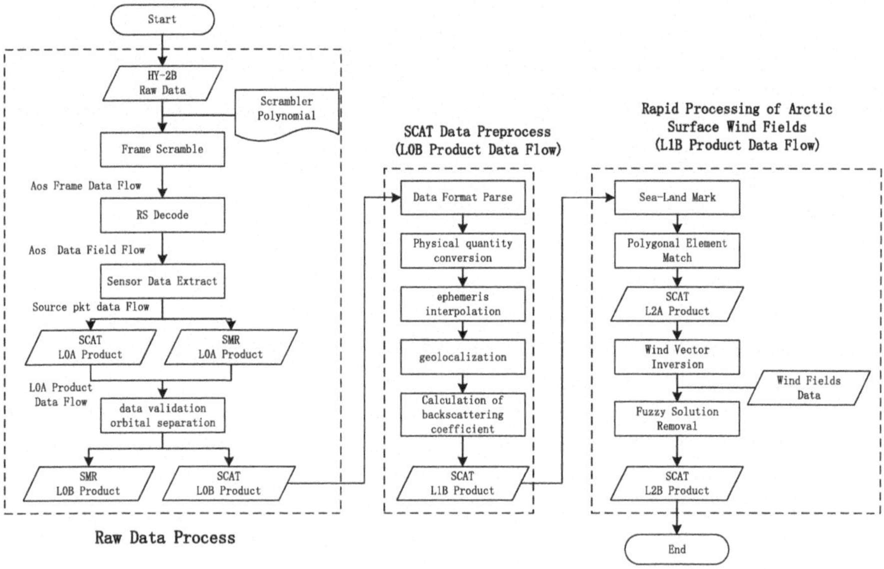

Fig. 2. Fast processing Flow of Arctic Surface Wind Fields.

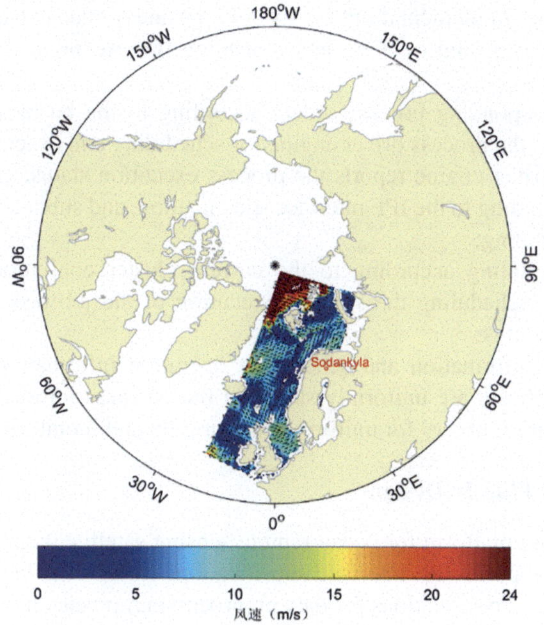

Fig. 3. Sea surface wind field map.

4 Streaming Scheduling Platform Design

This paper proposes a fast processing method for ocean remote sensing satellite data, based on streaming architecture for the data processing of ocean remote sensing satellites.

4.1 Platform Design

The construction of the rapid processing platform for ocean remote sensing satellite data is based on streaming processing architecture. Data pre-processing, product production and other software are plugged in. The platform provides middleware for data and information interaction and accomplishes the scheduling and management of plug-ins.

The technical process of the Fast processing platform is described as follows:

(1) According to the IPF (Instrument Processing Facility, the scheduling interface specification for data preprocessing, precision orbiting, data processing and product quality inspection algorithm components) interface specification, carry out the registration of the IPF components and saved to the business information base;
(2) Perform business process modeling according to business requirements and save it to the business information library;
(3) Real-time and timed business plans are formulated, and the plan information is saved to the business information base;
(4) Generating corresponding business plan instances and triggering corresponding business processes according to the received load data;

(5) Scheduling the implementation of specific business plans, triggering specific business processes, and tracking and managing the execution status of business plans;
(6) Generate corresponding process orders according to the business processes and submit them to the process driver engine for scheduling and execution;
(7) The process driver engine reports the process execution status, generates specific job orders according to the IPF interface specification, and submits the job orders to resource scheduling;
(8) Resource scheduling according to the current cluster configuration policy and resource load, scheduling the specific execution of the job tasks, and report the job execution status;
(9) Configuration information and scheduling operation information of the process scheduling software are uniformly stored in the business information library and business operation library for uniform configuration and monitoring management.

4.2 Professional Plug-In Design

The Fast processing platform for ocean remote sensing satellite data based on streaming architecture proposes IPF interface specifications (nstrument Processing Facility, scheduling interface specifications for data preprocessing, precision orbiting, data processing and product quality inspection algorithm components) for data recording, transmission, preprocessing, geo-positioning, radiometric calibration, and product generation of sea surface wind field plug-ins, which makes the platform It has become a common data processing platform for processing components.

The versatility of the processing components of the ocean remote sensing satellite data Fast processing platform based on streaming architecture includes the following aspects:

(1) Having a universal component integration platform, the initial stage of software construction takes into full consideration the versatility, cross-platform, and scalability of the software platform, and it is not a processing platform that is customized and developed for a certain or a certain type of component;
(2) Having a common integration framework, extracting and abstracting all kinds of common problems and forming a unified solution to common problems during the construction and design of various types of business processing processes, i.e., constructing a common integration framework to unify and solve the common problems of component integration and scheduling;
(3) Integration and invocation of components through the IPF component integration specification, and all kinds of processing components conforming to the IPF component integration specification can be integrated into the Fast processing platform.

5 Results and Analysis

In this paper, the data reception and fast processing of HY-2 satellites are tested through cooperative ground stations in the Arctic region. By checking the operational records and adding up the reception, preprocessing, processing and transmission times, the fast processing time is less than 30 min for the HY-2 satellite sea surface wind field, which is a great improvement compared with the 180-min processing time of the existing processing system.

During the experiment period, the HY2B satellite was arranged to receive 5 rails per day, with a total of 310 rails received as a result, and the processing was successful in generating scattero meter and radiometer level 2 products, with a raw data volume of more than 600 GB, and the full-processing time was improved to less than 30 min (Fig. 4).

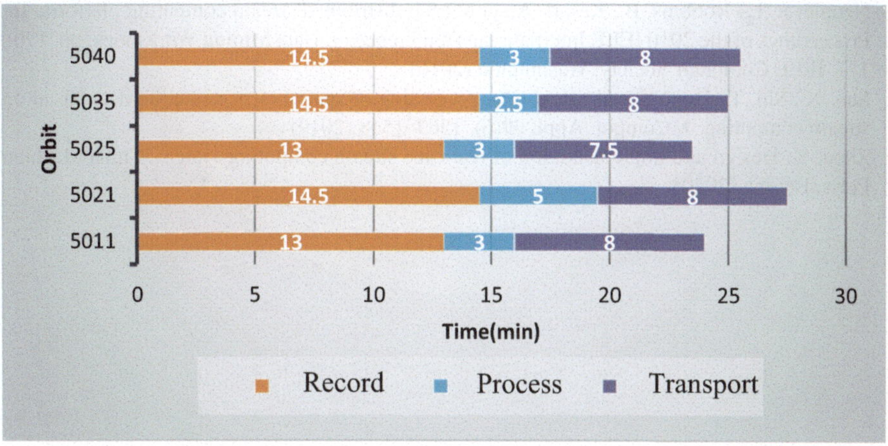

Fig. 4. HY-2B Satellite Full Process Test Timing.

6 Conclusion

The data processing platform based on the streaming processing framework is oriented towards the real-time Fast processing of Arctic regional data products, supports the incremental development of new processing plug-ins, and provides more flexible data processing workflow scheduling and more efficient management and control of data processing work status. It shortens the processing time for level 2 products of the surface wind field of Arctic data from 180 min to 30 min, forming a solution for satellite data polar downlinking, real-time transmission of raw data and fully automated streaming Fast processing technology.

By upgrading the processing timeliness of Arctic regional data sea surface wind field level 2 products, it can enhance China's Arctic environment operational monitoring and forecasting and service guarantee capability, improve China's deep participation in Arctic ocean governance level and international influence, and provide technical support and information services for the Silk Road on Ice.

Funding. This research was funded by the Science and Technology Innovation Foundation of Command Center of Integrated Natural Resources Survey (KC20240004, KC20230006).

References

1. Sun, D., Zhang, G., Zheng, W., et al.: Big data stream computing: technologies and instances. J. Softw. **25**(4), 839–862 (2014)
2. Simoncelli, D., Dusi, M., Gringoli, F., et al.: Scaling out the performance of service monitoring applications with BlockMon. In: Proceedings of the 2013 International Conference on Passive and Active Measurement, LNCS, vol. 7799, pp. 253–255. Springer, Berlin (2016). https://doi.org/10.1007/978-3-642-36516-4_26
3. Han, D., Chen, X., Lei, Y., et al.: Real-time data analysis system based on Spark Streaming and its application. J. Comput. Appl. **37**(5), 1263–1269 (2017)
4. Neumeyer, L., Robbins, B., Kesari, A., et al.: S4: distributed stream computing platform. In: Proceedings of the 2010 IEEE International Conference on Data Mining Workshops, pp. 170–177. IEEE Computer Society, Washington (2010)
5. Sun, X., Shi, T., Hu, Y., et al.: Real-time processing of space science satellite data based on stream computing. J. Comput. Appl. **39**(6), 1563–1568 (2019)
6. Zhou, S.: Design and implementation of real-time stream computing system. China Machine Press, Beijing (2020)

Fast-SISENet: A Fast Sea Ice SAR Image Segmentation Network for Sea Ice Disaster Response

Jiande Zhang[1], Wenyi Zhang[2(✉)], Xiao Zhou[1], Qingwei Chu[2], and Shuo Hu[1]

[1] QiLu Aerospace Information Research Institute, Jinan 250132, China
{zhangjd,zhouxiao,hushuo}@aircas.ac.cn

[2] Aerospace Information Research Institute, Chinese Academy of Sciences, Beijing 100094, China
{wyzhang,chuqw}@aircas.ac.cn

Abstract. The presence of Arctic sea ice increases navigation risks and introduces numerous instability factors, representing a major safety hazard for Arctic Route. This paper addresses the frequent sea ice hazards in Arctic Route by proposing a deep learning-based rapid SAR sea ice image segmentation algorithm, named Fast-SISENet (Fast-Sea Ice Segment Network), to achieve real-time acquisition and processing of Arctic ice conditions. This method improves upon the DeeplabV3+ backbone network by modifying the original ASPP module into a parallel dual-branch structure, allowing it to process features at different scales simultaneously. This modification reduces the number of data transmissions through the network and enhances computation speed. Additionally, a specialized adaptive feature fusion module and a lightweight channel attention mechanism are designed to integrate local and global features, providing more comprehensive feature information to the decoder. Compared to other mainstream segmentation methods, this approach achieves the fastest inference speed while maintaining high classification accuracy. The inference speed of this method can be integrated with SAR quick view systems to enable real-time acquisition and classification of SAR sea ice images, offering a reliable reference for real-time path planning and emergency response to sea ice hazards in Arctic Route.

Keywords: Sea Ice Segmentation · Synthetic Aperture Radar · Deep Learning · Arctic Route · Real-time Processing

1 Introduction

Global warming is accelerating Arctic sea ice transformation, presenting unprecedented challenges to navigation and safety in the region [1]. The risks posed by sea ice to maritime traffic necessitate enhanced planning and emergency protocols for Arctic shipping lanes. To tackle these issues, integrating Synthetic Aperture Radar (SAR) imagery with deep learning for sea ice classification is essential. SAR technology, unaffected by cloud cover or polar night darkness, enables reliable monitoring of sea ice. Combined with

G. Luo (Ed.): GFSISD 2024, SPPHY 417, pp. 241–250, 2025.
https://doi.org/10.1007/978-981-96-2490-4_17

deep learning, it automates the identification and classification of sea ice features like thickness and roughness, which are vital for navigation, safe shipping lane mapping, and emergency response strategy development.

In the field of sea ice classification, various strategies have been explored, including classifying ice by backscatter intensity. However, the similarity in scattering traits among different ice types can affect the method's reliability. To improve classification accuracy, researchers have incorporated advanced attributes like texture [2], scattering intensity [3], and polarimetric features [4] to form a comprehensive feature vector. However, the feature extraction process heavily depends on researchers' judgment and intuition, making it impractical to find an optimal solution by testing all feature combinations. This highlights the pressing need for innovation in sea ice classification techniques.

Convolutional Neural Networks (CNNs) have shown good performance in detection and classification tasks [5, 6], including the classification of sea ice. Initially, basic CNNs were used to differentiate sea ice through backscatter signal variance in Synthetic Aperture Radar (SAR) images, offering robust, automated feature extraction [7–9]. Advanced deep learning frameworks like ResNet [10], VGG-16 [11], LSTM [12], and MobileNet [13] have since been utilized to optimize classification models, outperforming traditional methods. Additionally, image segmentation techniques with classic deep learning methods such as Unet [14] and FFN (Feature Fusion Network) [15] have been explored for sea ice classification.

Current deep learning-based sea ice segmentation methods predominantly focus on optimizing network architectures to enhance classification accuracy. However, given the short acquisition time and large volumes of remote sensing imagery, real-time processing can only be achieved by improving the model's inference speed. In the field of semantic segmentation, innovative structures and strategies are often employed to reduce the model's parameters and computational complexity, thereby increasing inference speed. Examples include BiSeNet's parallel branch design [16], STDCNet's Short-Term Dense Concatenate (STDC) module and Detail Guidance module [17], and DDRNet's Dilated Parallel Paired Modules (DAPPM) [18].

This paper addresses the issue of rapid segmentation of Synthetic Aperture Radar (SAR) sea ice imagery by proposing the Fast-SISENet (Fast-Sea Ice Segmentation Network). The model is based on the Deeplabv3+ framework, with modifications made to the original network's Atrous Spatial Pyramid Pooling (ASPP) module. Atrous convolutions and PyramidPooling module are separated into two parallel pathways, reducing the number of times data is transmitted within the network and enhancing the model's computational speed. Additionally, a novel adaptive feature fusion module is introduced, which provides richer semantic information compared to the original simple channel-wise concatenation, thereby improving the model's final classification accuracy. The research findings of this paper will be applied to the rapid processing of sea ice SAR images in the Arctic Route, providing a reliable reference for the rapid response to sudden sea ice disasters in the Arctic Route.

2 Dataset

The fully polarimetric working mode is a significant feature of the Gaofen-3 (GF-3) satellite, allowing the satellite to acquire the reflective characteristics of ground objects in all possible polarimetric combination during a single imaging process. This paper selects eight scenes of GF-3 fully polarimetric strip mode SLC (Single Look Complex) complex image data, all captured within a time frame from 21:05 to 21:06 (UTC) on February 28, 2020, with a swath width of 30 km and a nominal resolution of 8 m. The imaging area is the Okhotsk Sea, as shown in Fig. 1a, where the red box represents the coverage area of the GF-3 fully polarimetric strip mode SAR sea ice images. The data preprocessing steps are generally similar to those described in the literature [13], including radiometric calibration, speckle noise removal, geocoding, and synthesis of pseudo-color images. Figure 1b–d correspond to the pseudo-color images for scenes numbered 4 to 6.

(a) (b) (c) (d)

Fig. 1. (a) The coverage area of nine GF-3 images used in this paper. The false color images corresponding to (b) number 6, (c) number 5, (d) number 4 (R = HH, G = HV, B = VV).

This paper combines the visual interpretation of sea ice characteristics as described in [4, 8, 13, 14], and classifies the sea ice into four categories: open water (OW), new ice (NI), young ice (YI), and first-year ice (FYI). In SAR images, OW and NI typically exhibit lower backscatter coefficients due to their smooth surfaces, appearing as darker or smoother areas. The scattering intensity of YI is the highest among these four types of sea ice, often presenting as the brightest areas in SAR images. FYI is more structurally complex, often containing a variety of scattering types, and its textural features are typically interwoven in SAR images. We extracted 512 × 512 patches from the aforementioned images using a sliding window and manually annotated them with the labelme software, obtaining a total of 600 sample images, with 500 used for training and 100 for testing. Sample examples are shown in Fig. 2.

3 Methodology

The overall structure of this method is depicted in Fig. 3. The feature extraction network of this method employs the relatively lightweight ResNet18, which undergoes three down-sampling operations during the feature extraction process, resulting in three feature maps of different scales: 256 × 256 × C (denoted as F1), 128 × 128 × C (denoted as

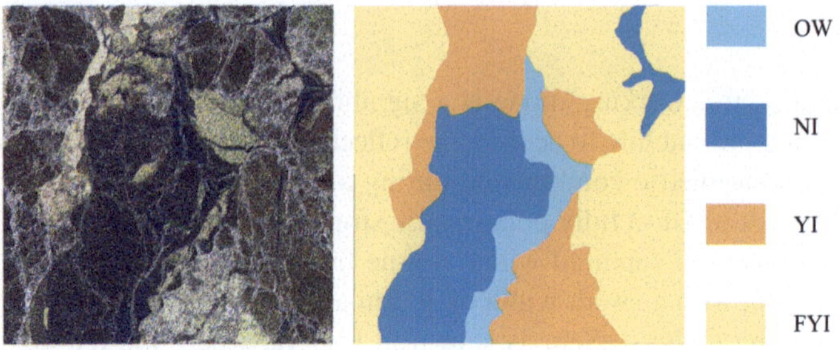

Fig. 2. A sea ice sample example and its' corresponding manual annotations used in this paper.

F2), and 64 × 64 × C (denoted as F3). The F3 feature map is subsequently input into the ASPP module to further extract richer semantic features. Thereafter, F1 and F3 are adjusted for the number of channels by a 1 × 1 convolutional module and are simultaneously input into the FFM module for feature fusion, followed by a lightweight channel attention mechanism module. Finally, a convolutional module and a four-time up-sampling process yield the ultimate output. The ASPP module, FFM module, and ECA (Efficient Channel Attention) module will be detailed in subsequent sections.

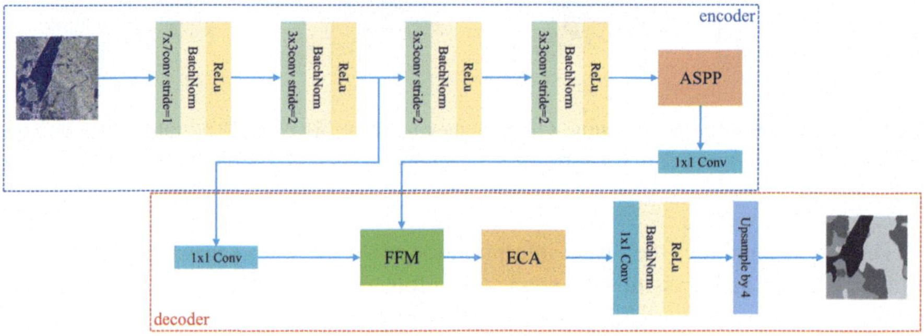

Fig. 3. Overall structure of Fast-SISENet.

3.1 ASPP Module

The modified ASPP module consists of two parts: the Atrous Convolution Module (ACM) and the Pyramid Pooling Module (PPM), as shown in Fig. 4. These two components together form the ASPP module found in the original DeeplabV3+ architecture. Atrous convolution expands the receptive field without losing resolution by employing various dilation rates, capturing more contextual semantic information. The pyramid pooling module integrates information from different sub-regions through pooling at multiple scales, which aids the network in capturing object features at various scales, thereby enhancing the recognition capability for objects of different sizes. Considering that a parallel structure can process features of different scales simultaneously, potentially increasing computational efficiency as it allows for concurrent operations and reduces the number of times data is transmitted within the network. Therefore, we designed the original ASPP module's atrous convolution and pyramid pooling modules as two parallel branches for processing, referred to as the ACM (Atrous Convolutional Module) branch and the PPM (Pyramid Pooling Module) branch. Each branch independently processes the input feature map, and upon completion, the outputs of these two branches are concatenated with the original input feature map. Ultimately, a 1×1 convolutional layer is applied to the concatenated feature map to compress it, reducing the number of channels and producing the final output of the ASPP module. This structure not only enhances the diversity of features but also improves the model's ability to integrate information at different scales.

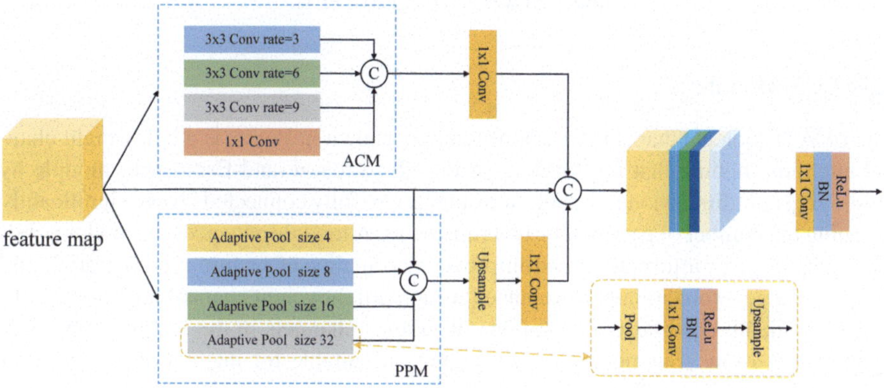

Fig. 4. Overall structure of modified ASPP module.

3.2 FFM

The Feature Fusion Module (FFM) innovates the feature fusion strategy in the DeeplabV3+ network by introducing an adaptive feature fusion mechanism, which is shown in Fig. 5. This module takes two feature maps of different sizes as input: one is a shallow feature map with higher resolution, and the other is a deep feature map with lower resolution. During the fusion process, the deep feature map is first up-sampled to match the resolution of the shallow feature map, and then the two are element-wise added together. The added data is then sent into two parallel paths: one path is for capturing global features, and the other is for extracting local features. The features extracted from these two paths are then merged and passed through a sigmoid activation function to calculate the fusion weights for each. Ultimately, the calculated weights are applied to the shallow and deep feature maps, respectively, and the outputs of the FFM module are obtained through a weighted summation, thereby achieving an effective fusion of the feature maps.

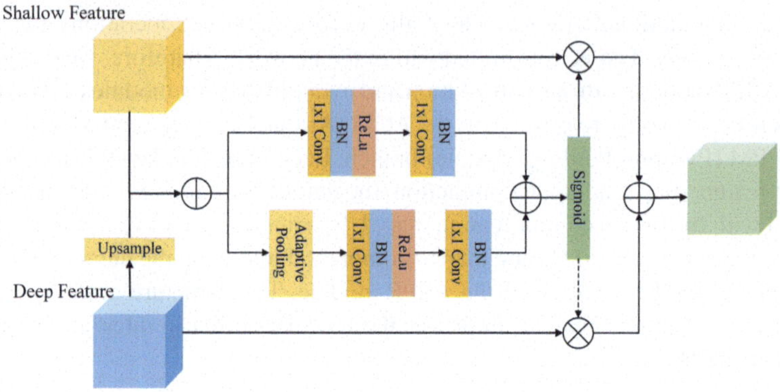

Fig. 5. Overall structure of FFM.

3.3 ECA Module

The ECA (Efficient Channel Attention) module, shown in Fig. 6, is a lightweight channel attention module that improves upon the SE (Squeeze-and-Excitation) module by employing one-dimensional convolutions in place of fully connected layers, significantly reducing the number of parameters. Overall, the ECA module enhances the model's ability to capture key information and improves the quality of feature representation and the model's generalization performance through a lightweight channel attention mechanism. It adaptively adjusts the weights of various channels in the feature map while maintaining a low computational cost, making it suitable for integration into various deep learning models to improve efficiency and performance.

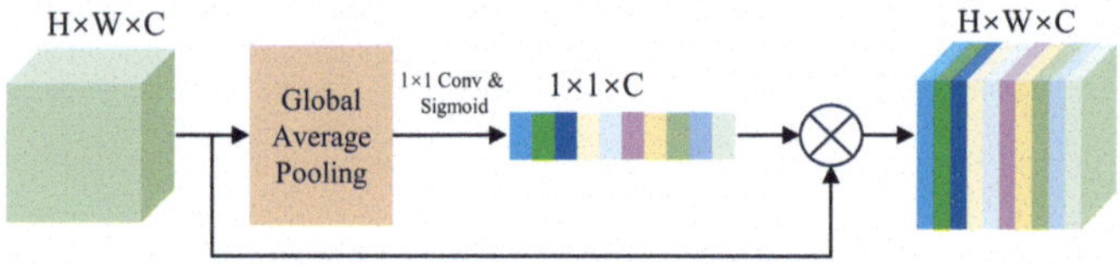

Fig. 6. Overall structure of ECA Module.

4 Experiments Result

4.1 Experiment Setup

The experiments were divided into two parts: one was a comparison of classification accuracy, and the other was a comparison of inference speed. The model training and accuracy testing were both conducted on an RTX 3090Ti graphics card. The following five metrics were selected for comparison with several common segmentation methods: PA (Pixel Accuracy), CPA (Class Pixel Accuracy), mPA (mean Pixel Accuracy), IoU (Intersection over Union), and mIoU (mean Intersection over Union). The inference speed testing was performed on an RTX 3060Ti graphics card, as this graphics card has a similar floating-point operation speed to the V100S graphics card used in the SAR quick-view system. The inference speed testing mainly compared the following three metrics: FLOPs (Floating-point Operations Per Second), Params (number of model parameters), and FPS (Frames Per Second).

4.2 Comparison with State-of-the-Art Methods on Classification Accuracy

The experimental results of the classification accuracy are shown in Table 1. PIDNet demonstrated the best performance in most metrics, particularly in mPA and mIoU, indicating its high accuracy in pixel-level prediction, especially when processing NI and YI. Our method also exhibited excellent performance in Cpa, IoU, and mPA, especially in overall and FYI, which may imply that our model is more effective in feature extraction and classification in these areas. DeeplabV3+ and DDRNet showed relatively balanced performance across all metrics, without significant strengths or weaknesses, demonstrating good stability, but with slightly lower classification accuracy. BiSeNet performed well in some metrics but showed some deficiency in NI. Unet achieved the best results in OW, but its performance in IoU and pixel accuracy in other areas was average, especially in NI and YI. Overall, our method showed the most outstanding performance in most key evaluation metrics, with Pa, mPA, and mIoU reaching the highest levels of 90.622%, 90.788%, and 83.266%, respectively.

To further demonstrate the effectiveness of our method, Fig. 7 compares the segmentation results of Fast-SISENet with other methods. Fast-SISENet achieves more precise and consistent segmentation, particularly in complex and detailed regions. It accurately captures fine structures and aligns closely with the ground truth, while minimizing errors in ambiguous areas. This highlights the robustness and superiority of Fast-SISENet in producing high-quality segmentation results.

Table 1. Quantitative comparison with state-of-the-art models. The best values in the columns are in bold. All scores are expressed as percentages (%).

Model	Cpa				IoU				Pa	Mpa	MIoU
	OW	FYI	NI	YI	OW	FYI	NI	YI			
Unet [20]	**95.24**	80.37	77.46	88.55	85.02	67.81	67.89	77.00	85.44	85.41	74.68
DeeplabV3+ [21]	94.16	82.75	85.42	87.97	88.26	71.01	73.04	79.71	87.36	87.58	78.01
DDRNet [18]	94.48	84.78	85.49	89.07	87.75	73.82	74.46	81.17	88.31	88.45	79.30
BiSeNet [16]	94.51	85.15	85.08	88.26	88.46	72.86	74.47	80.31	88.04	88.25	79.03
PIDNet [19]	94.50	86.42	87.03	**92.07**	89.96	76.91	77.79	**83.86**	90.06	90.00	82.13
Ours	94.21	**88.41**	**90.06**	90.46	**91.02**	**77.61**	**80.92**	83.52	**90.62**	**90.79**	**83.27**

4.3 Comparison with State-of-the-Art Methods on Infrence Speed

The experimental results of the inference speed of various models are shown in Table 2. Our model (Ours) demonstrates significant advantages in key performance

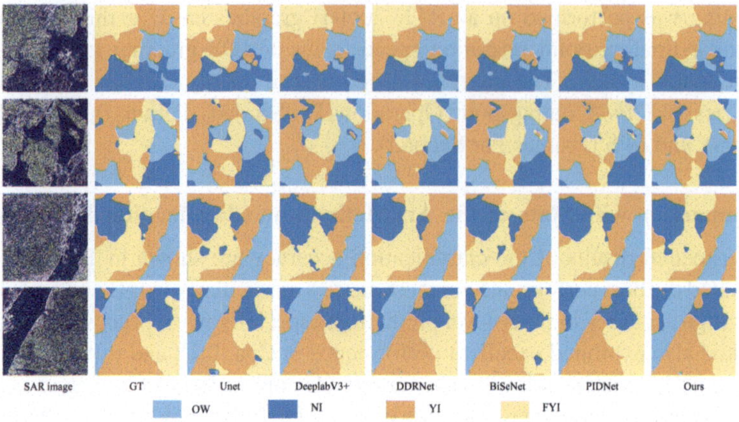

Fig. 7. Qualitative comparison of the visualization results of our method with other methods

metrics. Although Unet's high Floating Point Operations Per Second (FLOPS) limit its Frames Per Second (FPS), making it potentially unsuitable for real-time applications, DeeplabV3+ has achieved a good balance between reducing FLOPS and increasing FPS, but there is still room for improvement. DDRNet, by further reducing FLOPS, has successfully improved FPS, but it has a slightly higher parameter count. BiSeNet has well-controlled parameters and FLOPS, but its FPS does not reach the optimal level. PIDNet is closest to our method in terms of FPS, but there is room for improvement in the other two metrics. Our model (Ours) maintains a low FLOPS comparable to BiSeNet, significantly reduces the model parameter count, and most importantly, achieves the highest FPS among all models, reaching 271.1. This outstanding performance not only proves the computational efficiency of our model but also indicates its great potential in processing speed, especially in real-time semantic segmentation scenarios that require rapid response.

Table 2. Experimental results of inference speed for different segmentation methods.

Model	FLOPS(Gbps)	Params(Mb)	FPS
Unet [20]	438.003	31.038	27.716
DeeplabV3+ [21]	46.964	5.814	132.557
DDRNet [18]	35.944	20.147	186.895
BiSeNet [16]	10.932	**1.942**	136.663
PIDNet [19]	11.917	7.624	218.439
Ours	**10.266**	3.581	**271.1**

5 Conclusion

This paper presents a rapid segmentation method for SAR sea ice imagery, which improves upon the original DeeplabV3+ network's ASPP module and feature fusion components, enhancing inference speed while maintaining classification accuracy. The original ASPP module's dilated convolutions and pyramid pooling have been decoupled into two independent parallel branches, allowing the improved ASPP module to extract features from different perspectives simultaneously, reducing the number of times data is transmitted within the network and effectively increasing inference speed. The simple concatenation operation has been refined into an adaptive feature fusion module that integrates features from different hierarchical levels or sources. The feature fusion module consolidates rich information, enhances the expressiveness of features, and thereby improves classification accuracy. Experimental results indicate that our method achieves the best performance in both classification accuracy and inference speed tests. The inference speed of our method reaches 271.1 FPS, which has the potential to be combined with SAR quick-view systems to realize real-time acquisition and classification of SAR sea ice images, providing a reliable reference for real-time route planning in the Arctic shipping lanes and emergency response to sea ice disasters.

References

1. Lin, B., et al.: An overview of scholarly literature on navigation hazards in Arctic shipping routes. Environ. Sci. Pollut. Res. **31.28**, 40419–40435 (2024)
2. Zakhvatkina, N.Y., et al.: Classification of sea ice types in ENVISAT synthetic aperture radar images. IEEE Trans. Geosci. Remote Sens. **51.5**, 2587–2600 (2012)
3. Park, J.-W., et al.: Classification of sea ice types in Sentinel-1 synthetic aperture radar images. Cryosphere **14.8**, 2629–2645 (2020)
4. He, L., et al.: Investigation of polarimetric decomposition for Arctic summer sea ice classification using Gaofen-3 fully polarimetric SAR data. IEEE J. Select. Top. Appl. Earth Observ. Remote Sens. **15**, 3904–3915 (2022)
5. Yin, Q., et al.: Coherence matrix power model for scattering variation representation in multi-temporal PolSAR crop classification. IEEE J. Select. Top. Appl. Earth Observ. Remote Sens. (2024)
6. Liu, Y., et al.: SAR ship detection based on explainable evidence learning under intra-class imbalance. IEEE Trans. Geosci. Remote Sens. (2024)
7. Boulze, H., Korosov, A., Brajard, J.: Classification of sea ice types in Sentinel-1 SAR data using convolutional neural networks. Remote Sens. **12**(13), 2165 (2020)
8. Montpetit, B., et al.: Assessing the Parameterization of RADARSAT-2 Dual-polarized ScanSAR scenes on the accuracy of a convolutional neural network for sea ice classification: case study over coronation Gulf, Canada. Can. J. Remote Sens. **49.1**, 2247091.4 (2023)
9. Jiang, M., et al.: Semi-supervised sea ice classification of SAR imagery based on graph convolutional network. In: IGARSS 2022–2022 IEEE International Geoscience and Remote Sensing Symposium. IEEE (2022)
10. Song, W., Li, M., Gao, W., et al.: Automatic sea-ice classification of SAR images based on spatial and temporal features learning. IEEE Trans. Geosci. Remote Sens. **59**(12), 9887–9901 (2021)
11. Khaleghian, S., et al.: Sea ice classification of SAR imagery based on convolution neural networks. Remote Sens. **13.9**, 1734 (2021)

12. Song, W., et al.: Si-stsar-7: a large sar images dataset with spatial and temporal information for classification of winter sea ice in hudson bay. Remote Sens. **14.1**,168 (2021)
13. Zhang, J., et al.: An improved sea ice classification algorithm with Gaofen-3 dual-polarization SAR data based on deep convolutional neural networks. Remote Sens. **14.4**, 906 (2022)
14. Huang, Y., Ren, Y., Li, X.: Deep learning techniques for enhanced sea-ice types classification in the Beaufort Sea via SAR imagery. Remote Sens. Environ. **308**, 114204 (2024)
15. Wan, H., et al.: Multi-featured sea ice classification with SAR image based on convolutional neural network. Remote Sens. **15.16**, 4014 (2023)
16. Yu, C., et al.: Bisenet: bilateral segmentation network for real-time semantic segmentation. In: Proceedings of the European conference on computer vision (ECCV) (2018)
17. Fan, M., et al.: Rethinking bisenet for real-time semantic segmentation. In: Proceedings of the IEEE/CVF Conference on Computer Vision and Pattern Recognition (2021)
18. Hong, Y., et al.: Deep dual-resolution networks for real-time and accurate semantic segmentation of road scenes. arxiv preprint arxiv:2101.06085 (2021)
19. Xu, J., Zixiang, X., Bhattacharyya, S.P.: PIDNet: a real-time semantic segmentation network inspired by PID controllers. In: Proceedings of the IEEE/CVF Conference on Computer Vision and Pattern Recognition (2023)
20. Ronneberger, O., Fischer, P., Brox, T.: U-net: convolutional networks for biomedical image segmentation. Medical Image Computing and Computer-Assisted Intervention–MICCAI 2015: 18th International Conference, Munich, Germany, October 5–9, 2015, proceedings, part III 18. Springer International Publishing (2015)
21. Chen, L.-C., et al.: Encoder-decoder with atrous separable convolution for semantic image segmentation. In: Proceedings of the European Conference on Computer Vision (ECCV) (2018)

Application of Multi-source Remote Sensing Data for Sea Ice Classification in Arctic Shipping Route Emergency Response

Zhang Wenyi[1](✉), Zhang Jiande[2], Dai Xiangyu[1], Chu Qingwei[1], and Li Guangzuo[1]

[1] Aerospace Information Research Institute, Chinese Academy of Sciences, Beijing, China
wyzhang@aircas.ac.cn

[2] QiLu Aerospace Information Research Institute, No. 44 Industry North Road, Jinan, China

Abstract. The gradual opening of Arctic shipping routes will significantly impact the global shipping industry, but the severe sea ice conditions in the Arctic pose serious risks, with shipwreck incidents caused by collisions with sea ice occurring frequently. This paper focuses on the Yongxing shipwreck, a typical sea ice accident, and utilizes a deep learning model to classify sea ice images from the incident area, aiming to analyze the causes of the Yongxing accident. The model used in this study is based on the MobileNetV3 backbone, combined with the CBAM and FPN modules to enhance the network's feature extraction capabilities. Experimental results show that the overall classification accuracy of the proposed network reached 97.18%, with a kappa coefficient of 0.962. The classification accuracy for each of the four categories (seawater, new ice, gray-white ice, and thick ice) exceeded 95%. The analysis of the classification results indicates that the Yongxing ship mistakenly entered a thick ice zone in the Tatar Strait, leading to a collision with sea ice and subsequent sinking. The review suggests that acquiring high-accuracy sea ice maps through remote sensing can provide reliable decision-making support for Arctic vessels in planning safe routes to avoid sea ice hazards.

Keywords: Multi-Source Remote Sensing Data · Sea Ice Classification · Arctic Shipping Route

1 Introduction

As a major global power in trade and shipping, China places great importance on the development and utilization of Arctic shipping routes. Compared to the traditional route via the Suez Canal, the Arctic route would significantly shorten the maritime distance between Asia and Europe, while also mitigating the navigational risks associated with sensitive waters such as the Strait of Malacca and the Bab-el-Mandeb Strait. The opening of this route would provide a safer and more efficient maritime corridor for trade between Asia and Europe, potentially leading to a substantial transformation of the current shipping landscape and international trade patterns.

G. Luo (Ed.): GFSISD 2024, SPPHY 417, pp. 251–261, 2025.
https://doi.org/10.1007/978-981-96-2490-4_18

While the Arctic shipping routes offer the potential advantage of shortening maritime distances, their adverse impacts on navigation cannot be overlooked, particularly the significant threat posed by sea ice to navigational safety. The seasonal fluctuations and unpredictable spatial distribution of sea ice make the navigability of these routes highly unstable. On February 19, 2023, the cargo ship Yongxing 56 collided with sea ice during its voyage, resulting in hull damage. In the following days, the ship inadvertently entered an area with dense sea ice, worsening the damage and ultimately leading to the sinking on March 1. The sinking of the Yongxing 56 highlights that sea ice is a major safety hazard for maritime activities, underscoring the importance of real-time access to sea ice distribution maps for ensuring safe navigation.

2 Related Work

Early methods of sea ice detection primarily relied on fixed observation stations, aircraft, and ships, which were not only limited in coverage but also time-consuming and labor-intensive. The advent of satellite remote sensing technology has effectively addressed these challenges. With its advantages of wide coverage and high spatial and temporal resolution, satellite remote sensing has gradually become the mainstream method for sea ice observation. The resolution of optical satellite data is high, and the classification of ground objects is easy to distinguish; SAR satellite data has a wider range of applicable conditions and can be observed all day and all weather. Both types of satellite data above are suitable for sea ice classification. The data from Gaofen-1 satellite has been proven to be effective in sea ice detection, while Gaofen-3 satellite data has been used by more scholars for sea ice classification, achieving good experimental results [1–3].

Sea ice classification methods are mainly divided into traditional machine learning based classification methods and deep-learning based classification methods. The use of machine learning methods for sea ice classification usually requires manual extraction of features such as backscatter coefficient [4, 5], texture [6, 7], polarization [8], etc. from sea ice images. The extracted features are combined into feature vectors and input into traditional machine learning classifiers for training to obtain the final classification model. At present, the commonly used machine learning classifiers include support vector machines [4], decision trees and random forest [9], Bayesian networks [7], and other segmentation and clustering algorithms [10, 11]. The deep learning method places more emphasis on model construction and dataset production, enabling network models to autonomously learn sea ice features from massive data and improve classification accuracy. At present, the sea ice classification methods based on deep learning include convolutional neural network [12], residual neural network [3], Long short-term memory network [13], feature fusion network [14] and attention mechanism, and their classification effects are generally better than traditional machine learning methods.

3 Data and Methodology

3.1 Date

In this paper, we utilize a combination of multi-source satellite data and AIS data to analyze the sinking of the Yongxing 56. The optical satellite data is sourced from the Gaofen-1 series, while SAR satellite data is derived from the Gaofen-3 series. Based on the specific observation plan, we selected two images captured by the Gaofen-1 04 satellite (GF1D) and one image captured by the Gaofen-3 02 satellite (GF3B). The Automatic Identification System (AIS) is a system used for the automatic identification of vessels. It transmits both dynamic and static information, such as the ship's call sign, IMO number, speed, position, course, and destination, to shore-based AIS stations, satellite-based AIS detection systems, and nearby vessels via very high frequency (VHF) radio waves.

For specific data, we used 2-m resolution panchromatic images and 8-m resolution multispectral images from the GF1D satellite, 5-m resolution dual-polarization SAR images from the GF3B satellite in FSII mode, as well as the full AIS data. These datasets cover the entire voyage of the Yongxing 56 from February 12 to March 3, 2023. We selected three images that captured the trajectory of Yongxing 56, taken on February 17 and February 24, 2023, as shown in Fig. 1. Notably, the two images captured on February 24 clearly show the Yongxing 56 trapped in an ice floe area along with a nearby Russian rescue ship. Therefore, we will conduct an in-depth analysis of these two images in the subsequent sections.

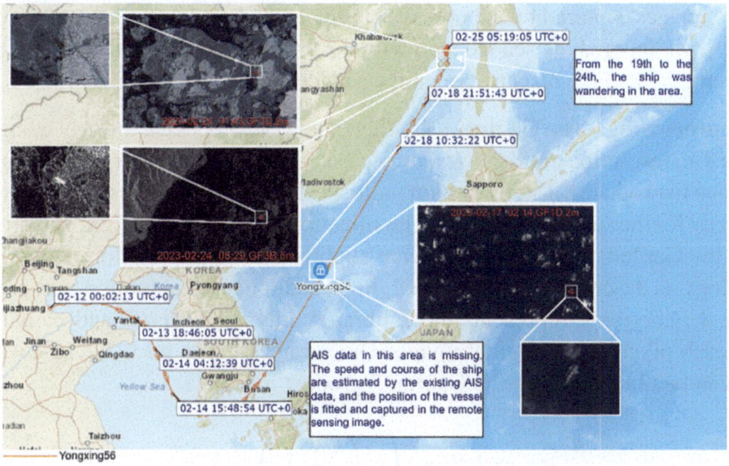

Fig. 1. The trajectory map of Yongxing 56

For data processing, we used ENVI software to fuse the 2-m resolution panchromatic image with the 8-m resolution multispectral image from the GF1D satellite, resulting in a multispectral image with a 2-m resolution. The fused image not only has higher spatial resolution but also retains rich color information, making it easier to highlight key features. Additionally, we used ENVI to convert the projection coordinate system of the dual-polarization SAR image from the GF3B satellite to match that of the GF1D multispectral image. Subsequently, the GDAL toolkit was employed to clip the overlapping regions of both datasets based on their geographic coordinates. This allows us to analyze the types of sea ice in the area and investigate the causes of the Yongxing 56 shipwreck.

3.2 Methodology

In general, optical satellite images display distinct differences in the color and texture characteristics of sea ice, making them suitable for sea ice observation and classification. However, they are susceptible to factors such as lighting, cloud cover, and weather conditions, which can degrade image quality. SAR images, on the other hand, are not affected by these factors and can provide all-weather, around-the-clock observation. Therefore, in this study, we use both the daytime panchromatic images from GF1D captured on February 24 and the dual-polarization SAR images from GF3B. The GF1D images provide a reference for identifying sea ice types in the SAR images during the day, while the GF3B data compensate for the limitations of optical satellites during nighttime, ensuring a comprehensive review of the incident by combining the strengths of both data.

To further analyze the incident, we employed deep learning methods to classify sea ice in the SAR images of the affected area, while using optical images for validation to obtain accurate and reliable sea ice maps. The sea ice classification process follows the method used in Reference 2. However, since no suitable ice condition map products are available for the region covered by the trajectory of Yongxing 56, we manually interpreted the sea ice images captured by the optical satellite and labeled different categories of sea ice. Finally, we merged the sea ice map with the trajectory data, overlaying the paths of the Yongxing 56 and the rescue ship on the sea ice map.

Data Preprocessing. The Gaofen-1 satellite data were processed directly using ENVI, where the 8-m resolution multispectral image was fused with the 2-m resolution panchromatic image to produce a 2-m resolution multispectral image. This fused image was then used directly for the subsequent sea ice classification, as shown in Fig. 2a. The processing steps for the Gaofen-3 satellite images are slightly more complex, following the same methodology as described in Reference 2. First, radiometric calibration was performed using Eq. 1, where σ_{db} represents the calibrated gray value in dB, and p^I is the gray value of each pixel in the original image. The parameter m is set to 32,767, while QV and K_{db} can be obtained from the Gaofen-3 product metadata.

$$\sigma_{db} = 10log_{10}\left(p^I\left(\frac{QV}{m}\right)^2\right) - K_{db} \tag{1}$$

Next, an enhanced Lee filter was applied to remove speckle noise. Finally, the dual-polarization SAR images were combined to create a false-color image. The specific method for this combination involved assigning the HH polarization image to the red (R) channel, the HV polarization image to the green (G) channel, and the average of both to the blue (B) channel. The resulting image is shown in Fig. 2b.

Principle of Classification. The World Meteorological Organization (WMO) has established various standards for classifying sea ice in polar and high-latitude regions, including classifications based on ice thickness, ice motion, and ice size. In this study, we adopt the classification based on ice thickness, as there is a general assumption that the longer the sea ice persists, the thicker it becomes. The WMO classifies sea ice into 13 categories based on thickness. In this paper, we use a classification system similar to the WMO

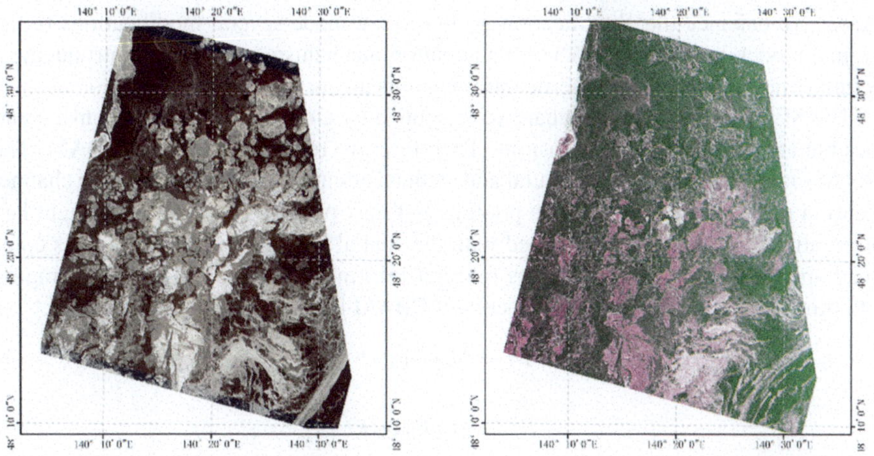

(a) (b)

Fig. 2. The remote sensing images of the area where Yongxing 56 was trapped, captured by GF1D and GF3B. (a) Multi-spectral image at 2m resolution for GF1D, (b) pseudo-colour image synthesized from dual-polarized SAR data at 5m resolution for GF3B.

standard but with a more simplified categorization, which is sufficient to meet the needs of this study, as shown in Table 1.

Table 1. The sea ice classification standards used in this paper.

type	thickness	code
open water	0 cm	1
new ice	0–15 cm	2
gray-white ice	15–30 cm	3
thick ice	>30 cm	4

Model. Our classification model is based on MobileNetV3 as the backbone network, enhanced with the integration of the Convolutional Block Attention Module(CBAM) and Feature Pyramid Network(FPN) modules. The MobileNet series is a lightweight network that can be deployed on mobile devices, with fewer training parameters and lower training complexity, while still maintaining high accuracy. MobileNetV3 is the current outstanding performance network architecture of the MobileNet series, achieving higher accuracy while reducing runtime [16].

The original MobileNetV3 backbone consists of a series of convolutional modules, with each module resembling a residual block and incorporating a Squeeze-and-Excitation (SE) module to introduce a channel-wise attention mechanism. Our model largely retains the key features of this feature extraction network, with only slight adjustments to the convolution kernel sizes and pooling layer positions to better fit our custom dataset. To enhance the classification accuracy, we made several modifications to the original model, including replacing the attention mechanism module and introducing a feature fusion module. The specific modifications are as follows.

The SE module introduces adaptive weights only along the channel dimension and does not account for other dimensions. Therefore, we replaced it with the CBAM. The CBAM module combines both spatial and channel attention mechanisms. In the channel attention mechanism, both average pooling and max pooling are used to highlight key information, which is then integrated with the spatial attention mechanism. This combined approach effectively preserves relevant information in the image by leveraging both dimensions. The calculation process of CBAM is generally as follows:

$$F\prime = M_c(F) \otimes F \tag{2}$$

$$F'' = M_s(F) \otimes F\prime$$

Where M_c represents the channel attention mechanism process, M_s represents the spatial attention mechanism process, F denotes the original feature map, $F\prime$ is the result obtained by applying the channel attention mechanism to F and then multiplying it with itself, F'' is the result obtained by applying the spatial attention mechanism to $F\prime$ and then multiplying it with itself. The computation processes for M_c and M_s are shown in Eq. 3.

$$M_c(F) = \sigma\left(W_1\left(W_0\left(F_{avg}^c\right)\right)\right) + W_1(W_0(F_{max}^c)) \tag{3}$$

$$M_s(F) = \sigma(f^{7\times7}([F_{avg}^c; F_{max}^c]))$$

Where W_0 and W_1 represent two fully connected layers, σ is the activation function, $f^{7\times7}$ denotes a convolutional layer with a kernel size of 7, F_{avg}^c and F_{max}^c represent the average pooling and max pooling operations, respectively.

Due to the limited size of the input image, only the convolutional strides of the second, eighth, and ninth layers are set to 2 to achieve downsampling. After three pooling operations, feature maps of three different sizes are produced, with the smallest feature map being 1/8 the dimension of the original image. For instance, if the input is a 32 × 32 slice, the smallest feature map would be only 4 × 4 pixels, which may not adequately reflect the overall features of the original image. Therefore, we introduce a feature fusion module, which upsamples the smaller feature maps to match the dimensions of the feature map from the previous layer. Then, a 1 × 1 convolution operation is applied to adjust the channel number of the current feature map to match that of the previous feature map. The two feature maps are then combined to produce a fused feature map.

The overall network structure is shown in Fig. 3. The first row illustrates the entire classification process: input data consists of sea ice images, which are processed through the classification network to produce a sea ice classification map, where different colors represent different sea ice categories. The classification network comprises four main modules: the preprocessing module, convolutional module, feature extraction module (backbone network), and reconstruction module. The backbone network structure is detailed below. We have incorporated the CBAM attention mechanism [17] and a feature fusion network into the feature extraction module to fully utilize features at different scales, adaptively learning the weights of various features to enhance classification accuracy. The reconstruction module, implemented with fully connected layers, takes the results from the feature extraction module and generates the final classification outcome.

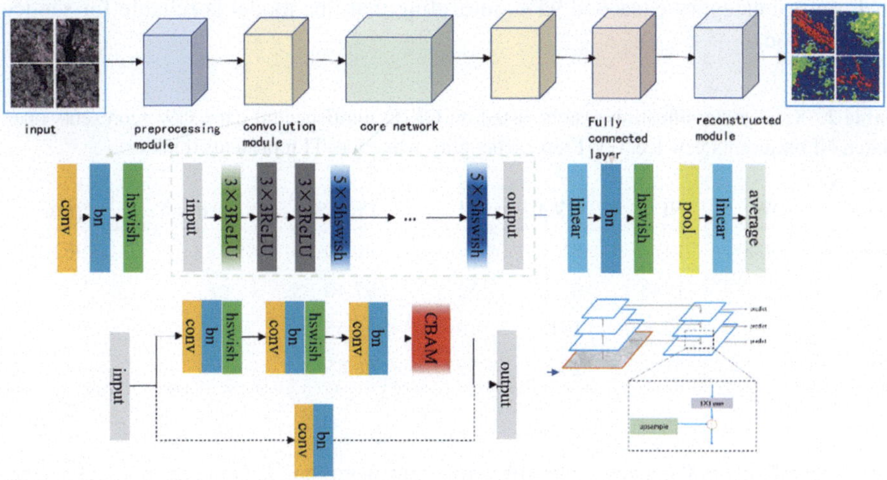

Fig. 3. The sea ice classification network used in this paper.

Evaluation Metrics. The confusion matrix, also known as an error matrix, is one of the standard formats for evaluating classification accuracy. By treating samples belonging to a given class as positive examples and those not belonging to the class as negative examples, the classification process can be viewed as a binary decision. Based on this, the following evaluation metrics can be introduced: overall accuracy (Accu), per-class precision (Prec), and the Kappa coefficient.

Overall accuracy (Accu) is defined as the ratio of correctly classified samples to the total number of samples. Per-class precision (Prec) is the ratio of correctly classified sea ice samples in a specific category to the total number of samples in that category. The Kappa coefficient is a measure of classification accuracy, calculated using Eq. 4.

$$kappa = \frac{p_o - p_e}{1 - p_e} \tag{4}$$

Where p_o represents the overall accuracy (Accu), and p_e is calculated using Eq. 5, Assuming the total number of samples is n, and the number of categories is C, a_k denotes the actual number of samples in each category, and b_k represents the predicted number of samples in each category, where $1 \leq k \leq C$.

$$p_e = (\sum\nolimits_{k=1}^{C} a_k b_k)/n \times n \tag{5}$$

4 Result

In the GF1D multispectral image, we classified the sea ice into four categories based on their different characteristics as seen in the image, forming a high-resolution ice map. The categories are open water, new ice, gray-white ice, and thick ice. The acquired samples were then split into training and test sets. After model training, the classification performance on the test set is shown in Table 2. The overall classification accuracy reached 97.18%, with a Kappa coefficient of 0.962. The classification accuracy for each sea ice category exceeded 95%, indicating that this model is reliable for sea ice classification.

Table 2. Sea ice classification results based on GF1D multispectral data, OW represents open water, NI represents new ice, GWI represents gray-white ice, TI represents thick ice.

type	OW	NI	GWI	TI	Prec(%)	Accu(%)	kappa
OW	1938	11	51	0	96.90	97.18	0.962
NI	0	1948	0	52	97.40		
GWI	17	32	1911	40	95.55		
TI	0	0	23	1977	98.85		

Figure 4 shows the sea ice classification map from the GF1D multispectral image. From the map, it is evident that along the route taken by the Yongxing 56, gray-white ice and thick ice were predominant, with a thickness of over 30 cm. Typically, ordinary cargo ships lack icebreaking capabilities, so based on the ice map, the Yongxing 56 was sailing through relatively thick ice floes both before and after its collision with the ice on February 19–20. On February 24, after becoming trapped in the ice floe region, the Yongxing 56 was stuck at the boundary between gray-white ice and thick ice. At that point, it should have avoided the thick ice and chosen a thinner ice route to attempt self-rescue. However, as seen in the images taken by the GF3B satellite, the Yongxing 56 does not appear to have successfully avoided the thick ice area. As shown in Fig. 5, because thin ice has little impact on ship navigation, we used the false-color composite image generated from GF3B dual-polarization data to classify the sea ice in this region into two categories: thin ice and thick ice. The GF3B image was taken nine hours after the GF1D image, during which time there was a slow relative movement between the sea ice and the ship. The results show that the Yongxing 56 had completely moved from the gray-white ice and thick ice boundary into the thick ice region, as illustrated in Fig. 6. This is consistent with media reports that the Yongxing 56 was in "Level 10 ice conditions." With the surrounding sea ice thickening, the difficulty of rescue increased. As the waves continued to batter the cargo ship and the bilge pumps failed to fully remove water from the hull, the Yongxing 56 sank at 5:46 a.m.

Russia is one of the earliest countries to begin sea ice research, and the sea ice products developed by its research institutions are now capable of guiding maritime navigation. This rescue event serves as a typical example. We fused the AIS data of the rescue ship on February 23 with the sea ice map, and the result is shown in Fig. 5. From the fusion map, it is clear that the rescue ship, when heading to the site, did not take the shortest route. Instead, it planned a route that circumvented areas of thick ice based on the sea ice conditions. Additionally, according to the AIS data, the rescue ship did not significantly reduce its speed after entering the ice-heavy region, which suggests

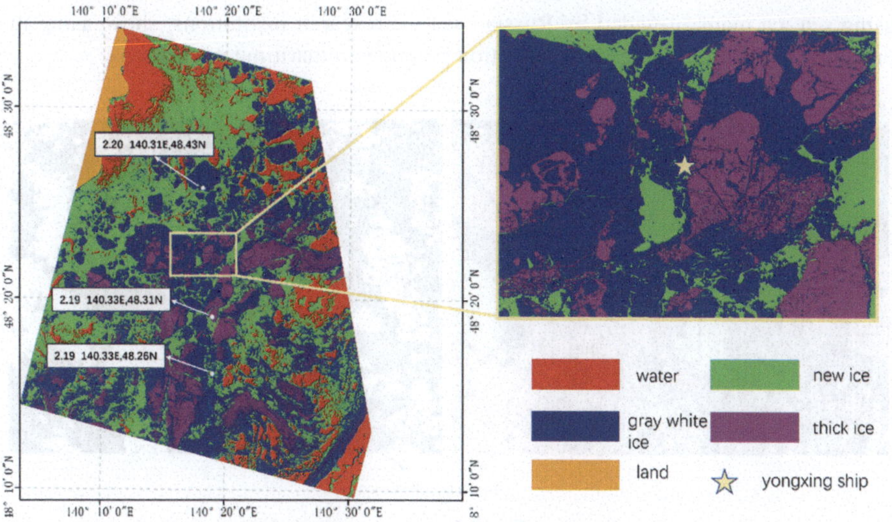

Fig. 4. Sea ice classification results based on GF1D multispectral images. The positions marked with pentagrams indicate the area where Yongxing 56 was trapped. White dots represent the positions of Yongxing 56 at corresponding times during its voyage.

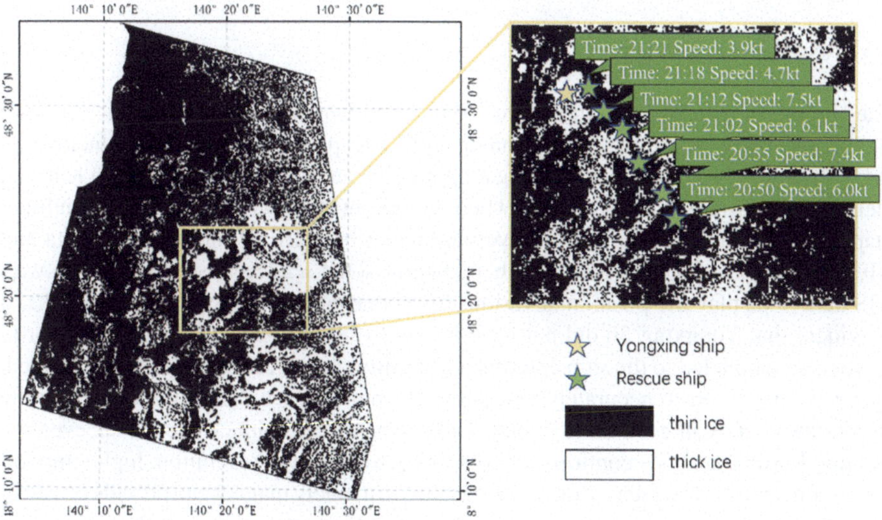

Fig. 5. Sea ice classification results based on GF3B FSII data, yellow pentagrams indicate the area where Yongxing 56 was trapped, green pentagrams represent the trajectory of the rescue ship.

that the ship may have already been aware of the sea ice conditions and had planned its route before entering the area. This indicates that even professional rescue ships avoid crossing regions with dense thick ice. The route of the rescue ship was likely planned using sea ice maps provided by Russian sea ice research institutions, showcasing the practical application of these ice condition products in maritime navigation.

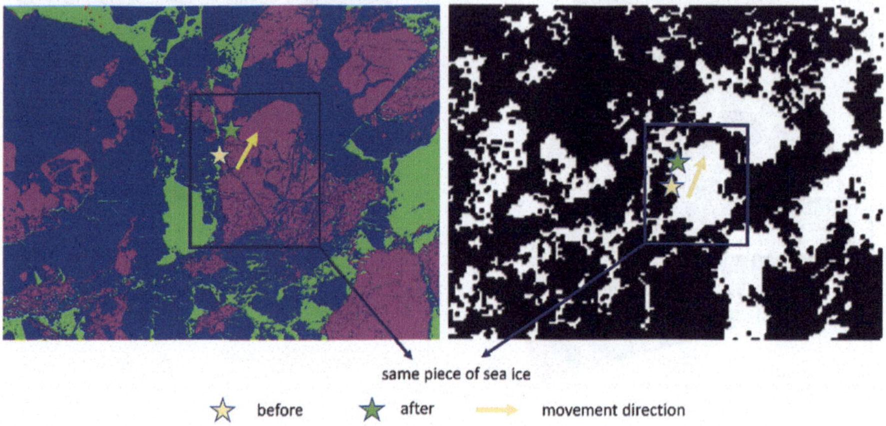

Fig. 6. By comparing the positions of Yongxing 56 in GF1D and GF3B, the movement direction of Yongxing 56 relative to the sea ice over the nine-hour period can be determined.

5 Result

The sinking of Yongxing 56 in a Level 10 ice floe area highlights that sea ice poses a significant safety hazard to maritime navigation. This paper uses the sinking incident to study the role of sea ice maps in navigating ice floe areas. We employed deep learning methods to classify sea ice in the area where Yongxing 56 sank, creating an ice condition map. The accuracy of this map was verified using both GF1D multispectral data and GF3B dual-polarized SAR data. We then integrated the sea ice classification map with AIS data to display the position at various times during its voyage. From the results, it is evident that Yongxing 56 did not have access to sea ice maps of the region prior to its voyage, which led to the ship entering a hazardous area. After the hull was damaged, the crew still failed to accurately assess the surrounding ice distribution, causing the ship to move into an even more complex ice area, which indirectly led to its eventual sinking. Further analysis confirms that obtaining high spatial resolution, high temporal resolution, and accurate sea ice distribution maps in a timely manner can provide effective safety guarantees for polar navigation. As Arctic routes continue to open, the widespread application of real-time sea ice classification maps is likely to reduce accidents in Arctic waters, minimizing casualties and financial losses.

References

1. He, L., He, X., Hui, F., et al.: Investigation of polarimetric decomposition for Arctic summer sea ice classification using Gaofen-3 fully polarimetric SAR data. IEEE J. Select. Top. Appl. Earth Observ. Remote Sens. **15**, 3904–3915 (2022)

2. Zhang, J., Zhang, W., Hu, Y., et al.: An improved sea ice classification algorithm with Gaofen-3 dual-polarization SAR data based on deep convolutional neural networks. Remote Sens. **14**(4), 906 (2022)
3. Zhang, T., Yang, Y., Shokr, M., et al.: Deep learning based sea ice classification with Gaofen-3 fully polarimetric SAR data. Remote Sens. **13**(8), 1452 (2021)
4. Liu, H., Guo, H., Zhang, L.: SVM-based sea ice classification using textural features and concentration from RADARSAT-2 dual-pol ScanSAR data. IEEE J. Select. Top. Appl. Earth Observ. Remote Sens. **8**(4), 1601–1613 (2014)
5. Shokr, M., Dabboor, M., Lacelle, M., et al.: Observations from C-band SAR fully polarimetric parameters of mobile sea ice based on radar scattering mechanisms to support operational sea ice monitoring. Can. J. Remote. Sens. **48**(2), 197–213 (2022)
6. Park, J.W., Korosov, A.A., Babiker, M., et al.: Classification of sea ice types in Sentinel-1 synthetic aperture radar images. Cryosphere **14**(8), 2629–2645 (2020)
7. Zakhvatkina, N.Y., Alexandrov, V.Y., Johannessen, O.M., et al.: Classification of sea ice types in ENVISAT synthetic aperture radar images. IEEE Trans. Geosci. Remote Sens. **51**(5), 2587–2600 (2012)
8. Ressel, R., Singha, S., Lehner, S., et al.: Investigation into different polarimetric features for sea ice classification using X-band synthetic aperture radar. IEEE J. Select. Top. Appl. Earth Observ. Remote Sens. **9**(7), 3131–3143 (2016)
9. Lohse, J., Doulgeris, A.P., Dierking, W.: An optimal decision-tree design strategy and its application to sea ice classification from SAR imagery. Remote Sens. **11**(13), 1574 (2019)
10. Yu, Q., Clausi, D.A.: SAR sea-ice image analysis based on iterative region growing using semantics. IEEE Trans. Geosci. Remote Sens. **45**(12), 3919–3931 (2007)
11. Ochilov, S., Clausi, D.A.: Operational SAR sea-ice image classification. IEEE Trans. Geosci. Remote Sens. **50**(11), 4397–4408 (2012)
12. Boulze, H., Korosov, A., Brajard, J.: Classification of sea ice types in Sentinel-1 SAR data using convolutional neural networks. Remote Sens. **12**(13), 2165 (2020)
13. Song, W., Li, M., Gao, W., et al.: Automatic sea-ice classification of SAR images based on spatial and temporal features learning. IEEE Trans. Geosci. Remote Sens. **59**(12), 9887–9901 (2021)
14. Han, Y., Liu, Y., Hong, Z., et al.: Sea ice image classification based on heterogeneous data fusion and deep learning. Remote Sens. **13**(4), 592 (2021)
15. Ren, Y., Li, X., Yang, X., et al.: Development of a dual-attention U-Net model for sea ice and open water classification on SAR images. IEEE Geosci. Remote Sens. Lett. **19**, 1–5 (2021)
16. Howard, A., Sandler, M., Chu, G., et al.: Searching for mobilenetv3. In: Proceedings of the IEEE/CVF International Conference on Computer Vision, pp. 1314–1324 (2019)
17. Woo, S., Park, J., Lee, J.Y., et al.: Cbam: convolutional block attention module. In: Proceedings of the European Conference on Computer Vision (ECCV), pp. 3–19 (2018)

Author Index

G. Luo (Ed.): GFSISD 2024, SPPHY 417, pp. 263–264, 2025.
https://doi.org/10.1007/978-981-96-2490-4

The manufacturer's authorised representative in the EU is Springer Nature Customer Service Centre GmbH, Europaplatz 3, 69115 Heidelberg, Germany. If you have any concerns regarding our products, please contact ProductSafety@springernature.com

Printed and bound by CPI Group (UK) Ltd, Croydon, CR0 4YY
01/10/2026
02225913-0012